C++ for Mathematicians

An Introduction for Students and Professionals

Edward Scheinerman

Chapman & Hall/CRC
Taylor & Francis Group
Boca Raton   London   New York

Chapman & Hall/CRC is an imprint of the
Taylor & Francis Group, an informa business

Cover photograph: Ira Scheinerman

Cover design concept: Jonah Scheinerman

Published in 2006 by
Chapman & Hall/CRC
Taylor & Francis Group
6000 Broken Sound Parkway NW, Suite 300
Boca Raton, FL 33487-2742

Taylor & Francis Group
is the Academic Division of Informa plc.

Visit the Taylor & Francis Web site at
http://www.taylorandfrancis.com

and the CRC Press Web site at
http://www.crcpress.com

In loving memory of Pauline and of Arnold

זכרונם לברכה

Contents

Programs

Figures

Preface

To my fellow students of mathematics

This book is written for you. This is the book that I wish someone had written for me. This is a book that introduces the C++ language for people who are interested in solving mathematical problems.

There is a dizzying selection of books on C++ written for a wide array of audiences. Visit your favorite bookseller and you can find C++ books for finance, numerics, computer security, game programming, embedded controllers, graphical user interfaces, network protocols, data and file structures, engineering, scientific computing, digital signal processing, simulation, neural nets, artists, virtual machine design, graphics, computational physics, cellular automata, cryptography, Web agents, business, aerospace and flight simulation, music and MIDI instruments, mobile phones, language translation, computer-aided design, speech recognition, database development, computer architecture, photographic imaging, fuzzy logic, hardware control, rigid body kinematics, real programmers, and—of course—for dummies.

We assume that none of the above applies to you. We approach C++ from the point of view of solving mathematical problems. We organize our discussion around the mathematics and bring in the relevant C++ ideas as we need them.

Why C++?

There is a plethora of computer tools available to the mathematical scientist. Many of these are suited for specific purposes. For example, if you need to perform extensive calculations in support of a number theory problem, the *Pari* package is perfect. There is a variety of commercial software packages that are excellent for mathematical work including Maple, MATLAB, and *Mathematica* to name but a few.

For many mathematical problems, these systems work perfectly. However, for problems that require extensive computation the speed of a compiled language such as C++ cannot be beat. A C++ program can work through billions of examples faster than most other computing choices.

The length of time it takes to solve a problem on a computer begins not when you run your program; it begins when you first start to write your program. The object-oriented nature of C++ enables you to create correct programs quickly. Furthermore, there is an extensive collection of C++ programs freely available on the Web that can be customized and used for your purposes (we discuss a few of these in Chapter 13).

In addition, C++ is available for free on most computer systems. See Appendix A for more information about different versions of C++ for various computing environments (Windows, UNIX, Macintosh).

What can my computer do for me?

Although the utility of computers in the sciences and engineering is unquestionable, it is not as clear that a computer is useful for mathematics. However, there are several arenas in which a computer can be a mathematician's best friend.

- *Symbolic computation.* Mathematical work often requires us to fuss with formidable formulas and solve elaborate equations. Computer algebra systems such as Maple and *Mathematica* are extremely useful for such work.

- *Visualization.* The computer can draw precise pictures and diagrams; often these images provide key insights to understanding problems.

- *Examples and counterexamples.* The computer can run through millions of examples and try myriad possibilities. It is a laboratory in which to perform experiments to see if ideas work and for discovering patterns.

- *Contribution to proof.* Sometimes, parts of proofs can be relegated to the computer for checking. A celebrated example of this is the 1970s-era proof of the Four Color Theorem by Appel and Haken. More recently, Tom Hales's announced proof of the Kepler Conjecture required extensive computation.

I have used the computer in all these ways in my research. Allow me to share an amusing anecdote in which the computer contributed to the proof of a theorem. I was working with two colleagues on a problem in discrete mathematics. We knew that we could complete one portion of the proof if we could find a graph with certain specific properties. We carefully wrote down these criteria on a blackboard and set out to find the elusive graph. Although we tried to create the required graph by hand, each example we tried took us several minutes to check. We realized that the criteria could be checked mechanically and so we wrote a program to generate graphs at random until one with the needed properties was found. The first time we ran the program, it asked us for the number of vertices. Because we were unsuccessful with small examples, we typed in 50. Nearly instantly we were rewarded when the computer printed out a few screenfuls of data specifying the graph.

Heartened by this (but not wanting to tangle with such a large graph), we ran the program again asking for a graph with only 20 vertices. Again, we were greeted with near instant success. We started to draw the graph on the blackboard, but it was a nasty random mess (no surprise—our program was designed to examine random graphs one after another). One more run. Shall we be optimistic? Certain this would not work, we ran the program again for graphs on 5 vertices. Success! The graph was actually quite simple and we had a good laugh over how we found our answer.

Using this book

This book is ideal either for self-study or as a text for a semester-long course in computer programming (with an emphasis on mathematics). It is important that undergraduate mathematics majors know how to use the computer effectively. This is a skill that will serve them well whether for applied scientific/engineering/financial work, or as a means for forming and testing conjectures in pure research.

We explain how to use C++ from the ground up, however, some experience in writing programs in any language will help. The reader does not need extensive experience in programming. Nor is a deep mathematical background needed to read this book. Our examples are drawn from standard mathematical topics such as Pythagorean triples [integers (a, b, c) such that $a^2 + b^2 = c^2$] and polynomials.

Whether you are reading this book on your own or in conjunction with a course, the most effective way to learn is to do. **Every chapter ends with a collection of exercises; do them all.** This is important for two reasons. First and foremost, mastery of any skill requires practice, and learning C++ is no exception. Second, additional ideas and subtle points are explored in the exercises. To assist you, we include complete solutions to nearly every exercise in Appendix D.

Organization

We organize our discussion around mathematical themes. For example, Chapters 3 to 5 cover many central ideas in C++ (from how to write procedures to dynamic allocation of arrays), but a single mathematical problem runs throughout, taking us from Euclid to Riemann.

The main body of the book is divided into three parts: *Procedures*, *Objects*, and *Topics*.

Procedures focuses on writing C++ procedures (often called *functions* by our computer science colleagues, but we have a different meaning for that word). *Objects* introduces the object-oriented method of programming. If you want to solve a problem that involves permutations, Chapter 11 shows how C++ can handle these structures nearly as comfortably as it handles integers. *Topics* discusses how to use freely available packages that you can use in your programs, more advanced input/output, visualization, and selected special features of the C++ language.

Four appendices provide (A) an overview of computing systems (including Integrated Development Environments), (B) the use of the *Doxygen* documentation system, (C) a quick reference to the C++ language and supporting libraries, and (D) answers to nearly every exercise.

No pointers! (almost)

The C++ concepts covered in this book are not exhaustive. There are aspects of the language that are relevant only to computer scientists and software engineers. For example, C++ provides a number of exotic casting operators (such as `reinterpret_cast`) that are not of interest to mathematicians; we omit those. Neither multiple inheritance nor pure virtual functions are for us. We do not explain how to make C++ programs work with other languages such as assembly language. We don't create our own namespaces. For these and other C++ topics that are useful to large software engineering projects, please refer to any of several excellent, comprehensive C++ reference books.

One topic that we touch only gently is the use of *pointers*. Mostly, we do not need them. We successfully avoid this confusing (and errorprone) concept through the use of call-by-reference and STL container classes. A mathematician shouldn't be worrying about exotic data structures such as red–black trees; we just want to insert and delete elements in a set and not to worry about how the computer manages the set.

There are, however, a few instances where a rudimentary understanding of pointers is necessary.

- The name of a C++ array is a pointer to its first element. We need to understand this when we pass an array to a procedure and when we dynamically allocate storage for an array.

- Sometimes an object needs to refer to itself (e.g., when a method such as `operator+=` returns the value of the object). In these instances, the `this` pointer is useful.

We do not provide an extensive discussion of string processing (but do cover the basics in Chapter 14). As mathematicians, we are interested in getting data into our programs and results out of them; we are not going to construct word processors.

We omit the exotic and focus on those aspects that make C++ a formidable weapon in the mathematician's arsenal. With your brain and this book, your problem doesn't stand a chance. Enjoy!

Additional resources

The CD-ROM that comes with this book includes the code for all the numbered programs (see the List of Programs on page xiii). The programs are free for you to use under the terms of the GNU General Purpose License. (See the CD-ROM for details.) The disk also includes solutions to some of the lengthier exercises.

Please visit the Web site for this book www.ams.jhu.edu/~ers/cpp4m/ where we maintain a list of errata and submissions for Exercise 1.1.5.

Acknowledgments

Many thanks to my editor Sunil Nair and his helpful staff at Taylor & Francis/CRC Press. They helped with everything from LATEX issues to copy editing to securing permissions.

Promit Roy is an undergraduate computer science major at Johns Hopkins University. He read through the entire manuscript checking for accuracy and compatibility with Microsoft Visual Studio. In addition, he prepared the MS Visual Studio project files for the accompanying CD-ROM. Thank you, Promit!

At various points in this book I recommend that the reader consult a "friendly computer science colleague." I am fortunate to have such a colleague. Thank you, Joanne Houlahan for your help (and patience) with getting me unstuck from computer woes.

I greatly appreciate my department chair, Daniel Naiman, for his support and encouragement for this project.

It gives me great joy to acknowledge the contributions of my father, Ira, and my son Jonah to the front cover. Grandpa took the cover photo and grandson provided the design concept.

Most important, thank you to my wife, Amy, and our children, Jonah, Naomi, Danny, and Rachel, for the world of love and happiness I share with them.

Ed Scheinerman
Baltimore
May 24, 2006

Part I

Procedures

Chapter 1

The Basics

1.1 What is C++?

C++ is a computer programming language.

Computers are electronic information-processing machines. Data and programs in these machines are saved, moved, and transformed in the form of electrical voltages. These electrical voltages can be interpreted as a zeros and ones. The zeros and ones can be aggregated and interpreted as words, numbers, images, sounds, and so on. Long ago, information—be it data or programs—could be entered into the

Figure 1.1:   A PDP-8 computer with front panel switches for entering instructions and data. (Image courtesy of the *Computer Museum* at the University of Stuttgart. Photograph by Klemens Krause.)

computer by manipulating switches on the front of the machine. Today, there are better methods. Computer programming languages convert text into the requisite binary instructions.

C++ is a compiled language. This means that before the program is run, it is first translated into a form that the machine can use directly. The C++ files (and a typical project contains several) are called the *source files*. You create the source files by typing them using a program called a *text editor*.

The translation of the source files into a program proceeds in two phases. First, the individual source files are translated by a program called the *compiler* into so-called *object files*. Next, a second program, called a *linker* (or *loader*) combines the individual object files into an *executable file*, that is, a program you can execute (run).

The precise manner in which is all this is done (source file creation/editing, compiling, linking, and execution) varies among different computing platforms. In Appendix A we discuss how this is done for some common computing platforms (Windows, Macintosh, UNIX).

Ideally, you have already done some programming, say in C, and so you are familiar with the general flow of this process. If not, your best option is to have someone show you how to perform these basic steps. In theory, your computer contains documentation that explains all this. However, such documentation is most useful to someone who already knows what to do, but needs reminders on specifics.

With the help of Appendix A, a friendly neighbor knowledgeable in programming, and documentation (if any) you will get past this first, often frustrating hurdle. Rest assured that the process is simple once you know what it is. The hard part is knowing just which menu to select or what command to type to set the process in motion. For example, on many computers you translate your C++ files into a working program with the single command:

```
g++ *.cc
```

and then run your program by typing this:

```
./a.out
```

If you are using an integrated development environment (also called an IDE) compiling your C++ files may be as simple as clicking on a button that says "build" and then running your program by clicking on a button that says "run".

1.2 Hello C++

To begin we need to write a program; it's time to present our first example. It is traditional to start with a program that does nothing more than type the words Hello, world onto the computer screen. Instead, we opt for a bit of (bad) poetry.

Program 1.1: The classic "Hello, world" program updated for the mathematics world.

```
1   #include <iostream>
2   using namespace std;
3
4   /**
5    * A simple program for demonstrating the basics of a C++ project.
6    *
7    * It does a good job of demonstrating C++ fundamentals, but a
8    * terrible job with the poetry.
9    */
10
11  int main() {
12    cout << "Don't you just feel like a louse";
13    cout << endl;
14
15    cout << "To learn that your \"new\" theorem was proved by Gauss?";
16    cout << endl;
17
18    return 0;
19  }
```

There are naming conventions for C++ program files. The name of the file that contains this code is `poem.cc`. The name has two parts: `poem` is called the base name and `.cc` is called the extension. The base name can be more or less anything you like, but the extension should be `.cc`. Other extensions that are permissible for C++ programs are `.C`, `.cpp`, and `.cxx`. The extension is important because this is what the compiler uses to recognize that your file contains C++ code.

Let's analyze `poem.cc` to understand the purpose of its various parts.

The core of this program is in lines 12–16, so we begin our discussion there. Line 12 contains four important components.

- The `cout` object: This is an object to which we send information that we want written on the computer screen. The word `cout` is an abbreviation of "console output."

- The `<<` operation: This is an operation that acts on the objects immediately to its left and right: in this case, the `cout` object on its left and the character array (described next) on its right.

- The character array enclosed in quotation marks: These are the words `Don't you just feel like a louse`. The quotation marks mark the beginning and end of the character array. Note that the single quote (apostrophe) is a valid character.

- The statement terminator: The semicolon at the end of the line denotes the end of the statement.

As mentioned, the << operation takes two arguments: the cout object on its left and the character array on its right. The effect of this operation is that the character array is typed onto the computer's screen.

The semicolon at the end of the line is mandatory; it marks the end of the statement. We could have broken this statement into two (or more lines), like this,

```
cout
    <<
        "Don't you just feel like a louse"
    ;
```

Line 13 is similar to line 12, but in place of the character array, we have the endl object. The endl stands for *end of line*; it causes the computer's output to move to a new line in the output. Without this statement, the next output would begin immediately after the e in the word louse.

Lines 15 and 16 write the second line of the poem to the computer screen. The only interesting thing here is that we want quotation marks to be part of the output (surrounding the word new). Because quotation marks signal the beginning and end of the character array, we need a way to indicate that the quotes around new are not delimiters, but part of the text message. This is done by preceding the " mark with a backslash, \.

Lines 12 through 16 comprise four separate statements. These could be combined into a single statement without loss of clarity. The various objects to be printed can be fed to cout by repeated use of the << operation.

```
cout << "Don't you just feel like a louse" << endl
     << "To learn that your \"new\" theorem was proved by Gauss?"
     << endl;
```

Here is the output of the program.

```
Don't you just feel like a louse
To learn that your "new" theorem was proved by Gauss?
```

The other parts of the program are important, too. Let's examine them one by one.

- The first line of the program is #include <iostream>. This line is necessary to incorporate the definitions of various input/output objects and operations. The objects cout and endl are not in the main part of C++, and need to be read into the program.

  Ironically, this first line of our first C++ program is not actually C++ code. It is a request to the C++ compiler to read a file called iostream and include the contents of that file in this program. The file iostream is called a *header file*. This header file is part of the C++ system, and so is called a *system header file*. Soon, you will write your own header files, and those are known as *user header files*.

  The compiler knows where to find iostream and before it does anything else with your program, it inserts the full contents of the file iostream at the start of your program.

Because this line is not a C++ statement, no semicolon is needed at its end. Instructions that begin with the # sign are *preprocessing* directives. We examine a few other such commands later.

Any time you write a C++ program that reads or writes data from the computer screen, you need to include the iostream header.

- Line 2 is particularly inscrutable. For now, just know that if you use #include to read in any standard system header files, then you must have the statement using namespace std; as well.

  You may safely skip the rest of this explanation. The objects cout and endl are not core parts of C++, but standard additions to the language. It is possible that a software developer—let's call her Sophie—might want a different version of cout that is somehow different from the standard version.

  Sophie also wants to call her console output object cout; this is possible in C++. Sophie creates (don't worry about how, you are not a software developer) a separate namespace, which she calls, say, sophie. The full name of Sophie's cout is sophie::cout. The full name of the standard cout is std::cout, where std stands for the "standard" namespace.

  In the iostream file, cout and endl are defined to have the full names std::cout and std::endl. It is possible to delete line 2 from the program, but then it would be necessary to replace cout with std::cout (and likewise for endl). However, for our purposes, it is much easier to declare: we are going to be using the standard namespace, so we will use the short forms of the names. That's what the statement using namespace std; does.

- Lines 4–9 are a comment. Comments are important features of computer programs. This comment immediately precedes the main part of the program and explains what the program does.

  Comments in C++ are of two forms: multiple-line comments and single-line comments.

  Multiple-line comments may span one or more lines. The comment begins with the pair of characters /* and ends with */. Everything between these is ignored by the compiler; the comments are meant for a human reader. The comment spanning lines 4 through 9 is of this sort. The single asterisks on lines 5 through 8 are present only to make it clear that this block of text is a comment. Clean formatting like this makes the program more readable. The extra * on line 4 serves a purpose that is revealed later (see Appendix B); for the present, you can think of it as optional.

  Single-line comments begin with a double slash // and continue to the end of the line. They are useful for short comments. For example, to explain line 12 of the program, I could have written this:

```
cout << "Don't you feel like a louse"; // rhymes with Gauss
```

- Line 11 and lines 18–19 surround the main part of the program. As we create more elaborate C++ programs, we separate the program into multiple files, with each file containing various procedures. In each program, there must be exactly one procedure whose name is `main`. We examine procedures in more detail later, but for now here is what you need to know.

  - Procedures take various values as arguments, execute instructions, and return a value. In this case, the procedure named `main` returns an integer value; this is signified by the word `int` before the name of the procedure. The value returned is zero, and this is accomplished at the end of the procedure with the statement `return 0;` which returns the value 0.

  - The `main` procedure does not take any arguments; this is signified by the empty pair of parentheses following the word `main`. There is an alternative way to create a `main` procedure that does take arguments; we explore that later (see Chapter 14).

  - Thus the beginning of line 11, `int main()`, says that this is a procedure that takes no arguments and returns an integer-valued answer.

    After `int main()`, the statements of the procedure are enclosed in curly braces. The opening curly brace is on line 11 and its matching closing brace appears alone on line 19.

    Notice that lines 12–18 are indented from the left margin. This indenting helps enormously with readability.

This completes our analysis of Program 1.1.

1.3 Exercises

1.1 The following comment in a C++ program would cause the compiler to issue an error message.

```
/*
 * In C++ comments are enclosed between /* and */
 */
```

What's wrong?

1.2 The following program is typed into the computer and saved as `bad.cc`.

```
#include <iostream>

int main() {
  cout << "What is wrong with this program?"
  cout << endl;
```

```
    return 0;
}
```

When this program is compiled, the following output is produced.

```
bad.cc: In function 'int main()':
bad.cc:4: error: 'cout' undeclared (first use this function)
bad.cc:4: error: (Each undeclared identifier is reported only
    once for each function it appears in.)
bad.cc:5: error: parse error before '<<' token
```

There are two errors in the program. Use the output to figure out what those errors are.

1.3 A programmer included the following line in a program.

```
cout << "My favorite web site is http://www.google.com";
```

Do the slashes // start a comment?

1.4 To include a quotation symbol in output, we use the sequence \". Write a program to see what the sequences \n, \t, and \\ do when part of a character sequence.

For example, your program should contain a line such as this:

```
cout << "What does \n do?";
```

1.5 Write your own bad math poem. There's great potential for a terrible rhyme with *Euler* including *under the broiler* and *stop or you're goin' to spoil her* and *how long does an egg boil fer?*

Send your worst to the author at ers@jhu.edu and we'll post the best/worst on this book's Web site www.ams.jhu.edu/~ers/cpp4m. Use math poem as the subject line of your message.

Chapter 2

Numbers

Numbers are the building blocks of mathematics and, of course, computers are extremely adept with numbers. There are several ways in which numbers are handled in C++. In this chapter, we explore the different representations of numbers and how to convert between representations. We also catalogue the various operations we can perform with numbers.

Numbers in C++ are divided into two broad categories: integers and reals. Each of these categories is refined further. Of course, to a mathematician every integer is a real number, so these categories may seem spurious. The reasons for having many different ways to represent numbers are efficiency and accuracy; if the quantities with which we are computing are known to be integral, then using an integer representation is not subject to roundoff error and the computations are faster.

2.1   The integer types

Every variable in a C++ program must be given a *type*. The type of a variable is a specification of the kind of data the variable can store. The most basic type is `int`. An `int` represents an integer quantity. Consider the following program.

Program 2.1: Introducing the `int` type.

```
 1  #include <iostream>
 2  using namespace std;
 3
 4  int main() {
 5    int x;
 6    int y;
 7
 8    x = 3;
 9    y = 4;
10
11    cout << x+y << endl;
12
13    return 0;
14  }
```

In this program, two variables, x and y, are *declared* to be of type int. These declarations occur on lines 5 and 6. C++ requires that we specify the type of every variable, and the declaration statement is the manner by which this is accomplished. Variables may be declared anywhere in the program so long as they are declared before they are used.

Subsequently (lines 8 and 9) the variables are *assigned* values. In this case x is assigned the value 3 and y the value 4. The equal sign = is an operation (called *assignment*) that stores the value of the expression to its right into the variable on its left.

It is possible to combine declarations on a single line; in place of lines 5 and 6, we could have this:

```
int x,y;
```

It is possible to combine declaration with assignment. Lines 5–9 can be replaced by these two:

```
int x = 3;
int y = 4;
```

It's easy to see what the rest of Program 2.1 does. In line 11, the contents of x and y are added, and the result is written on the computer's screen.

Variables of type int can store a finite range of integer values. The size of that range depends on your specific computer. On many computers an int is held in 4 bytes of memory (one byte is 8 bits, so this is 32 bits). With 32 bits one can store 2^{32} different values. In this case, the 2^{32} values are from -2^{31} to $2^{31} - 1$ inclusive.

There is an easy way to learn the size of an int on your computer in C++ with the sizeof operator. Evaluating sizeof(int) gives the number of bytes used to store variables of type int. If an int on your computer is b bytes long, then the minimum and maximum values an int may hold are -2^{b-1} and $2^{b-1} - 1$, respectively. See also Program 2.3 later in this chapter (page 15).

There are a few variations of the fundamental int type.

- A short is an integer type that is either the same as int, or else a smaller size than int. That is, sizeof(short) cannot exceed sizeof(int).

- A long is an integer type that is either the same size as int, or else a larger size than int. In other words, sizeof(long) is at least sizeof(int).

- Some compilers provide a long long type. This is an integer type that is at least the size of a long. That is, sizeof(long long) cannot be less than sizeof(long).

  Other compilers may provide an equivalent alternative. For example, in Microsoft's Visual Studio, use the type __int64 (the name begins with two underscore characters) in place of long long.

In summary,

```
sizeof(short) ≤ sizeof(int) ≤ sizeof(long) ≤ sizeof(long long).
```

Finally, each of the integer types can be modified for holding only nonnegative integers using the `unsigned` keyword. For example, a variable can be declared

```
unsigned int x;
```

In this case, x may take on only nonnegative values. If `sizeof(unsigned int)` is 4, then x can be any value from 0 to $2^{32} - 1$ inclusive.

Because integer variables are of finite capacity, if a calculation exceeds the limits incorrect results emerge. Consider the following program that we run on a computer on which `sizeof(int)` equals 4. Please note the operation ∗ in the code which is used for multiplication.

Program 2.2: A program to illustrate integer overflow.

```
1   #include <iostream>
2   using namespace std;
3
4   /**
5    * A  program  to illustrate  what  happens  when  large integers
6    * are multiplied together.
7    */
8
9   int main() {
10      int million = 1000000;
11      int trillion = million * million;
12
13      cout << "According to this computer, " << million << " squared is "
14           << trillion << "." << endl;
15  }
```

Line 10 defines a variable `million` equal to 10^6. Because 2^{31} is roughly 2 billion, there is no problem assigning x this value. However, when we define (line 11) y to be x∗x, the result of squaring x is greater than 2^{31}. What happens? Let's run the program and see.

```
According to this computer, 1000000 squared is -727379968.
```

The result is obviously wrong! Unfortunately, such errors are rarely so easy to detect. Here is some advice on how to handle this situation. First, know the `sizeof` of the various integer types on your machine (see Program 2.3 later in this chapter). Second, use `long` or `long long` routinely unless for reasons of speed or space you need to use a smaller size. Finally, if the `long` or `long long` type is not big enough for your calculations (and you need to keep the full precision) you can learn about arbitrary precision arithmetic packages (see Section 13.1).

2.2   The real number types

C++ can handle more than integers. Real numbers are represented using floating point approximations. There are two principal types of floating point types: `float` and `double`. Both hold real values of limited precision. Variables of type `float` use less memory and have less precision than those of type `double`.

Some computers may have a `long double` type that has even greater precision than `double`.

Unless you have special needs (for increased speed or decreased memory), use the `double` type for all your real number calculations.

One may be tempted to use `double` for all numbers. For example, if we replace `int` by `double` in Program 2.2, we have the following output.

```
According to this computer, 1e+06 squared is 1e+12.
```

Notice that the value of x is reported as `1e+06`; this is simply scientific notation for 1×10^6. The value of y is `1e+12` and this is correct: 10^{12}.

So why bother with the integer types at all? First, if the problem you are solving deals with integers, using `int` or `long` is more efficient than `float` or `double`. Second, the integer types are not subject to roundoff errors. Finally, there are certain instances in which an integer type is required by C++ (e.g., when accessing elements of an array).

2.3   The `bool` and `char` types

There are other basic data types available in C++. Two that we encounter frequently in C++ are designed for handling Boolean and character data.

The data type `bool` is used to represent the logical values TRUE and FALSE. In C++, these are represented as integers: 1 for TRUE and 0 for FALSE. Indeed, `bool` is considered to be an integer type.

Boolean values emerge as the result of comparison operations. In C++, to see if two variables hold equal values we use the == operator. The result of x == y is either 1 (TRUE) in the case where x and y hold equal values and 0 (FALSE) otherwise. If your program contains the statement

```
cout << (x==y) << endl;
```

either a 0 or a 1 is typed on the screen.

Individual characters have the type `char`. One can create variables to hold single characters by declaring them to be of type `char` as in this example.

```
char x;
x = 'A';
cout << x << endl;
```

This code causes the letter A to appear on the computer's screen.

Notice that a single character is enclosed in single quotes. Arrays of characters are enclosed in double quotes. It is incorrect to write char x = "A"; because x is of type char whereas "A" is an array of elements of type char.

C++ has two principal ways of handling textual data: arrays of characters and objects of type string. We discuss these later (see Chapter 14). However, programs whose purpose is to solve mathematical problems rarely have much need for extensive processing of textual data.

2.4 Checking the size and capacity of the different types

Earlier we mentioned the sizeof operator that is used to determine the number of bytes a given data type occupies in memory. The following program reports on the sizes of the various basic data types we have discussed.

Program 2.3: A program to show the sizes of the fundamental data types.

```
1   #include <iostream>
2   using namespace std;
3
4   /**
5    * Report on the size of various C++ data types.
6    *
7    * This program may give different results when run on different
8    * computers depending on how each of the fundamental data types is
9    * defined on those platforms.
10   */
11
12  int main() {
13      // Integer types:
14      cout << "The size of short is " << sizeof(short)
15          << " bytes" << endl;
16      cout << "The size of int is " << sizeof(int)
17          << " bytes" << endl;
18      cout << "The size of long is " << sizeof(long)
19          << " bytes" << endl;
20
21      // long long might not exist on all computers
22      cout << "The size of long long is " << sizeof(long long)
23          << " bytes" << endl;
24
25      // Character and boolean types:
26      cout << "The size of char is " << sizeof(char) << " bytes" << endl;
27      cout << "The size of bool is " << sizeof(bool) << " bytes" << endl;
```

```
28
29    // Floating point types
30    cout << "The size of float is " << sizeof(float)
31         << " bytes" << endl;
32    cout << "The size of double is " << sizeof(double)
33         << " bytes" << endl;
34
35    // long double might not exist on all computers
36    cout << "The size of long double is " << sizeof(long double)
37         << " bytes" << endl;
38
39    return 0;
40  }
```

The output of this program depends on the computer on which it is run. The following show the results of running this code on two different machines.

```
The size of short is 2 bytes
The size of int is 4 bytes
The size of long is 4 bytes
The size of long long is 8 bytes
The size of char is 1 bytes
The size of bool is 4 bytes
The size of float is 4 bytes
The size of double is 8 bytes
The size of long double is 8 bytes
```

```
The size of short is 2 bytes
The size of int is 4 bytes
The size of long is 4 bytes
The size of long long is 8 bytes
The size of char is 1 bytes
The size of bool is 1 bytes
The size of float is 4 bytes
The size of double is 8 bytes
The size of long double is 12 bytes
```

Notice that an int is 4 bytes on both machines and so, in both cases, int variables can hold values from -2^{31} to $2^{31} - 1$.

Note also that double variables on both machines are 8 bytes, but this does not tell us the range of values that double values can take.

For the float and double data types, there are three quantities of interest: the largest value, the smallest positive value, and the difference between 1 and the smallest value greater than 1. This latter value is known as the epsilon value of the data type. Knowing that a double is 8 bytes does not immediately reveal these three quantities.

C++ provides header files with this information. The header file climits gives the minimum and maximum value of various integer types. It defines symbols such as INT_MIN and INT_MAX that are equal to the smallest and largest value an int may hold. Similarly, the header file cfloat gives the minimum positive, maximum, and epsilon values for float, double, and long double (if available) data types.

Running the following program reports all this information for the data types we have discussed.

Program 2.4: A program to show the maximum and minimum values of various data types.

```
1   #include <iostream>
2   #include <climits>    // max & min size of integer types
3   #include <cfloat>     // max & min size of real types
4   using namespace std;
5
6   /**
7    * Print out the extreme values of various integer types.
8    */
9
10  int main() {
11      cout << "The maximum size of a short is " << SHRT_MAX << endl;
12      cout << "The minimum size of a short is " << SHRT_MIN << endl;
13
14      cout << "The maximum size of an int is " << INT_MAX << endl;
15      cout << "The minimum size of an int is " << INT_MIN << endl;
16
17      cout << "The maximum size of a long is " << LONG_MAX << endl;
18      cout << "The minimum size of a long is " << LONG_MIN << endl;
19
20      // long long values might not exist on some computers
21      cout << "The maximum size of a long long is " << LLONG_MAX << endl;
22      cout << "The minimum size of a long long is " << LLONG_MIN << endl;
23
24
25      cout << "The minimum positive value of a float is "
26          << FLT_MIN << endl;
27      cout << "The minimum epsilon value of a float is "
28          << FLT_EPSILON << endl;
29      cout << "The maximum value of a float is "
30          << FLT_MAX << endl;
31
32      cout << "The minimum positive value of a double is "
33          << DBL_MIN<< endl;
34      cout << "The minimum epsilon value of a double is "
35          << DBL_EPSILON << endl;
36      cout << "The maximum value of a double is "
37          << DBL_MAX << endl;
38
39      // long double might not be defined on some systems
40      cout << "The minimum positive value of a long double is "
41          << LDBL_MIN<< endl;
42      cout << "The minimum epsilon value of a long double is "
43          << LDBL_EPSILON << endl;
44      cout << "The maximum value of a long double is "
45          << LDBL_MAX << endl;
46
47      return 0;
48  }
```

Here is the output of this program on a particular computer. The result on other computers may be different.

```
The maximum size of a short is 32767
The minimum size of a short is -32768
The maximum size of an int is 2147483647
The minimum size of an int is -2147483648
The maximum size of a long is 2147483647
The minimum size of a long is -2147483648
The maximum size of a long long is 9223372036854775807
The minimum size of a long long is -9223372036854775808
The minimum positive value of a float is 1.17549e-38
The minimum epsilon value of a float is 1.19209e-07
The maximum value of a float is 3.40282e+38
The minimum positive value of a double is 2.22507e-308
The minimum epsilon value of a double is 2.22045e-16
The maximum value of a double is 1.79769e+308
The minimum positive value of a long double is 2.22507e-308
The minimum epsilon value of a long double is 2.22045e-16
The maximum value of a long double is 1.79769e+308
```

2.5   Standard operations

C++ provides the familiar arithmetic operations. The expressions x+y, x-y, x*y, and x/y are the usual sum, difference, product, and quotient of x and y.

In these expressions, x and y may be any of the numeric types we discussed. However, mixing types can sometimes cause problems and confusion. If x and y are of the same type, then the result of the operation is also that type. For example, consider the following code.

```
int     numerator   = 13;
int     denominator =  5;
double quotient;

quotient = numerator/denominator;

cout << quotient << endl;
```

We might expect this code to print the value 2.6 on the computer's screen. The surprise is that the computer prints 2. To understand why, we have to remember that when two int values are divided, the result is an int. In this case, C++ divides the integers 13 and 5 to give the result 2 (the fractional part is lost) and then assigns that value to quotient. Because quotient is of type double there is a silent, behind-the-scenes conversion of the int quantity into a double quantity.

We can coerce C++ to give us the answer 2.6 by converting 13 or 5 (or both) into double variables. We replace quotient = numerator/denominator with this:

```
quotient = double(numerator) / double(denominator);
```

The expression `double(numerator)` converts the integer value in `numerator` into a `double` quantity. This conversion process is known as *casting*. Note that the variables `numerator` and `denominator` are unaffected by this; they remain type `int` and their type cannot be changed. What `double(numerator)` does is create a temporary `double` number to be used in the division.

We do not have to cast both `numerator` and `denominator` to type `double`. Once we cast one of them to `double`, we have an expression involving a `double` and an `int`. In this case, C++ automatically converts the other value as well.

Algebraic expressions may contain explicit numbers in addition to variables, such as `x = 2*x+3;`. This expression means that we take the value of `x`, multiply that by 2, then add 3, and then save the result of the calculation back into `x`. So if the initial value of `x` is 10, after this statement `x` would hold the value 23.

Numbers written without a decimal point are assumed to be integers, and numbers with decimal points are assumed to be floating point values. Consider the following code.

```
cout << 13/5    << endl;
cout << 13./5   << endl;
cout << 13./5.  << endl;
cout << 13/5.   << endl;
```

The first of these prints 2 and the other three print 2.6.

For integer variables, the percent sign denotes the mod operation. The expression `237 % 100` evaluates to `37`.

Unfortunately, the `%` operation is not exactly the same as the mathematician's mod. For us, when a and b are integers with $b > 0$, we have that a mod b is the remainder r in the division of a by b where r must satisfy $0 \le r < b$. Even if a is negative, the result of a mod b is nonnegative. However, in C++, if `a` is negative and `b` is positive, `a%b` is negative. Furthermore, C++ allows `b` to be negative, or even zero! Use the following program to explore the result of `a%b` various choices of a and b.

Program 2.5: A program to explore C++'s mod operation.

```
1  #include <iostream>
2  using namespace std;
3
4  /**
5   * A program to investigate the behavior of the mod (%) operation.
6   */
7
8  int main() {
9     int a,b;
10
11    cout << "Enter the first number --> ";
12    cin >> a;
13
14    cout << "Enter the second number --> ";
15    cin >> b;
```

```
16
17      cout << a << " % " << b << " = " << a%b << endl;
18
19      return 0;
20  }
```

Line 11 types the words `Enter the first number -->` onto the screen. Notice that we did not include an `endl`, so the computer does not move the cursor to a new line.

Line 12 introduces the `cin` object. The statement `cin >> a` means read a value from the computer's keyboard and save the result in the variable a.

C++ provides an interesting combination of arithmetic operations and assignment. The statement `x += 4;` is an abbreviation for `x = x+4;`—increase the value held in x by 4. All of the usual arithmetic operations can be used in this combined syntax.

Statement	Meaning
x += y;	x = x+y;
x -= y;	x = x-y;
x *= y;	x = x*y;
x /= y;	x = x/y;
x %= y;	x = x%y;

In computer programs, one frequently wishes to increase or decrease an integer variable by 1; this happens when we want to consider successive elements of an array or when counting. A special syntax can be used in place of `x += 1` and `x -= 1`. The expressions `x++` and `++x` both increase x by one, and `x--` and `--x` both decrease x by one. We call `++` and `--` the *increment* and *decrement* operators, respectively.

The expressions `x++` and `++x` are not exactly the same. We explain the difference here because you might need to understand someone else's program that relies on this difference. You should not take advantage of the difference between `++x` and `x++`. Doing so makes your code more confusing, and the most likely victim of that confusion will be you. So read the next portion if you are curious, or you can safely skip ahead (past the material enclosed between the double lines) to the discussion on exponentiation.

With that admonishment firmly in place, here is the difference between `x++` and `++x`. Although both of these expressions increase the *value* of x by one, the *result* of this expression is, in the first case the old value of x and in the latter case, the new value of x. For example, consider this code.

```
int x,y;
x = 10;
y = x++;
cout << y << endl;

int a,b;
a = 10;
b = ++a;
cout << b << endl;
```

The statement y = x++; assigns the old value of x to y, so cout << y << endl; types 10 on the console. On the other hand, the statement b = ++a; assigns the new value of a to b, so the statement cout << b << endl; types 11 on the console.

The two versions of ++ are called *preincrement* (for ++x) and *postincrement* (for x++). Similarly, the two versions of -- are called *predecrement* and *postdecrement*. And, as you might suspect, the name of the language is inspired by the ++ operator as C++ is an extension of the C language.

Do yourself a big favor. Do not write code that depends on this difference! Instead, write your program this way.

```
int x,y;
x = 10;
y = x;
x++;
cout << y << endl;

int a,b;
a = 10;
a++;
b = a;
cout << b << endl;
```

Missing thus far from our discussion is exponentiation. There is no exponentiation operator in C++. Rather, one may use the pow procedure. To do so, one needs to load the cmath header file using the following preprocessor directive.

```
#include <cmath>
```

With this in place, you can compute a^b with the expression pow(a,b).

The cmath header defines a number of standard mathematical functions and constants. For example, exp is the usual e^x function and M_PI gives the value[1] of π. A classic problem (that one ought to solve without a computational aid) is to determine which is greater: e^π or π^e. The following program settles the issue.

Program 2.6: A program to calculate e^π and π^e.

```
 1  #include <iostream>
 2  #include <cmath>
 3  using namespace std;
 4
 5  /**
 6   * Which is larger, pi to the e or e to the pi? We calculate both to
 7   * find out.
 8   */
 9
10  int main() {
11    double e = exp(1.);
12    double pi = M_PI;
13
```

---

[1]The symbol M_PI is not completely standardized. If your compiler complains that M_PI is undefined, add the line const double M_PI = 3.14159265358979; to the beginning of your program.

```
14    cout << "e to the pi is " << exp(pi) << endl;
15    cout << "pi to the e is " << pow(pi,e) << endl;
16  }
```

Here is the output of the program.

```
e to the pi is 23.1407
pi to the e is 22.4592
```

2.6   Comparisons and Boolean operations

C++ provides comparison operators for testing numbers for equality and order. Each of these operators results in a `bool`. In this chart, x and y are variables, and x and y are their respective values.

Expression	Result
x == y	true iff $x = y$
x != y	true iff $x \neq y$
x < y	true iff $x < y$
x <= y	true iff $x \leq y$
x > y	true iff $x > y$
x >= y	true iff $x \geq y$

Boolean values can be combined with the standard logical operations. In the following chart x and y are of type `bool`, and x and y are their values.

Expression	Description
!x	$\neg x$, i.e., not x
x && y	$x \vee y$, i.e., logical and
x || y	$x \wedge y$, i.e., logical or
x ^ y	$x \veebar y$, i.e., exclusive or

These logical connectives are often used in conjunction with comparison operators in an expression such as

```
( (x == y) && (y <= z) )
```

which evaluates to true if and only if $x = y \leq z$. Although the following is syntactically correct,

```
x == y <= z
```

it is not equivalent to $x = y \leq z$. Here one of x == y or y <= z is evaluated first and then that results in a `bool` value. Unfortunately, `bool` values are convertible to integers, so the next comparison can take place. The expression x == y <= z is poor programming because it is not clear to the reader which comparison is evaluated first. (C++ has specific rules for this, but you should not rely on them.) Furthermore, the use of a `bool` as a numeric type is a sneaky trick, and sneaky tricks (regardless of how clever they may be) make code confusing and hard to understand.

2.7 Complex numbers

C++ handles complex numbers nearly as easily as real numbers. Although the real types such as `double` are part of the C++ core, the complex type requires the inclusion of a header file and a slightly different syntax for variable declaration.

First, because complex numbers are not part of the C++ core, we need to issue the preprocessor directive `#include <complex>`.

Second, we have some choices on what sort of complex numbers we want. For example, we may want complex numbers in which the real and imaginary parts are of type `float`, or for greater accuracy, both of type `double`. On the other hand, we may be interested in working with Gaussian integers, in which case we want the real and imaginary parts to be of type `int` or `long`. Fortunately, all these choices are available to us. Here is how we declare variables:

```
complex<double> z;   // z's real and imaginary parts are of type double
complex<long> w;     // w is a Gaussian integer
```

Next, we need to be able to assign complex values to these newly declared complex variables. We do that as follows.

```
z = complex<double> (4., -0.5);   // z is set to 4-0.5i
w = complex<long> (6,2);          // w is set to 6+2i
```

Recall that in order to convert (cast) an `int` into a `double`, we used an expression of the form `double(x)` (where x was a variable of type `int`). The same idea applies here. To create a `complex<double>` we use the type name, but this time with two arguments, as in `complex<double>(4.,-0.5)`.

Typing `complex<double>` repeatedly can be annoying. Fortunately, C++ enables us to create an abbreviation. The statement

```
typedef complex<double> C;
```

defines the symbol `C` as a substitute for `complex<double>`. Once we have created this abbreviation, to declare a variable to be of type `complex<double>` we only need to type `C z;` and z has the desired type. To assign z a value, we can simply write `z = C(6,-3);` and now z has the value $6 - 3i$.

Complex variables are printed on the screen as ordered pairs. If z has value $6 - 3i$, then `cout << z << endl;` prints `(6,-3)` on the computer's screen.

The following program illustrates these ideas. Note the unusual way in which we extract the real and imaginary parts of a complex variable (lines 26–27). It is too soon for us to explain this unusual syntax, but all is revealed in time.

Program 2.7: A program to demonstrate C++'s ability to handle complex numbers.

```
1  #include <iostream>
2  #include <complex>
3  using namespace std;
4
```

```
5    /**
6     * A program to illustrate the use of complex numbers.
7     */
8
9    typedef complex<double> C;
10
11   int main() {
12     C x(3,4);        // define x = 3+4i
13     C z;             // declare z to be complex
14     z = C(2,7);      // assign z = 2+7i
15     C i(0,1);        // define i = sqrt(-1)
16
17     cout << "z = " << z << endl;
18     cout << "x = " << x << endl;
19     cout << "z+x = " << z+x << endl;
20     cout << "z*x = " << z*x << endl;
21     cout << "z/x = " << z/x << endl;
22
23     z = 5. - 4.*i;
24     cout << "Now z = " << z << endl;
25
26     cout << "The real part of z is " << z.real()
27          << " and the imaginary part is " << z.imag() << endl;
28
29     return 0;
30   }
```

The output of this program follows.

```
z = (2,7)
x = (3,4)
z+x = (5,11)
z*x = (-22,29)
z/x = (1.36,0.52)
Now z = (5,-4)
The real part of z is 5 and the imaginary part is -4
```

If you plan to work with complex numbers, the typedef abbreviation is convenient, as is defining the variable i to be $\sqrt{-1}$. A convenient way to do this is to create your own header file to #include at the start of your program. The name of a header file that you create is some base name of your choosing plus a .h extension. For example, you could name the header file my_complex.h, but my preference is complexx.h with the extra x for extension.

Here is the complexx.h header file.

Program 2.8: A header file, complexx.h, containing convenient definitions for working with complex numbers.

```
1    /**
2     * @file complexx.h
3     * @brief A header file that adds convenient extensions for working
4     * with complex numbers.
5     */
6
```

```
7   #ifndef COMPLEXX_H
8   #define COMPLEXX_H
9
10  #include <complex>
11  using namespace std;
12
13  /**
14   * We define C to be an abbreviation for complex<double>.
15   */
16  typedef complex<double> C;
17
18  /**
19   * We define i to be a constant equal to sqrt(-1), i.e., C(0.,1.).
20   */
21  const C i = C(0., 1.);
22
23  #endif
```

When we want to use this header file, we employ a slightly different version of the `#include` directive. We write this at the start of our program:

```
#include "complexx.h"
```

We can now use `C` and `i` as desired.

There's a fair amount going on in the `complexx.h` file, but the heart of the matter can be found on lines 16 and 21 where we define `C` to be an abbreviation for `complex<double>` and `i` to be a constant equal to $\sqrt{-1}$.

Line 21 contains a new C++ keyword, `const`. By declaring `i` to be of type `const C` we are saying two things. First, `i` represents a quantity of type `C` (which, in turn, means `complex<double>`). Second, `i` cannot be changed later in our program.

Because `i` is declared outside any procedure (e.g., it is not enclosed between the curly braces of `main()`), it is a global constant. The use of global constants is good. If your work makes extensive use of the golden mean, $\phi = \frac{1}{2}(1 + \sqrt{5})$, you could create a header file containing the following line.

```
const double phi = (1. + sqrt(5.))/2.;
```

Side comment: It is possible to create global *variables*, but this is a bad practice that should be avoided if at all possible. The problem is that one part of your program might modify a global variable causing unexpected behavior later in another part of your program. A common source of confusion and bugs in computer programs are the side effects procedures might have. We try to rein these in as much as possible. Global variables are antithetical to this effort.

There are a number of other parts of `complexx.h` that we need to explain. Let's work through them one at a time.

- Lines 1–5 are a comment. Without going into detail, these comments give a description of what this file does. The `@file` and `@brief` are instructions for a program called Doxygen that we describe later. Suffice it to say that by including these markers here, the Doxygen program can use comments like this to create beautiful Web pages that you can use to look up what your program does. (The extra ∗ on line 1 is a signal to Doxygen that this is a comment it should read.)

- Lines 7, 8, and 23 are a safeguard against accidental double inclusion. It is easy to #include header files more than once. How can this happen? Some day you might create a header file for handling Möbius transformations, that is, functions of a complex variable of the form $f(z) = (az + b)/(cz + d)$. In your header file mobius.h you need to define various quantities as complex, so you start with the directive #include "complexx.h" to be sure that C and i are defined before you get started.

  Now in the main program you might write this:

  ```
  #include "complexx.h"
  #include "mobius.h"
  ```

  Without the double-inclusion safeguards, this would cause the two statements typedef complex<double> C; and const C i = C(0.,1.); to be inserted twice into your program. The compiler would then complain about this and refuse to compile your program.

  There are two solutions to this problem. A bad solution is to require you to remember which of your various header files already includes which other and make the programmer (you) responsible for avoiding double inclusion.

  The better solution is to build in a mechanism in the header file that prevents double inclusion. Here is how this mechanism works.

  Line 7 begins with the directive #ifndef. This is not a C++ statement, but rather an instruction to the preprocessor; it stands for "if not defined." If what is not defined? If the symbol COMPLEXX_H is not defined, then we should do what follows up to the matching #endif on line 24 (at the end of the file).

  If the symbol COMPLEXX_H is not yet defined, we include everything in the file. But if the symbol COMPLEXX_H is already defined, we skip everything through to the #endif.

  Suppose COMPLEXX_H is not defined (and it won't be when we first set out), and we next come to line 8 which reads #define COMPLEXX_H. This line defines the symbol COMPLEXX_H, although it does not specify any particular value for the symbol. (We don't care whether COMPLEXX_H has a value; we just want to know whether it is defined.)

  Suppose we try to #include "complexx.h" a second time. On the first pass through complex.h, we executed the directive #define COMPLEXX_H. Thus, on this second pass, when we reencounter #ifndef COMPLEXX_H the preprocessor notes that COMPLEXX_H is defined and then skips everything in the file until the matching #endif.

- The rest of the file is easier to understand. We need to include the C++ header file complex in order to use complex<double> later. We need the statement using namespace std; so the definitions in <complex> are available.

The comments before lines 16 and 21 document how C and i are defined. These comments can be processed by Doxygen.

2.8 Naming variables

C++ allows you to name your variables more or less whatever you like. The name should begin with one of the 26 letters (lower or upper case). After that, it may contain letters, digits, or the underscore symbol _.

Pick short, easily remembered names for your variables. For example, if your variables are named center_x, center_y, and radius, it is instantly clear to anyone (especially yourself) what these variables hold.

There are some obvious restrictions on your choice of names. You cannot use a C++ keyword as a variable name. So you cannot name your variables if or for or long; these already have meanings in C++. It's not worth memorizing all of C++'s keywords. If you accidentally try to name a variable with one of C++'s more obscure keywords, the compiler will complain. Unfortunately, it won't complain in a way that you like. For example, imagine you tried to declare a long variable to be named volatile (perhaps you are working with a chemist trying to model some nasty substance) like this:

```
long volatile;
```

It turns out the volatile is a C++ keyword (one that you do not need to know about for mathematical work). What you would like is an error message from the compiler that this is an illegal variable name. Here is what the compiler on my computer says about this line:

```
warning: declaration does not declare anything
```

Not particularly helpful, but at least the compiler did complain about the offending line.

Try to develop a consistent style of variable names. I prefer to name variables beginning with lowercase letters. (Later, when we create our own types, I name these beginning with uppercase letters.) Multiword names either can use the underscore as a space (you cannot have a space or hyphen in a variable name) or use capital letters to show the start of each word. Some examples:

```
left_end_point
rightEndPoint
geometric_mean
upperBound
```

Some people like to use all capitals for constants, for example, GOLDEN_MEAN.

Variables may be declared anywhere you like as long as they are declared before they are used.

2.9   Exercises

2.1 Consider the following program.

```
#include <iostream>
using namespace std;
int main() {
   float x;
   int y;
   double z;
   x = 3.5;
   y = x;
   z = y;
   cout << z << endl;
   return 0;
}
```

What is printed to the screen (and why)?

2.2 What is the output of the following program and which (if any) of the lines gives a result equal to $\frac{40}{3}$?

```
#include <iostream>
using namespace std;

int main() {
   cout << (4/3)*10 << endl;
   cout << 4*(10/3) << endl;
   cout << (4*10)/3 << endl;
   cout << (4/3)*10. << endl;
   cout << (4./3)*10 << endl;
   return 0;
}
```

2.3 What is the output of the following program? How does C++'s % operator differ from the mathematical mod operation?

```
#include <iostream>
using namespace std;

int main() {
   cout << (-5) % 3 << endl;
   cout <<   5 % 3 << endl;

   cout << (-5) % (-3) << endl;
   cout <<   5 % (-3) << endl;
   return 0;
}
```

2.4 Explain why the expression '3'==3 is legal C++ and what it means. What value does this expression have?

2.5 Consider this program.

```
#include <iostream>
using namespace std;
int main() {
  int a, b;
  a = 5;
  b = 10;
  cout << (a==b) << endl;
  cout << (a=b) << endl;
  cout << (a==b) << endl;
  return 0;
}
```

This program gives the following output.

```
0
10
1
```

Explain.

2.6 Write a program to explore division by zero. Include the cases $\frac{0}{0}$ and $\frac{1}{0}$. Consider the variants in which the numerator and denominator are `float` or `int` (or one of each).

2.7 What is the output of this program?

```
#include <iostream>
using namespace std;
int main() {
  int a = 10;
  a += a;
  a *= a;
  cout << a << endl;
  return 0;
}
```

2.8 All of the following are illegal names for variables in a C++ program. Explain what is wrong with each.

(a) `2nd_coord`

(b) `y-val`

(c) `double`

(d) `x:y_ratio`

(e) `purple&orange`

(f) `1e-2`

Chapter 3

Greatest Common Divisor

3.1 The problem

For integers a and b, the *greatest common divisor* of a and b is the largest integer d such that d is a factor of both a and b. The greatest common divisor is denoted $\gcd(a, b)$. We say a and b are relatively prime provided $\gcd(a, b) = 1$. In this chapter we develop many C++ concepts by studying a particular problem involving the gcd of integers.

Here is the classic problem: Let n be a positive integer. Two integers, a and b, are selected independently and uniformly from the set $\{1, 2, \ldots, n\}$. Let p_n be the probability that a and b are relatively prime. Does $\lim_{n \to \infty} p_n$ exist and, if so, what is its value?

The computer cannot solve this problem for us, but it can help us to formulate a conjecture. We try a number of approaches including exhaustive enumeration, generating pairs of numbers at random and recording the results, and the use of Euler's totient.

The first order of business, however, is to develop a procedure to compute the greatest common divisor of two integers.

3.2 A first approach

In this section we develop a correct, but highly inefficient, procedure for calculating the greatest common divisor of two integers. Our goal is to introduce a number of C++ concepts as well as to create the gcd procedure. This procedure takes two integers and returns their greatest common divisor. Later, we replace this inefficient procedure with a much more efficient method.

Before we begin, however, we need to address a bit of terminology. Mathematicians and computer programmers use the word *function* differently. A mathematician's function assigns to each value x a unique value $y = f(x)$. Suppose we calculate, say $f(8)$ and the result is 17. Then if we calculate $f(8)$ a few minutes later, we are guaranteed that the result is still 17.

By contrast, for a C++ programmer, it is natural that a function might return different values (even with the identical arguments) at different times! This is because C++ functions can access data beyond their arguments; for example, there is a C++ function that reports the current time. Clearly, the value of this function changes from one minute to the next.

As a mathematician, my bias is, of course, for the mathematical use of the word. Therefore, in this book, we use a different word for C++ functions; we call them *procedures*. This nomenclature ought not upset our computer science colleagues, but when you read other books or documentation about C++ procedures, be aware that they are likely to be called functions. (Some books may refer to *methods* and we introduce that terminology later.)

With issues of nomenclature behind us, we now develop a procedure to compute greatest common divisors.

We need to name the procedure. We could name it `greatest_common_divisor`, but there is no loss of clarity in simply naming it `gcd`. We want `gcd` to accept two arguments (inputs) of type `long` and return an answer that is also of type `long`.

The code for the `gcd` procedure is written in two files named `gcd.h` and `gcd.cc`. The header file, `gcd.h`, is used to describe the procedure (both in C++ and in English comments). The file `gcd.cc` contains the actual instructions for calculating the greatest common divisor.

This organization is similar to declaring versus assignment variables. The declaration announces the variable's type and the assignment gives the variable a value. Likewise, the header file announces the type of the procedure (two `long` inputs and return a `long`) and the `.cc` file gives the actual instructions to be carried out.

The file `gcd.cc` does not have a `main()` procedure; the `main()` is found in another file. The latter file includes the directive `#include "gcd.h"`.

The first `gcd` algorithm we present is terribly inefficient. When we develop a better algorithm, we replace the file `gcd.cc`, but the file `gcd.h` does not change. The file `gcd.h` looks like this.

Program 3.1: The header file `gcd.h`.

```
1   #ifndef GCD_H
2   #define GCD_H
3
4   /**
5    * Calculate the greatest common divisor of two integers.
6    * @param a the first integer
7    * @param b the second integer
8    * @return the greatest common divisor of a and b
9    */
10
11  long gcd(long a, long b);
12
13  #endif
```

Line 11 is the most important line of this file. The statement

```
long gcd(long a, long b);
```

declares gcd to be a procedure. This procedure takes two input arguments (named a and b) that are both of type long. The first long on this line (to the left of gcd) is the return type of the procedure; this indicates that the procedure returns a value of type long.

Other features of this file: lines 1, 2, and 13 are the mechanism to prevent double inclusion (see the discussion on page 26). Lines 4–9 are a description of the gcd procedure. This description includes a sentence that explains what the procedure does, an explanation of the parameters passed to the procedure, and an explanation of the value returned by the procedure. The tags @param and @return are read by Doxygen to produce a nice Web page for this procedure.

We are ready to get to work on the file gcd.cc. This file has the following structure.

```
#include "gcd.h"

long gcd(long a, long b) {
   .......
   return d;
}
```

The definition of the gcd procedure looks nearly identical to the declaration in the header file. The semicolon on line 11 of gcd.h is replaced by an open brace. The open brace is followed by the instructions to calculate the greatest common divisor of a and b, and that value eventually ends up in a variable named d. The return d; statement causes the value in d to be the "answer" returned by this procedure.

Our strategy is to test successive integers to see if they are divisors of a and b, and keep track of the largest value that divides both.

There are a few things we need to worry about first.

- What happens if the gcd procedure is given negative values for a or b?

  There is nothing wrong with allowing a or b to be negative. After all,

  $$\gcd(a,b) = \gcd(-a,b) = \gcd(a,-b) = \gcd(-a,-b).$$

- What happens if one (or both) of a or b is zero? If only one of these is zero, then there is no mathematical problem because $\gcd(a,0) = |a|$ provided $a \neq 0$.

  However, $\gcd(0,0)$ is undefined. We need to decide what to do in this case. We could have the program immediately stop (this is done by the statement exit(1);). A better solution, however, is to print a warning message and return a value, say zero.

  We need to revise the documentation in gcd.h to reflect this.

Program 3.2: Revised documentation for `gcd` in the header file `gcd.h`.

```
4    /**
5     * Calculate the greatest common divisor of two integers.
6     * Note: gcd(0,0) will return 0 and print an error message.
7     * @param a the first integer
8     * @param b the second integer
9     * @return the greatest common divisor of a and b.
10    */
```

Now we work on `gcd.cc`. The file begins as follows.

Program 3.3: Beginning of the file `gcd.cc`.

```
1    #include "gcd.h"
2    #include <iostream>
3    using namespace std;
4
5    long gcd(long a, long b) {
6
7      // if a and b are both zero, print an error and return 0
8      if ( (a==0) && (b==0) ) {
9        cerr << "WARNING: gcd called with both arguments equal to zero."
10            << endl;
11       return 0;
12     }
```

We require `#include <iostream>` on line 2 because we may need to write an error message (in case `gcd(0,0)` is invoked). Line 5 starts the definition of the `gcd` procedure.

The first thing we check is if both arguments are equal to zero; this occurs at line 8. The general structure of an `if` statement is this:

```
if ( condition ) {
  statements;
}
```

If the condition (an expression that evaluates to a `bool`) is true, then the statements in the enclosing braces are executed. Otherwise (the condition is false), all the statements in the enclosing braces are skipped.

For our program, if both `a` and `b` are zero, then the condition is true and the two enclosing statements are executed. The first writes a warning message to an object named `cerr`. The `cerr` object is similar to the `cout` object. It would not have been a mistake to use `cout` here instead. However, computers provide two output streams, `cout` and `cerr`, and both write to the screen. The `cout` is usually used for standard output and `cerr` for error messages.

The second statement controlled by this `if` is `return 0;`. When this statement is executed, the procedure ends and the value 0 is returned; the rest of the program is not executed.

Next, we ensure that `a` and `b` are nonnegative.

Program 3.4: Ensuring a and b are nonnegative in `gcd.cc`.

```
14    // Make sure a and b are both nonnegative
15    if (a<0) {
16        a = -a;
17    }
18    if (b<0) {
19        b = -b;
20    }
```

The code is reasonably straightforward. If a is negative, it is replaced by -a; and likewise for b. However, there is something to worry about. Do these statement have a side effect? We are changing the arguments to the gcd procedure. Does this change the values of a and b in the procedure that called gcd?

The answer is that no change is made to any values outside the gcd procedure; there are no side effects. The reason is that when another procedure (say, main()) calls gcd, the arguments are *copied* to a and b. We say that C++ procedures *call by value*; the arguments are copies of the originals. For example, suppose the main() contains this code:

```
long x = -10;
long y = 15;
cout << gcd(x,y) << endl;
```

When gcd is invoked, the computer sets a equal to -10 and b equal to 15; the values a and b are private copies of these values. Eventually gcd reaches line 16 where it replaces a with -a (i.e., sets a to 10). However, the original x in main is unaffected by this.

Next we get to the heart of the matter. We test all possible divisors from 1 to a and see which divides both a and b. There's one slight mistake, though. If a is zero, then the answer should be b. We treat that as a special case. Here is the last part of the program.

Program 3.5: The last part of the gcd.cc program.

```
22    // if a is zero, the answer is b
23    if (a==0) {
24        return b;
25    }
26
27    // otherwise, we check all possibilities from 1 to a
28
29    long d;   // d will hold the answer
30
31    for (long t=1; t<=a; t++) {
32        if ( (a%t==0) && (b%t==0) ) {
33            d = t;
34        }
35    }
36
```

```
37    return d;
38  }
```

Lines 23–25 handle the special case in which a is zero. After that, we declare a variable d to hold the answer.

Lines 31–35 do the bulk of the work of this program. We begin with a new C++ keyword: for. The general form of the for statement is this:

```
for( starting statement ; test condition ; advancing statement) {
    statements to be done on each iteration;
}
```

The *starting statement* is executed the first time the for statement is reached. In our case, the starting statement is long t=1;. This declares a new long integer variable named t and assigns it the value 1.

Next the *test condition* is evaluated; if this condition evaluates to TRUE, the statements enclosed in the curly braces are executed. In our case, the test condition is t<=a;. As long as t is less than or equal to a, we do the statements enclosed in the curly braces.

After the enclosed statements are executed, the *advancing statement* is executed; in our case, that statement is t++;. This means that t is increased by 1.

Now the entire process repeats. For our example, t now holds the value 2. As long as this is less than or equal to a, the enclosed statements are executed. Then t is advanced to 3, then to 4, and so forth, until the test condition is FALSE. At that point, the for loop is exhausted and we proceed to the next statement after the close brace; in our example that's at line 37 and the statement is return d;.

In other words, the code

```
for (long t=1; t<=a; t++) {
    statements;
}
```

executes the statements between the curly braces a times with t equal to 1, then 2, then 3, and so on, until t equals a.

This style of for statement is common in programs. Of course, we can use the for statement to step down through values as in this code:

```
for (long s=n; s > 0; s--) {
    statements;
}
```

Alternatively, we could step only through odd values of the index:

```
for (long j=1; j <= n; j += 2) {
    statements;
}
```

For our gcd procedure, letting t take the values, 1, 2, 3, and so on, until a is precisely what we want. For each value of t, we check if t is a divisor of both a and b. We do this with the conditional if((a%t==0) && (b%t==0)). If this condition is satisfied, we update d to the current value of t, and this is what happens on line 33.

After the loop is finished, the value held in d is returned.

Now that our `gcd` program is finished, it's time to try it out. In a separate file, that we name `gcd-tester.cc`, we write a simple `main()` procedure to try out our code.

Program 3.6: A program to test the `gcd` procedure.

```
1   #include "gcd.h"
2   #include <iostream>
3   using namespace std;
4
5   /**
6    * A program to test the gcd procedure.
7    */
8   int main() {
9     long a,b;
10
11     cout << "Enter the first number --> ";
12     cin >> a;
13     cout << "Enter the second number --> ";
14     cin >> b;
15
16     cout << "The gcd of " << a << " and " << b << " is "
17          << gcd(a,b) << endl;
18     return 0;
19  }
```

3.3 Euclid's method

The program we developed in Section 3.2 works, but it is a slow, inefficient algorithm. Suppose we want to calculate the greatest common divisor of numbers that are around one billion? This is not too large for a `long` integer, but in order to find the answer, the trial-division algorithm runs for billions of iterations. There is a much better way that was developed by Euclid.

The key idea is the following result.

Proposition 3.1. *Let a,b be positive integers and let $c = a$ mod b. Then $\gcd(a,b) = \gcd(b,c)$.*

Proof. Let a,b be positive integers and let $c = a$ mod b; that is, $a = qb + c$ where $q,c \in \mathbb{Z}$ and $0 \le c < b$.

Note that if d is a common divisor of a and b, then d is also a divisor of c because $c = a - qb$. Conversely, if d is a common divisor of b and c, then (because $a = qb + c$) d is also divisor of a. Hence the set of common divisors of a and b is the same as the set of common divisors of b and c. Therefore $\gcd(a,b) = \gcd(b,c)$. $\square$

We use this to develop an algorithm. Suppose we want to find $\gcd(80,25)$. By Proposition 3.1, we calculate 80 mod $25 = 5$ and so $\gcd(80,25) = \gcd(25,5)$. To find gcd(25,5), we again apply Proposition 3.1. Because 25 mod 5 = 0, we have gcd(25,5) = gcd(5,0). At this point Proposition 3.1 does not apply because 0 is not positive. However, we know that gcd(5,0) = 5 and so we have

$$gcd(80,25) = gcd(25,5) = gcd(5,0) = 5$$

and we only needed to do two divisions (not 25 or 80).

In this section we present two programs for computing the greatest common divisor by this method. Here is the first.

Program 3.7: A recursive procedure for `gcd`.

```
1   #include "gcd.h"
2   #include <iostream>
3   using namespace std;
4
5   long gcd(long a, long b) {
6     // Make sure a and b are both nonnegative
7     if (a<0) a = -a;
8     if (b<0) b = -b;
9
10    // if a and b are both zero, print an error and return 0
11    if ( (a==0) && (b==0) ) {
12      cerr << "WARNING: gcd called with both arguments equal to zero."
13           << endl;
14      return 0;
15    }
16
17    // If b is zero, the answer is a
18    if (b==0) return a;
19    // If a is zero, the answer is b
20    if (a==0) return b;
21
22    long c = a%b;
23
24    return gcd(b,c);
25  }
```

Lines 7 and 8 ensure that `a` and `b` are nonnegative. We use a slightly different syntax for these `if` statements. When an `if` is followed by exactly one statement, the curly braces may be omitted. The single-line statement `if (a<0) a = -a;` is equivalent to this:

```
if (a<0) {
  a = -a;
}
```

Line 11 checks to see if the arguments are both zero; if they are, we issue a warning and return zero.

Lines 18 and 20 check if one of the arguments is zero; if so, the other argument is the answer we desire.

The real work of this procedure is on lines 22 and 24. If the program reaches line 22, then we know that both `a` and `b` are positive integers and Proposition 3.1

applies. We calculate c to be a mod b, so the answer to this problem is gcd(b,c) and we return that.

The question is: How is gcd(b,c) computed? The answer is that the gcd procedure calls itself. This is known as *recursion*. If we call gcd(80,25), then when we reach line 24, c holds the value 5. At this point we issue a call to gcd(25,5). The previous call to gcd(80,25) goes "on hold" pending the result of gcd(25,5). When this second invocation of gcd reaches line 24 we have a equal to 25, b equal to 5, and c equal to 0. A third call to gcd is generated at this point requesting gcd(5,0) and the second call is also placed on hold. During the call to gcd(5,0), we come to line 18 (because b==0 evaluates to TRUE) and so 5 is returned. This is passed back to the second and then to the first call to gcd and the final answer, 5, is returned.

Recursion is a powerful idea. When it applies, such as in this case, it can make for particularly simple code. The primary danger in using this technique is *infinite recursion*: if the program does not check adequate boundary conditions, it may run forever without returning an answer. This is akin to neglecting the basis case in a proof by induction.

A classic example of recursion is a program to calculate factorials. Here is the general idea presented in an incorrect program.

```
long factorial(long n) {
   return n*factorial(n-1);
}
```

To calculate factorial(5) the computer makes a call to factorial(4) which (if all goes well) returns the value 24. We then multiply this by 5 to get our answer.

The mistake, however, is that factorial(4) calls factorial(3) which calls factorial(2), and so on forever. We need to catch this process someplace, and the right place is when n is zero.

Here's a better version.

```
long factorial(long n) {
   if (n==0) return 1;
   return n*factorial(n-1);
}
```

This gives the correct result for factorial(5), but is not entirely free of the infinite recursion trap. Figure out for yourself the problem and what you can do to address it. (See Exercise 3.2.)

Program 3.7 is a perfectly good, efficient gcd procedure. At this point, it would be proper to move on to the problem at hand (described at the beginning of this chapter). However, it is possible to make the program slightly more efficient and, in so doing, we can study another C++ feature: while loops.

There are some minor inefficiencies in Program 3.7. The tests to ensure that a and b are nonnegative and not both zero are run at every iteration. One can prove, however, that we do not need to worry about this in the embedded calls to gcd. The extra tests are performed unnecessarily. Also, the computer needs to do a modest amount of work every time a new procedure is called. To calculate the greatest common divisor of two large numbers may result in dozens of calls to gcd. This overhead is not a serious problem, but if we plan to call gcd repeatedly, small improvements are worthwhile.

We now present another version of the gcd procedure that does not use recursion.

Program 3.8: An iterative procedure for gcd.

```
1   #include "gcd.h"
2   #include <iostream>
3   using namespace std;
4
5   long gcd(long a, long b) {
6     // Make sure a and b are both nonnegative
7     if (a<0) a = -a;
8     if (b<0) b = -b;
9
10    // if a and b are both zero, print an error and return 0
11    if ( (a==0) && (b==0) ) {
12      cerr << "WARNING: gcd called with both arguments equal to zero."
13           << endl;
14      return 0;
15    }
16
17    long new_a, new_b; // place to hold new versions of a and b
18
19    /*
20     * We use the fact that gcd(a,b) = gcd(b,c) where c = a%b.
21     * Note that if b is zero, gcd(a,b) = gcd(a,0) = a. If a is zero,
22     * and b is not, we get a%b equal to zero, so new_b will be zero,
23     * hence b will be zero and the loop will exit with a == 0, which
24     * is what we want.
25     */
26
27    while (b != 0) {
28      new_a = b;
29      new_b = a%b;
30
31      a = new_a;
32      b = new_b;
33    }
34
35    return a;
36  }
```

The beginning of this program is the same as Program 3.7; we make sure the arguments are nonnegative and not both zero.

At line 17 we declare two new variables: new_a and new_b. The idea is to let

$$a' \leftarrow b \qquad \text{and} \qquad b' \leftarrow a \bmod b$$

and to continue with a' and b' in lieu of a and b. We keep doing this until b reaches zero, and then a will hold the answer. For example, starting with $a = 80$ and $b = 25$, we have this:

Iteration	1	2	3
value of a	80	25	5
value of b	25	5	0

These steps take place on lines 27–33. The control statement is `while`. The general form of a `while` statement is this:

```
while (condition) {
  statements to perform;
}
```

When a program encounters a `while` statement, it checks to see if *condition* is TRUE or FALSE. If *condition* is FALSE, it skips all the statements enclosed in the curly braces. However, if *condition* is TRUE, then the statements inside the braces are executed and then we start the loop over again, checking to see if *condition* is TRUE or FALSE.

In this manner, the loop is executed over and over again until such time as the condition specified in the `while` statement is FALSE.

This is precisely what we need here. As long as b is not zero, we replace a and b with b and a%b, respectively. We enlist the help of the temporary variables new_a and new_b to do this.

When the `while` statement terminates, a holds the greatest common divisor of the original a and b, and so we return that at line 35.

It is interesting to note that we do not need two temporary variables for the `while` loop. The following works as well.

```
while (b != 0) {
  new_b = a%b;
  a = b;
  b = new_b;
}
```

Although this is correct, it is more difficult to understand than the `while` loop in Program 3.8. The slightly more verbose version shows clearly how a and b are updated from the previous values of a and b. Clarity is often preferred to cleverness because clarity is more likely to be correct.

3.4 Looping with `for`, `while`, and `do`

We have introduced two of C++'s looping statements: `for` and `while`. Here we introduce you to a third: `do`.

Recall that a `while` loop is structured like this:

```
while (condition) {
  statements to execute;
}
```

The *condition* is tested before the enclosed statements are ever executed. If *condition* is FALSE, the statements will never be executed. Incidentally, the `while` structure can be replaced by a `for` loop as follows.

```
for(; condition; ) {
  statements to execute;
}
```

The starting and advancing statements are missing, so they have no effect. We mention this mostly as a curiosity; the `while` version is preferable because it is clearer.

Sometimes you may find that the condition you want to check doesn't make sense until some series of instructions has been performed at least once. For example, a program might read in data from a file and should stop reading when a negative value is encountered. We need to read at least one data value before the condition makes sense.

For such situations, the following structure can be used.

```
do {
  statements to execute;
} while (condition);
```

The *statements to execute* are performed at least once. At the end of the loop, the condition is checked. If it is TRUE, then we return to the start of the loop for another round; if the condition is FALSE, the loop is finished and control passes to the next statement after the loop. The `do-while` loop structure is not used as frequently as `while` and `for` loops.

There are two special statements that can be used to modify the execution of a loop: `break` and `continue`.

The `break` command causes the loop to exit immediately with control passing to the next statement after the loop. Consider this code.

```
for (long a=0; a<100; a++) {
  statement1;
  statement2;
  if (x > 0) break;
  statement3;
  statement4;
}
statement5;
```

If after `statement2` the variable x holds a positive quantity, we skip `statement3` and `statement4`, and go directly to `statement5`.

The `break` command can be used with `while` and `do` loops, too.

The `continue` command directs the computer to go to the end of the loop and then do what is natural at that point. Consider this code.

```
for (long a=0; a<100; a++) {
  statement1;
  statement2;
  if (x > 0) continue;
```

```
  statement3;
  statement4;
}
statement5;
```

In this case, all 100 iterations of the loop take place. However, if during some iteration, after we execute `statement2`, we find `x` to hold a positive quantity, we skip both `statement3` and `statement4`. At this point we increase `a` (because the advancing statement is `a++`). If `a` is less than 100, another pass through the loop is performed.

3.5 An exhaustive approach to the GCD problem

It is now time to tackle the problem from the beginning of this chapter. Let p_n be the probability that two integers, chosen independently and uniformly from $\{1, 2, \ldots, n\}$, are relatively prime.

The following program counts the number of pairs (a, b) with $1 \le a, b \le n$ that satisfy $\gcd(a, b) = 1$ and divides by n^2.

Program 3.9: A program to calculate p_n.

```
1   #include <iostream>
2   #include "gcd.h"
3   using namespace std;
4
5   /**
6    * Find the probability that two integers in {1,...,n} are relatively
7    * prime.
8    */
9
10  int main() {
11    long n;
12    cout << "Enter n --> ";
13    cin >> n;
14
15    long count = 0;
16
17    for (long a=1; a<=n; a++) {
18      for (long b=1; b<=n; b++) {
19        if (gcd(a,b) == 1) {
20          count++;
21        }
22      }
23    }
24
25    cout << double(count) / (double(n) * double(n)) << endl;
26    return 0;
27  }
```

The code is straightforward. We ask the user to enter n and set a counter (named count) to zero. The main action takes place in lines 17–23. Two nested `for` loops run through all possible values of a and b with $1 \leq a, b \leq n$. On each iteration, if a and b are found to be relatively prime, the counter is increased by one.

At the end, we divide the counter by n^2 to get the probability. Because we want a floating point answer, we cast the appropriate terms into type `double` (see line 25).

To run this program, we need to use three files: `gcd.h`, `gcd.cc`, and this file (let's call it `exhaust.cc`). We can either load all three into an IDE or (on a UNIX system) save all three in a directory and compile with a command such as

```
g++ gcd.cc exhaust.cc -o exhaust
```

and the program is saved in a file named `exhaust`. (See Appendix A for more details.)

Here is a typical run of the program.

```
Enter n --> 100
0.6087
```

Thus, $p_{100} = 0.6087$.

If we run the program for various values of n, we generate the following results.

n	p_n
100	0.6087
500	0.608924
1000	0.608383
5000	0.608037
10000	0.60795
50000	0.607939

It appears that $\lim_{n \to \infty} p_n$ exists and is converging to a value around 0.6079. It would be useful to extend this table further. However, the last entry in this table took a long time for my computer to calculate. Is there a more efficient method? There is one modest modification we can make to the program. Notice that we are computing the same values twice in most instances. That is, we compute both $\gcd(10, 15)$ and $\gcd(15, 10)$. We can make our program twice as fast by calculating only one of these. Also, we do not have to bother calculating $\gcd(a, a)$; the only instance in which $\gcd(a, a)$ equals one is when $a = 1$.

Here is the modified version of the program.

Program 3.10: A slightly better program to calculate p_n.

```
1  #include <iostream>
2  #include "gcd.h"
3  using namespace std;
4
5  /**
6   * Find the probability that two integers in {1,...,n} are relatively
7   * prime.
8   */
```

```
 9
10   int main() {
11      long n;
12      cout << "Enter n --> ";
13      cin >> n;
14
15      long count = 0;
16
17      for (long a=1; a<=n; a++) {
18        for (long b=a+1; b<=n; b++) {
19          if (gcd(a,b) == 1) {
20            count++;
21          }
22        }
23      }
24      count = 2*count + 1;
25
26      cout << double(count) / (double(n) * double(n)) << endl;
27      return 0;
28   }
```

Notice that the second for loop (line 18) begins with b=a+1, so the inner loop runs from a+1 up to n. Thus the program calculates gcd(5, 10), but not gcd(10, 5). We need to double count at the end to correct. Also, we do not calculate gcd(*t*, *t*) for any *t*, so we add 1 to (the doubled) count at the end to correct for that. These corrections are on line 24.

In the following chapters, we approach the problem using two other methods: a Monte Carlo randomized algorithm (giving us an opportunity to learn about random numbers in C++) and a more intelligent exhaust using Euler's totient φ (giving us an opportunity to learn about arrays). However, before we move on to those other approaches, we make our gcd procedure better still and use that opportunity to learn additional C++ features.

3.6   Extended gcd, call by reference, and overloading

The Euclidean algorithm can be used not only to find the greatest common divisor of two integers *a* and *b*, but also to express the gcd as an integer linear combination of *a* and *b*. That is, if *a* and *b* are integers (not both zero) then there exist $x, y \in \mathbb{Z}$ such that $ax + by = \gcd(a, b)$.

Our goal is a procedure that can be called like this:

```
d = gcd(a,b,x,y);
```

where *a*, *b* are the numbers whose gcd we desire, *d* is $\gcd(a, b)$, and *x*, *y* have the property that $d = ax + by$. At first glance this appears impossible for two reasons. First, we already have a procedure named gcd. Shouldn't we name this something

else (such as `extended_gcd`)? Second, we can pass values to a procedure in its list
of arguments, but the procedure cannot change the value of the arguments. This fact
is known as *call by value* and was discussed earlier in this chapter (see Section 3.2).
The good news is that neither of these is a significant hurdle.

First, two different procedures may have the same name, but there is a caveat: The
procedures must have different types of arguments. For example, suppose we want
to create procedures for (a) finding the slope of a line segment from the origin to a
given point and (b) finding the slope of a line segment joining two given points. In
some programming languages, you would be required to give these different names;
in C++, we may name them both `slope`. The declarations for these procedures
(which we put in a header file named, say, `slope.h`) are these.

```
double slope(double x, double y);
double slope(double x1, double y1, double x2, double y2);
```

The first is for the origin-to-slope version of the procedure and the second for the
point-to-point version. Because the first takes two `double` arguments and the second takes four `double` arguments, the C++ compiler recognizes these as different
procedures.

The definitions of the procedures (which we place in a file named `slope.cc`) look
like this:

```
double slope(double x, double y) {
   return y/x;
}

double slope(double x1, double y1, double x2, double y2) {
   return (y1-y2)/(x1-x2);
}
```

When we use the same name for different procedures, these procedures ought to
be closely related and serve essentially the same purpose.

We mathematicians often use the same symbol for two closely related (but differ-
ent) objects. For example, if we have a function $f : \mathbb{R} \to \mathbb{R}$, then $f(x)$ only makes
sense when x is a real number. However, we often extend this notation. If $A \subseteq \mathbb{R}$,
then $f(A)$ (same f) means $\{f(x) : x \in A\}$.

In C++, we refer to this ability to name different procedures with the same name
as *overloading*.

Second, it is true that C++ passes data to procedures using call by value; this
means that the arguments to the procedure are copies of the original, and the invoked
procedure cannot change the original values. For example, consider this procedure.

```
void swap(long a, long b) {
   long tmp;
   tmp = a;
   a = b;
   b = tmp;
}
```

The procedure's return type is `void`. This means that the procedure does not return
any value. The code in `swap` takes the values held in `a` and `b` and exchanges them.

If, when the procedure is called a holds 5 and b holds 17, then at the end, a holds 17 and b holds 5. However, this procedure has no effect whatsoever on the calling procedure. Suppose the calling procedure contained this code:

```
long a = 5;
long b = 17;
swap(a,b);
cout << "a = " << a << endl;
cout << "b = " << b << endl;
```

The output of this would be as follows.

```
a = 5
b = 17
```

There is a way to modify the procedure swap so that it is capable of modifying its arguments. Instead of passing the *value* of a to the procedure, we pass a *reference* to a. To do this, we add the character & to the end of the type name, like so:

```
void swap(long& a, long& b) {
   long tmp;
   tmp = a;
   a = b;
   b = tmp;
}
```

The & does this: The argument a is a reference to a variable of type long. Not only is a of type long, it is more than a copy of the long. This syntax means: instead of sending a *copy* of a to this procedure, send the variable itself (and not a clone). When a procedure receives a reference to a variable, it is working on the original and not a copy.

With the rewritten swap procedure, the code

```
long a = 5;
long b = 17;
swap(a,b);
cout << "a = " << a << endl;
cout << "b = " << b << endl;
```

produces this output:

```
a = 17
b = 5
```

Call by reference is the mechanism we need so that the new gcd procedure can deliver three answers: the gcd returned via a return statement and the values x and y. The declaration for the procedure (which we add to gcd.h) looks like this:

```
long gcd(long a, long b, long& x, long& y);
```

A recursive approach to solving the extended gcd problem works well. Suppose we want to solve the extended gcd problem for positive integers a and b. We start by solving the extended gcd problem for b and c where $c = a \bmod b$. Let's say we have that solution so

$$d = \gcd(b, c) = bx_0 + cy_0$$

for some integers x_0, y_0. We know that c is the remainder when we divide a by b; that is,

$$a = qb + c$$

where $0 \le c < b$. Writing $c = a - qb$ we have

$$\begin{aligned} d &= bx_0 + cy_0 \\ &= bx_0 + (a - qb)y_0 \\ &= ay_0 + b(x_0 - qy_0) \\ &= ax + by \end{aligned}$$

where $x = y_0$ and $y = x_0 - qy_0$.

These ideas form the heart of the recursion. To construct the procedure, we need to check special cases (a or b might be zero or negative). The code follows.

Program 3.11: Code for the extended `gcd` procedure.

```
1   long gcd(long a, long b, long &x, long &y) {
2     long d; // place to hold final gcd
3
4     // in case b = 0, we have d=|a|, x=1 or -1, y arbitrary (say, 0)
5     if (b==0) {
6       if (a<0) {
7         d = -a;
8         x = -1;
9         y = 0;
10      }
11      else {
12        d = a;
13        x = 1;
14        y = 0;
15      }
16      return d;
17    }
18
19    // if b is negative, here is a workaround
20    if (b<0) {
21      d = gcd(a,-b,x,y);
22      y = -y;
23      return d;
24    }
25
26    // if a is negative, here is a workaround
27    if (a<0) {
28      d = gcd(-a,b,x,y);
29      x = -x;
30      return d;
31    }
```

```
32
33      // set up recursion
34      long aa = b;
35      long bb = a%b;
36      long qq = a/b;
37      long xx,yy;
38
39      d = gcd(aa,bb,xx,yy);
40
41      x = yy;
42      y = xx - qq*yy;
43
44      return d;
45    }
```

3.7   Exercises

3.1 Use the Euclidean algorithm to find $d = \gcd(51,289)$ and to find integers x and y so that $d = 51x + 289y$.

3.2 On page 39 we presented the following mostly correct procedure for computing factorials.

```
long factorial(long n) {
    if (n==0) return 1;
    return n*factorial(n-1);
}
```

This procedure, however, still has a bug. What is it?

3.3 Write a procedure to calculate the sign of its argument (the signum function). That is, your procedure should calculate

$$\operatorname{sgn} x = \begin{cases} +1 & \text{if } x > 0, \\ 0 & \text{if } x = 0, \quad \text{and} \\ -1 & \text{if } x < 0. \end{cases}$$

3.4 Write two procedures for producing the nth Fibonacci number when the number n is given as input. In one procedure, use a `for` loop and produce the answer using iteration. In the second version, use recursion. The input argument and return value should be type `long`.

Use the following definition of Fibonacci numbers: $F_0 = F_1 = 1$, and $F_n = F_{n-1} + F_{n-2}$ for all $n \geq 2$.

Have your procedure return -1 if the input argument is negative. Don't worry about overflow.

3.5 Write a `main` to test the Fibonacci procedures you created in Exercise 3.4. Use it to evaluate F_{20}, F_{30}, and F_{40}.

You should find that one version is much faster than the other. Explain why. Include in your explanation an analysis of how many times the recursive version of your procedure gets called in order to compute F_n.

3.6 Write two procedures to calculate

$$f(N) = \sum_{k=1}^{N} \frac{1}{k^2}.$$

Note that for N large, this approaches $\zeta(2) = \pi^2/6$.

The first procedure should calculate the sum in the usual order

$$1 + \frac{1}{4} + \frac{1}{9} + \cdots + \frac{1}{N^2}$$

and the second should calculate the sum in the reverse order

$$\frac{1}{N^2} + \frac{1}{(N-1)^2} + \cdots + \frac{1}{4} + 1.$$

In both cases, use `float` variables for all real values.

Evaluate these sums for $N = 10^6$ and report which gives the better result. ($\pi^2/6 \approx 1.6449340668482264365$.)

Explain.

3.7 Write procedures for converting between rectangular and polar coordinates named `xy2polar` and `polar2xy`. Invoking `xy2polar(x,y,r,t);` should take the rectangular coordinates (x,y) and save the resulting polar coordinates; (r,θ) are `r` and `t`, respectively. The procedure `polar2xy(r,t,x,y);` should have the reverse effect. Be sure to handle the origin in a sensible manner.

Of course, you need trigonometry functions to accomplish this task; consult Appendix C (especially Appendix C.6.1) where you can find useful functions such as `atan2`.

3.8 Every positive integer has a multiple that, when expressed in base ten, is comprised entirely of zeros and ones. Write a procedure to find the least multiple of n, the input parameter, of the required form.

Warnings: If the `long long` type[1] is available on your computer, use it. The least multiple of the required form may be much larger than n. For example, what is the least multiple of 9 that contains only zeros and ones? What is the least such multiple of 99?

Design your procedure to detect overflow and return -1 if it is unable to find the required multiple.

----

[1] Or `__int64`.

3.9 Write a procedure to find the gcd of three integers of the form

```
long gcd(long a, long b, long c);
```

and an extended version of the form

```
long gcd(long a, long b, long c, long& x, long& y, long& z);
```

that returns $d = \gcd(a,b,c)$ and populates the last three arguments with values x, y, z such that $ax + by + cz = d$.

You may use the gcd procedures developed in this chapter as part of your solution.

Chapter 4

Random Numbers

In this chapter we discuss the generation random numbers in C++. The motivation is the problem introduced in Section 3.1: let p_n be the probability that two numbers, sampled independently and uniformly from $\{1, 2, \ldots, n\}$ are relatively prime. What can we say about p_n as $n \to \infty$?

In Chapter 3 we used direct enumeration to calculate p_n for various values of n. However, as n approaches 100,000, the time for the computer to complete the calculation becomes excessive. This motivates us to find another attack. The approach we take is to sample pairs of integers in $\{1, 2, \ldots, n\}$ at random and keep track of how often we find that they are relatively prime. To do this, however, we require a mechanism to generate (pseudo) random numbers.

4.1 Pseudo random number generation

We need a procedure that produces uniform random values in the set $\{1, 2, \ldots, n\}$. In this section, we develop such a procedure as well as a procedure to produce random real values in an interval.

Before we begin, however, we need to understand that the computer does not produce random values. Instead, it provides a *pseudo random number generator*. This is a procedure whose output looks random, but is actually deterministic. The simplest type of random number generator is a *linear congruential generator* or LCG for short. An LCG produces a series of integers $x_0, x_1, x_2, \ldots$ by the following calculation,

$$x_{n+1} = (ax_n + b) \bmod c$$

where a, b, c are fixed positive integers. The method is specified fully once we choose a value for x_0. This first value is known as the *seed*.

Pseudo random numbers produced by an LCG behave in some ways as uniform random values from $\{0, 1, \ldots, c-1\}$, but there are problems. For example, by the pigeonhole principle, the sequence $x_0, x_1, x_2, \ldots$ must repeat itself over and over. The hope is that by taking c sufficiently large, this is not an issue.

A variety of more sophisticated pseudo random number generators have been proposed and implemented each with various advantages and disadvantages. The important point we need to keep in mind is that pseudo random numbers are not random, so we need to be mildly skeptical of the results they suggest and careful in how we use them.

Standard C++ includes the `rand` procedure for producing pseudo random numbers. To use `rand` it is necessary to include the `cstdlib` library. A call to `rand()` returns an integer value between 0 and a constant named `RAND_MAX` (inclusive). The value of `RAND_MAX` is defined in the `cstdlib` header file. A typical value is $2^{31} - 1$.

4.2   Uniform random values

It is possible to write programs so they call `rand` directly whenever a random value is required. However, it is a good idea to create our own procedures that serve as intermediaries between the program that requires a random value and `rand`. Why do we need a middle man? Here are three good reasons.

- First, `rand` produces random values in a large discrete set. We might want a random integer in a smaller set (such as $\{0, 1\}$ if we want to simulate a coin flip) or we might want a uniform continuous value in the interval $[0, 1]$. It is convenient to have procedures that do each of these.

- Second, there are preferred methods to extract random values from `rand`. For example, if we want to simulate a coin flip, it is tempting to write code such as this: `flip = rand()%2;`. The problem is that the lowest-order bit of random values from a pseudo random number generator might not behave as you would expect. In the worst case, this last bit might simply oscillate between 0 and 1, and such coin flips do not look random at all. We create a better alternative one time and use that procedure henceforth.

- Some day you might decide that you prefer a different random number generator. Rather than editing all your programs to replace `rand` with the new procedure, you only need to rewrite the intermediaries.

What do we want these procedures to do for us? We want to produce a random real value uniformly in the interval $[0, 1]$, or more generally in an interval $[a, b]$ where $a, b \in \mathbb{R}$. And we want to produce a random integer chosen uniformly from a finite set of the form $\{1, 2, \ldots, n\}$.

Because all of these return a random value that is uniformly distributed over its domain, we name all three of these `unif`. This procedure name overloading is permissible because the three versions have different type arguments.

Program 4.1: Header file, `uniform.h` for procedures to generate uniform random values.

```
1   #ifndef UNIFORM_H
2   #define UNIFORM_H
```

```
3
4   /**
5    * Generate a random number between 0 and 1.
6    * @return a uniform random number in [0,1].
7    */
8   double unif();
9
10  /**
11   * Generate a random number in a real interval.
12   * @param a one end point of the interval.
13   * @param b the other end point of the interval.
14   * @return a uniform random number in [0,1].
15   */
16  double unif(double a, double b);
17
18  /**
19   * Generate a random integer between 1 and a given value.
20   * @param n the largest value this procedure can produce.
21   * @return a uniform random value in {1,2,...,n}.
22   */
23  long unif(long a);
24
25  /**
26   * Reset the random number generator based on the system clock.
27   */
28  void seed();
29
30  #endif
```

The three flavors of `unif` are declared on lines 8, 16, and 23. The first produces a uniform real value in the interval $[0,1]$, the second generalizes this and produces a real value in an arbitrary interval $[a,b]$, and the third produces an integer value uniformly in $\{1,2,\ldots,n\}$.

In addition, we have declared a procedure named `seed` on line 28. The purpose of `seed` is to initialize the random number generator from the system clock (more on this later).

Let's turn to implementing each of these starting with `double unif()`. A call to `rand()` returns an integer between 0 and `RAND_MAX`. To convert this to a real value in $[0,1]$ we simply compute `rand() / double(RAND_MAX);`. (See lines 7–9 of the file `uniform.cc` in Program 4.2 below.)

Once we have the `double unif()` version written, we use it to write the second version. We simply multiply `unif()` by `(b-a)` and add `a`. See lines 11–13 in `uniform.cc`.

For the integer version, we multiply the continuous `unif()` by `a` and convert to an integer. This gives a value in $\{0,\ldots,a-1\}$, so we add 1 to place the value into the desired set. What if the user gives a negative or zero value for `a`? We have a choice as to handling these undefined situations. See lines 15–19 of `uniform.cc` to see our decision.

Now we consider the `seed` procedure. Because `rand` (and the procedures that we wrote based on `rand`) returns an unpredictable stream of values, we might expect that the programs we write would behave differently every time we run them. Interestingly (and with good reason) this is not the case. Every time a program containing `rand` is run, the `rand` procedure gives the same sequence of pseudo random values. The reason is that the first time `rand` is invoked, it contains a fixed starting value called the *seed* of the random number generator. The reason this behavior is desirable is reproducibility. If you perform a computational experiment and wish to report it in a journal, it is important that others can run the same program as you and see the same results.

However, there are times when this reproducible behavior is undesirable. For example, the motivating purpose of this chapter is to write code to generate many pairs of integers in a set $\{1,\ldots,n\}$ to see how frequently they are relatively prime. We might want to run our program a few times with different streams of random values so we can compare results.

The standard library (`cstdlib`) provides the procedure `srand` that is used to set the seed used by `rand`. Calling `srand(s)`, where s is a `long` integer, resets to the seed to s. We could ask the user to provide a seed value like this:

```
long s;
cout << "Enter a seed for the random number generator --> ";
cin >> s;
srand(s);
```

Another solution, that is easier for the user, is to use the computer's clock to provide the seed. As long as we don't run the program twice in the same second, a different value is used for the seed.

The header `ctime` defines the procedure `time`. Calling `time(0)` returns the number of seconds that have elapsed since a specific date and time (on many computers, that date and time is January 1, 1970 at 12:00 A.M. UTC). Our `seed` procedure simply takes the value[1] returned by `time(0)` as input to `srand`. See lines 21–23 of `uniform.cc`.

Program 4.2: Definitions of the `unif` procedures in `uniform.cc`.

```
1   #include "uniform.h"
2   #include <cstdlib>
3   #include <ctime>
4   #include <cmath>
5   using namespace std;
6
7   double unif() {
8     return rand() / double(RAND_MAX);
9   }
10
11  double unif(double a, double b) {
12    return (b-a)*unif() + a;
13  }
```

---

[1] To be precise, the procedure `time` returns a value of type `time_t`. Your compiler might require you to convert this to an unsigned integer before using it as an argument to `srand`. Replace line 22 with `srand(unsigned(time(0)));` in this case.

```
14
15   long unif(long a) {
16       if (a < 0) a = -a;
17       if (a==0) return  0;
18       return long(unif()*a) + 1;
19   }
20
21   void seed() {
22       srand(time(0));
23   }
```

4.3   More on pseudo random number generation

The integer version of `unif` used the expression `long(unif()*a)+1` to produce a random value in $\{1,\ldots,a\}$. This is mildly convoluted because we have the additional call to the continuous version of `unif` that calls `rand` and divides by `RAND_MAX`. It is both simpler and more efficient simply to calculate `1 + rand()%a`, but this latter approach is less reliable.

The problem is that some older versions of the `rand` pseudo random number generator are purported to be unreliable, and the lower-order bits of the values produced by `rand` do not behave well. One way to rectify this situation is to replace `rand` by a better procedure.

To understand why the low-order bits of a random number generator might not be good approximations of randomness, we develop our own linear congruential generator (LCG) here. (And this gives us an opportunity to introduce additional C++ concepts.)

Recall that an LCG produces a stream of values $x_0, x_1, x_2, \ldots$ where

$$x_{n+1} = (ax_n + b) \bmod c.$$

The LCG is fully specified once we select values for a, b, c, and x_0. For our example, we take

$$a = 17, \quad b = 3, \quad c = 64, \quad \text{and} \quad x_0 = 0.$$

Suppose we call our procedure `lcg`. Each time we call `lcg()` it should return the next x_j value in the sequence. However, for this the procedure `lcg` needs to remember the previous x value. The following code does not work.

```
int lcg() {
    int state = 0;
    return (17*state + 3) % 64;
}
```

Every time this procedure is called, it returns the value 3. We need to indicate that the variable `state` should only be initialized to zero the first time `lcg` is called.

To do this, we declare the variable `state` to be `static`. The declaration looks like this instead: `static int state = 0;`. With this declaration, the variable `state` retains its value after the procedure `lcg` ends. (Without the `static` modifier, the usual behavior is for variables to cease to exist once the procedure ends.) The variable `state` is initialized to zero only the first time `lcg` is called. Henceforth, it retains the value it held when `lcg` last terminated.

The `return` statement is fine as written (`return (17*state + 3) % 64`) but there is a better way. The constants 17, 3, and 64 should be given names and declared at the beginning of the procedure. For a short simple procedure such as `lcg`, this is not an important issue. However, for more complicated procedures, giving constants specific names makes the program easier to read and easier to modify. Imagine that we write a program in which we are often reducing numbers modulo 64. Rather than explicitly typing 64 repeatedly in the code, we can define a variable named `theMod` set equal to 64 instead. Although typing `theMod` is longer than typing 64, if we ever want to change the program so that `theMod` is now, say, 128, we only have to change one line. The declaration of `theMod` looks like this:

```
const int theMod = 64;
```

The `const` modifier means that the procedure does not modify the value of `theMod`. This enables the C++ compiler to generate more efficient object code and prevents you from mistakenly putting `theCode` on the left-hand side of an assignment statement.

Returning to the `lcg` procedure, we declare three variables (named a, b, and c) to be type `const int` (see lines 12–14 of Program 4.3). The return statement is then `return (a*state+b) % c;` which is easier to read. Later, if we wish to modify the `lcg` program, we simply need to change the values we ascribe to a, b, or c at the top of the procedure.

With the `lcg` procedure written, we create a short `main()` to test it out. The `main()` calls `lcg` 20 times, reducing each return value modulo two. It then calls `lcg` another 20 times, reducing each of those results mod four. Here is the full program.

Program 4.3: A program to illustrate the problem with lower-order bits in an LCG.

```
1   #include <iostream>
2   using namespace std;
3
4   /**
5    * A sample linear congruential pseudo random number generator that
6    * returns values in {0,1,...,63}.
7    */
8   int lcg() {
9       static int state = 0;
10      const long a = 17;
11      const long b = 3;
12      const long c = 64;
13
```

```
14      state = (a*state+b) % c;
15      return state;
16  }
17
18  /**
19   * This main calls lcg twenty times and prints out the value modulo
20   * two, and then prints twenty more values taken modulo four.
21   */
22  int main() {
23    cout << "Values mod 2: ";
24    for (int k=0; k<20; k++) {
25      cout << lcg()%2 << " ";
26    }
27    cout << endl;
28
29    cout << "Values mod 4: ";
30    for (int k=0; k<20; k++) {
31      cout << lcg()%4 << " ";
32    }
33    cout << endl;
34
35    return 0;
36  }
```

When this program is run, the following output is printed on the screen.

```
Values mod 2: 1 0 1 0 1 0 1 0 1 0 1 0 1 0 1 0 1 0 1 0
Values mod 4: 3 2 1 0 3 2 1 0 3 2 1 0 3 2 1 0 3 2 1 0
```

Clearly, the sequence of values we produce in this manner is far from random! However, we can change the way we extract zeros and ones from lcg. Consider the following alternative main.

```
int main() {
  cout << "Values mod 2: ";
  for (int k=0; k<20; k++) {
    double x = lcg() / 64.;
    cout << int(2*x) << " ";
  }
  cout << endl;
  return 0;
}
```

Here we first produce a double value x in the range $[0, 1)$ by dividing the output of lcg by 64. We then multiply x by 2 (so we are now somewhere in $[0, 2)$) and then cast to type int (so the value is now either 0 or 1). Here's the result.

```
Values mod 2: 0 1 0 1 1 1 0 0 0 0 1 0 0 1 0 1 1 1 0 0
```

As you can see, the output appears, at least superficially, more random.

4.4   A Monte Carlo program for the GCD problem

We now present a simple program to estimate p_n. The user selects n (thereby specifying the set $\{1, 2, \ldots, n\}$ from which pairs of integers are to be drawn) and the number of repetitions. The program generates the pairs, counts the number that are relatively prime, and reports the frequency.

Program 4.4: A Monte Carlo approach to calculating p_n.

```
 1  #include "uniform.h"
 2  #include "gcd.h"
 3  #include <iostream>
 4  using namespace std;
 5
 6  /**
 7   * This main generates many pairs of values from the set {1,2,...,n}
 8   * and reports how often the pairs are relatively prime. The value n
 9   * and the number of pairs are specified by the user.
10   */
11
12  int main() {
13      long n;          // max el't in the set {1,2,...,n}
14      long reps;       // number of times we perform the experiment
15      long a,b;        // values chosen from {1,2,...,n}
16      long count;      // number of pairs that are relatively prime
17
18      count = 0;
19
20      cout << "Enter n (maximum el't of the set) --> ";
21      cin >> n;
22
23      cout << "Enter the number of pairs to sample --> ";
24      cin >> reps;
25
26      for (long k=1; k<=reps; k++) {
27          a = unif(n);
28          b = unif(n);
29          if (gcd(a,b) == 1) ++count;
30      }
31
32      cout << count / (double(reps));
33
34      return 0;
35  }
```

To begin, let's run the program with $n = 1000$ for 10,000 repetitions. The session looks like this:

```
Enter n (maximum el't of the set) --> 1000
Enter the number of pairs to sample --> 10000
0.6117
```

The program estimates $p_{1000} \approx 0.6117$ which is reasonably close to the actual value 0.608383. If, however, we run the code for a million repetitions we get $p_{1000} \approx 0.608932$ which is correct to three decimal places.

We can use the program to estimate p_n for $n = 10^6$. Running the program for 10^8 repetitions, we find $p_n \approx 0.607979$. How good is this estimate? When we perform r independent repetitions of an experiment whose probability of success is p, the expected number of successes is rp, but the standard deviation is on the order of $\sqrt{r}$. In this case, $r = 10^8$, so $\sqrt{r} = 10^4$. So it is reasonable to suppose that the value of p_n we found is correct to four decimal places. To obtain greater accuracy, we need either to increase the number of repetitions or (as we do in the next chapter) find a better method for calculating p_n.

One hundred million repetitions take a modest amount of time. It would be feasible to run for a billion repetitions, but that would only give us another "half" digit of accuracy. We might consider running for a trillion repetitions, but that would take too long.

Here's a reasonable rule of thumb for today's computers. Programs that do millions of operations using millions of bytes of memory are quick. Programs that do billions of operations take some time, but are feasible to run. Holding billions of bytes in memory, however, is at the limit of today's personal computers. However, trillions of operations takes too long and most computers cannot hold trillions of bytes of data in memory.

4.5 Normal random values

Before we leave the subject of random numbers, we expand our repertoire of the types of random variables we can simulate. In Section 4.2 we developed code to produce discrete and continuous uniform random variables. Here, we produce normal (Gaussian) random values.

The distribution of a normal random variable is based on the classic bell curve, $f(t) = \exp(-t^2)$. We need to modify the bell curve slightly so that the total area under the curve is 1, the mean of the Gaussian is 0, and the standard deviation is 1. To that end, we use the following for the density.

$$f(t) = \frac{1}{\sqrt{2\pi}} \exp\left(-t^2/2\right).$$

From this it follows (with a bit of work) that $\int f(t)\,dt = 1$, $\int t f(t)\,dt = 0$, and $\int t^2 f(t)\,dt = 1$ (integration over all of $\mathbb{R}$).

We therefore define the Gaussian random variable X using f as its density, or equivalently

$$\Phi(x) = P[X \leq x] = \int_{-\infty}^{x} f(t)\,dt.$$

The integrals at the end of the previous paragraph imply that X has mean zero and standard deviation one; for this reason, X is also known as $N(0,1)$—a normal random variable with mean zero and standard deviation one.

There is an efficient algorithm for producing Gaussian random values known as the polar or Box–Muller method. One begins by generating a point (x,y) uniformly at random in the unit disc. We then let

$$r = x^2 + y^2, \qquad \mu = \sqrt{\frac{-2\log r}{r}}, \qquad Z_1 = \mu x, \qquad \text{and} \qquad Z_2 = \mu y.$$

Then Z_1 and Z_2 are independent $N(0,1)$ Gaussian random variables. Here is a C++ program based on this method:

```
#include <cmath>
#include "uniform.h"
using namespace std;

double randn() {
  double x,y,r;
  do {
    x = unif(-1.,1.);
    y = unif(-1.,1.);
    r = x*x + y*y;
  } while (r >= 1.);

  double mu = sqrt(-2.0 * log(r) / r);

  return mu*x;
}
```

The do/while loop generates points (x,y) uniformly in the square $[-1,1]^2$ until one that is interior to the unit disk is found. Each pass through the loop has a $\pi/4$ chance of succeeding, so after just a few iterations we are assured of finding a point chosen uniformly from the unit disk. Once the point (x,y) has been found, the rest of the algorithm follows the Box–Muller method.

There is an inefficiency in this implementation. The algorithm is capable of producing two independent normal random values. In our implementation we find one of these values and simply ignore the other. We could make this procedure twice as fast if there were a way to preserve that second normal value for the next call to this procedure. To do this, we enlist the help of static variables.

The header file `randn.h` looks like this:

```
#ifndef RANDN_H
#define RANDN_H

#include "uniform.h"

double randn();

#endif
```

Here is the program file (`randn.cc`).

Program 4.5: A program to generate Gaussian random values.

```
1   #include "randn.h"
2   #include <cmath>
3   using namespace std;
4
5   double randn() {
6      static bool has_saved = false;
7      static double saved;
8
9      if (has_saved) {
10        has_saved = false;
11        return saved;
12     }
13
14     double x,y,r;
15     do {
16        x = unif(-1.,1.);
17        y = unif(-1.,1.);
18        r = x*x + y*y;
19     } while (r >= 1.);
20
21     double mu = sqrt(-2.0 * log(r) / r);
22
23     saved = mu*y;
24     has_saved = true;
25
26     return mu*x;
27  }
```

The new version includes two static variables: a Boolean value has_saved and a real value saved. The has_saved variable is set to false to show that the procedure does not currently hold a saved Gaussian value (in saved). See lines 6–7.

At line 9 we check if there is a saved Gaussian value that we have not used yet. If so, then we change has_saved to false and return the value in saved.

Otherwise (starting at line 14), we generate the two Gaussian values Z_1 and Z_2 by the polar method. We save Z_2 in saved and set the flag has_saved to true. Finally, we return Z_1.

The statements on lines 14–21 are the time-consuming part of this procedure. By saving the second Gaussian value generated by the polar method, the slow part is only executed on every other invocation of the procedure. This makes the procedure nearly twice as fast.

4.6 Exercises

4.1 Suppose two points are chosen uniformly at random within the unit square $[0, 1]^2$. Write a program to estimate the expected (average) length of the segment joining such points.

4.2 *Buffon's Needle.* Imagine a needle is dropped at random onto a floor painted with equally spaced parallel lines. The length of the needle is the same as the distance between the lines.

Write a program that simulates dropping the needle and that counts the number of times the needle crosses one of the lines drawn on the floor.

If you wish to use standard trigonometry functions, such as `cos`, or the `floor` function, insert a `#include <cmath>` directive at the start of your program. See Appendix C.6.1.

4.3 *Sylvester's Four-Point Problem.* Let K be a compact convex subset of the plane with nonempty interior. Let $P(K)$ denote the probability that when four points are chosen independently and uniformly in K, then they lie at the vertices of a convex quadrilateral (as in the left portion of the illustration but not the right).

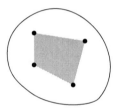 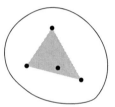

Write procedures to (a) generate a point uniformly at random inside a circle, (b) generate a point uniformly at random inside a triangle, and (c) test whether four points determine the vertices of a convex quadrilateral.

Use your procedures to estimate $P(K)$ for the cases where K is a circle or a triangle.

4.4 *Random point on a circle.* Suppose you wish to generate a point uniformly at random on the unit circle, $\{(x,y) : x^2 + y^2 = 1\}$. Suppose the procedure is declared like this:

```
void point_on_circle(double& x, double& y);
```

Here are two ways you might implement this procedure.

First, you can generate a uniform $[0,1]$ value that you multiply by 2π; call the result θ. Then set x and y to be $\cos\theta$ and $\sin\theta$.

Alternatively, you can generate a point in the interior of the unit ball by the rejection method. [That is, pick points $(x,y) \in [0,1]^2$ until you find one that satisfies $x^2 + y^2 \le 1$.] Then rescale by dividing by $\sqrt{x^2 + y^2}$.

Which do you think is faster? Create two versions of the procedure and time how long it takes each to generate 100 million points.

4.5 *Random point on a sphere.* Continuing Exercise 4.4, we consider the issue of generating a point at random on the surface of a sphere in $\mathbb{R}^3$ and in higher dimensions.

One way to generate a point on the unit sphere in $\mathbb{R}^3$ is to generate x, y, z uniformly in $[-1, 1]$. If $x^2 + y^2 + z^2 \le 1$, then we scale by $1/\sqrt{x^2 + y^2 + z^2}$ to generate the point. Otherwise, we generate another triple and try again.

What is the probability of successfully generating a point during a given iteration?

Next, explain why this procedure is woefully inefficient in high-dimensional space. Try to create an alternative method.

4.6 Create a procedure `int random_walk();` that simulates a random walk on the integers. That is, `random_walk()` returns the position of a particle whose initial position is 0. Each time `random_walk()` is called, the particle moves one step left or right, each with probability 50%. The return value is the new position of the particle.

For example, calling `random_walk` 20 times might produce the following values.

```
1 0 1 2 3 4 5 4 5 4 3 4 3 4 3 4 3 2 3 2
```

4.7 Write a pair of procedures `int up()` and `int down()` that behave as follows. When `up()` is called, a certain value is increased by 1 and the new value is returned. Similarly, when `down()` is invoked, the value is decreased by 1 and that new value is returned. At the start of the program, the value is zero.

For example, suppose we run the following code.

```
cout << up() << " ";
cout << up() << " ";
cout << up() << " ";
cout << down() << " ";
cout << up() << " ";
cout << down() << endl;
```

Then the following output is produced.

```
1 2 3 2 3 2
```

Chapter 5

Arrays

In Chapter 3 we introduced the following problem. Let p_n be the probability that two integers, chosen uniformly and independently from $\{1, 2, \ldots, n\}$, are relatively prime. What can we say about p_n as $n \to \infty$? We computed p_n for various values of n by direct enumeration. As n approached 10^5, the program became slow and so we sought another approach.

In Chapter 4 we tried a Monte Carlo approach. Although our program enables us to estimate p_n for n equal to one million (or larger), to get decent accuracy we need to run the program for too many iterations.

In this chapter we take another approach using Euler's totient.

5.1 Euler's totient

Euler's totient is a function φ defined on the positive integers. For $n \in \mathbb{Z}^+$, we define $\varphi(n)$ to be the number of elements of $\{1, 2, \ldots, n\}$ that are relatively prime to n. For example, $\varphi(10) = 4$ because the only integers between 1 and 10 (inclusive) that are relatively prime to 10 are 1, 3, 7, and 9.

Euler's totient has immediate relevance to our problem. We want to count the number of pairs (a, b) with $1 \le a, b \le n$ and $\gcd(a, b) = 1$. Alternatively, we can count the number of such pairs with $a \le b$. Of course, we miss the pairs with $a > b$, so we would just double the result and subtract one (for double-counting the pair $(1, 1)$). Symbolically, we have

$$p_n = \frac{1}{n^2} \left[-1 + 2 \sum_{k=1}^{n} \varphi(k) \right].$$

We can calculate $\varphi(k)$ by considering all the integers from 1 to k and check which are relatively prime to k. This, however, would lead to an algorithm that is no different from the one in Chapter 3.

Fortunately, there are more efficient ways to calculate φ.

We begin our analysis of Euler's totient with a few special cases.

If p is a prime, then all but one member of the set $\{1, 2, \ldots, p\}$ are relatively prime to p. Therefore $\varphi(p) = p - 1$.

Next consider $\varphi(p^2)$ for a prime p. In the set $\{1,2,\ldots,p^2\}$ the only numbers that are not relatively prime to p^2 are the multiples of p, and there are p of those. Therefore $\varphi(p^2) = p^2 - p = p(p-1)$.

More generally, consider $\varphi(p^n)$ where p is prime and $n \in \mathbb{Z}^+$. In $\{1,2,\ldots,p^n\}$ only the multiples of p are not relatively prime to p^n, and there are p^{n-1} of those. This gives the following.

Proposition 5.1. *Let p be a prime and let n be a positive integer. Then $\varphi(p^n) = p^n - p^{n-1} = p^{n-1}(p-1)$.* □

Now that we have examined φ for powers of primes, let us examine the special case $\varphi(77)$. Note that for $1 \le n \le 77$, we have

$$\gcd(n,77) = 1 \quad \Longleftrightarrow \quad \gcd(n,7) = 1 \quad \text{and} \quad \gcd(n,11) = 1.$$

By Proposition 3.1,

$$\gcd(n,7) = \gcd(7, n \bmod 7) \quad \text{and} \quad \gcd(n,11) = \gcd(11, n \bmod 11).$$

Let $n_1 = n \bmod 7$ and $n_2 = n \bmod 11$. If n is relatively prime to 77 and

$$x \equiv n_1 \pmod 7 \quad \text{and} \quad x \equiv n_2 \pmod{11}$$

then x is also relatively prime to 77. Furthermore, there is a unique such x in $\{1,2,\ldots,77\}$; this is a consequence of the Chinese Remainder Theorem.

Theorem 5.2 (Chinese Remainder). *Let a and b be relatively prime positive integers, and let $c,d \in \mathbb{Z}$. Then the system of congruences*

$$x \equiv c \bmod a$$
$$x \equiv d \bmod b$$

has a unique solution in $\{1,2,\ldots,ab\}$. □

So, every $x \in \{1,2,\ldots,77\}$ that is relatively prime to 77 satisfies a pair of congruences of the form

$$x \equiv n_1 \pmod 7 \quad \text{and} \quad x \equiv n_2 \pmod{11}$$

where n_1 is relatively prime to 7 and n_2 is relatively prime to 11. Conversely, for every pair (n_1,n_2) with (a) $1 \le n_1 \le 7$, (b) $1 \le n_2 \le 11$, (c) $\gcd(n_1,7) = 1$, and (d) $\gcd(n_2,11) = 1$, there is a unique x between 1 and 77 that satisfies the above congruences and is relatively prime to 77.

Therefore $\varphi(77)$ equals the number of choices for (n_1,n_2), and that is precisely $\varphi(7) \times \varphi(11) = 6 \times 10 = 60$.

A careful reading of the analysis of $\varphi(77)$ reveals that the only facts we used about 7 and 11 is that they are relatively prime. Thus the argument can be rewritten to give a proof of the following.

Proposition 5.3. *If a and b are relatively prime, then $\varphi(ab) = \varphi(a)\varphi(b)$.* $\qquad\square$

From these propositions we derive the following formula for $\varphi(n)$.

Theorem 5.4. *Let n be a positive integer and let $p_1, p_2, \ldots, p_t$ be the distinct prime divisors of n. Then*

$$\varphi(n) = n\left(1 - \frac{1}{p_1}\right)\left(1 - \frac{1}{p_2}\right)\cdots\left(1 - \frac{1}{p_t}\right).$$

Proof. We factor n into primes as

$$n = p_1^{e_1} p_2^{e_2} \cdots p_t^{e_t}$$

where the p_j are distinct primes and the e_j are positive integers. By Propositions 5.3 and 5.1 we have

$$\begin{aligned}
\varphi(n) &= \varphi(p_1^{e_1})\varphi(p_2^{e_2})\cdots\varphi(p_t^{e_t}) \\
&= \left(p_1^{e_1-1}(p_1 - 1)\right)\left(p_2^{e_2-1}(p_2 - 1)\right)\cdots\left(p_t^{e_t-1}(p_t - 1)\right) \\
&= p_1^{e_1}\left(1 - \frac{1}{p_1}\right) p_2^{e_2}\left(1 - \frac{1}{p_2}\right)\cdots p_t^{e_t}\left(1 - \frac{1}{p_t}\right) \\
&= n\left(1 - \frac{1}{p_1}\right)\left(1 - \frac{1}{p_2}\right)\cdots\left(1 - \frac{1}{p_t}\right).
\end{aligned}$$
$\qquad\square$

Thus, if we know the prime factors of n, we can calculate $\varphi(n)$. This results steers the discussion for the rest of this chapter. We begin by developing algorithms to factor `long` integers. We create a `factor` procedure that produces the full prime factorization of an integer; we use an *array* to hold the result.

5.2   Array fundamentals

The first goal of this chapter is to create a procedure that takes an integer and gives us a list of the prime factors of that integer. For example, if the integer is 120, then the list of prime factors is $(2, 2, 2, 3, 5)$.

There are several ways to hold lists in C++. The most primitive method is to use an *array*. An array is simply a list of values of a given type. To declare an array of, say, `long` integers, we use a statement such as this:

```
int vals[10];
```

This declares `val` to be an array of ten[1] integers. The elements of the array are indexed starting from zero. That is, the ten elements of `val` are these:

```
val[0]   val[1]   val[2]   val[3]   val[4]
val[5]   val[6]   val[7]   val[8]   val[9]
```

Each of these behaves exactly as does a `long` variable. One can have statements such as `val[3]=2*val[0]+val[5];` or `++val[7];`. Elements of an array are accessed using square brackets, not parentheses. If an array is declared to have n elements, the first is always indexed by subscript 0 and the last by subscript $n - 1$. Element n of such an array is not defined.

To print out all the members of an array, one can use a `for` loop:

```
for (int k=0; k<10; k++) {
  cout << val[k] << endl;
}
```

However, the following does not work: `cout << val << endl;`. This does not print the values of the `val` array, even though it is legal C++. Furthermore, arithmetic operations on elements of an array must be specified element by element. If you wish to increase every element of an array by one, you need a statement such as this: `for(int k=0; k<10; k++) val[k]++;`. Unfortunately, the statement `val++;` is not illegal, but it does not do what you might want or expect. If `alpha` and `beta` are two arrays, the expression `alpha==beta` does not determine if corresponding elements in the arrays are equal, but the compiler might not complain because it is a legal C++ expression.

Worse yet, if `val` is an array of ten elements, then only `val[0]` through `val[9]` are legitimate array elements. Surprisingly, using `val[10]` or `val[17]` is not prevented, but severe problems may result from accessing elements beyond the normal bounds of an array.

The C++ array is the most computationally efficient mechanism for storing a list, but it is not the only manner. We explore several alternatives in Chapter 8.

It is helpful to have a basic understanding the underlying mechanism by which arrays work. To do that, we need to mention the concept of a *pointer*. In this book we scrupulously avoid dealing with pointers; they are confusing and easily lead to subtle programming errors. However, we do need to be at least vaguely aware that C++ uses pointers. What is a pointer? A pointer is a variable whose value is a location in the computer's memory. Each byte of a computer's memory is numbered, and a pointer holds such a number. In the case of arrays, the name of the array is actually a pointer to the location of the first element of the array (e.g., the memory location that holds `val[0]`). So a statement of the form `val++;` changes the pointer `val` so that it now points to a different location in memory. (In fact, it would now point to the location of `val[1]` but this is more than we want or need to know right now.) A statement of the form `cout << val << endl;` prints out the location

---

[1]Of course, we can declare an array to hold hundreds or thousands of elements. However, some compilers place a limit on the maximum size of an array. There is a simple way to exceed that limit and this is explained in the footnote accompanying Program 5.8 on page 82.

number where the first element (with subscript zero) of `val` is housed. Try it! You should get a result that looks something like this: `0xbffffa20`. The leading `0x` is C++'s way of indicating that the number that follows is in base-16. The remainder is the number in standard base-16 (where `a` is a digit equal to ten, `b` is a digit equal to eleven, and so on).

When you declare an array with a statement such as `int val[10];` the computer sets aside a block of memory to hold ten `int`s. When you refer to, say, `val[3]`, the computer figures out where in memory this quantity sits by adding `3*sizeof(int)` to the base address `val`.

You may be thinking: Why do I need to know all this stuff?!? Mostly, you do not need to worry about this. The important things you need to know are these.

- Array elements are indexed starting from 0.

- Operations cannot be performed on arrays as a whole; you need to operate on the elements of arrays individually.

- The name of the array is a valid C++ entity. Once declared, the only information that the name of the array carries is the location in memory of the start of the array and the type of elements held in the array.

The third point is important when we write procedures that have array arguments. In the next section we develop a procedure called `factor` that takes two arguments. The first is the integer we want to factor. The second is an array to hold the prime factors.

5.3 A procedure to factor integers

In this section we develop a procedure to factor integers. The input to the procedure is a `long` integer, `n`. The procedure gives us two results: an array holding the prime factors and an integer telling us how many prime factors (multiplicities counted) we found. For example, if the number to be factored is 20, the procedure finds that the factors are $(2, 2, 5)$ and that there are three prime factors.

We use C++'s `return` statement to report on the number of factors found. However, the `return` mechanism does not work for arrays. Instead, we give the procedure an array in the argument list.

The declaration of the `factor` procedure is this:

```
long factor(long n, long* flist);
```

The first argument, `n`, is the number we want to factor. The second argument, `flist`, is an array to hold the answer (the factors of `n`). The type of `flist` is `long*`. The star indicates that `flist` is an array. (Technically, the star signifies that `flist` is a pointer.) The `long` to the left of the word `factor` signifies that `factor` returns a `long` integer—the number of prime factors found.

The technicality that `flist` is a pointer is relevant to us only in this regard. Although C++ uses call by value as its standard way to pass arguments to procedures, the array `flist` is not copied to the `factor` procedure. Instead, the location (i.e., pointer) of the array is sent. Therefore the `factor` procedure is capable of modifying the elements of `flist` in the procedure that called `flist`.

The `factor` procedure has no mechanism to ensure that the array `flist` contains enough elements to hold the primes found. It is the responsibility of the user to be sure that the array is large enough to hold the answers. If the size of your computer's `long` integers is 4 bytes, then these integers are less than 2^{32}. Therefore, no `long` can have more than 32 factors. Perhaps you have a computer in which `long` integers are 64 bits; in that case, if `flist` has at least 64 elements, no problem can arise. This means the procedure that calls `factor` should look something like this:

```
long prime_factors[200];
long nfactors;

nfactors = factor(60, prime_factors);
```

After this code runs `nfactors` holds 4 and the array `prime_factors` holds the values 2, 2, 3, and 5 in positions 0 through 3, respectively. Positions 4 through 199 are unaffected by this call to `factor`.

It is time to design the factoring algorithm. We begin by handling exceptional cases. What if the user sends a negative number or zero to be factored? For negative values, we can simply replace the argument by its absolute value. Asking to factor zero is asking for trouble. Returning a value of 0 is not quite right, because that is what we would return when we are asked to factor 1. Returning a positive value is also misleading. We settle for the unhappy choice of returning the value -1 in case the user requests a prime factorization of 0.

If we are asked to factor 1, we simply report there are no prime factors (i.e., `return 0;`).

So we suppose that $n \geq 2$. The method we use to factor n is this. We check if n is divisible by 2. If so, we record 2 in the first element (index zero) of `flist` and replace n by $n/2$. We keep track of where we last wrote into the `flist` array with a variable we name `idx`. Initially `idx` equals zero. Once we record a prime in `flist[idx]` we then increase `idx` by 1. We keep doing this until n is no longer divisible by 2.

At that point, we check n for divisibility by 3. We continue dividing until no factors of 3 remain. With each factor of 3 that we find, we record a 3 in `flist[idx]`, advance `idx` by one, and replace n by $n/3$.

It would be logical to next try divisibility by 5, but C++ does not "know" that 5 is the next prime. Instead, we try divisibility by 4 which, of course, fails because we have already divided out all of n's factors of 2.

We continue in this manner, trying a divisor until it is exhausted and then advancing to the next divisor. When do we stop? Once we have divided out all of n's prime factors, it equals 1; that's when we stop.

The algorithm is depicted in the flowchart in Figure 5.1.

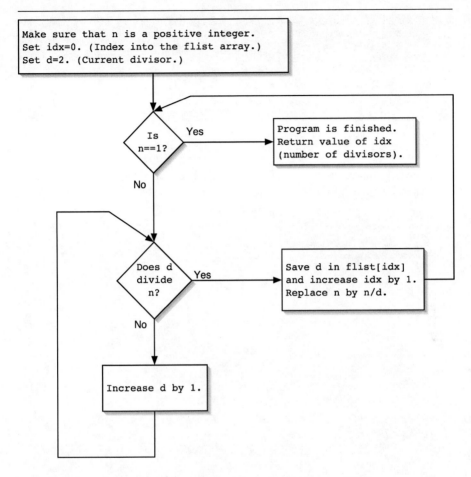

Figure 5.1: A flowchart for the factoring algorithm.

We are now ready to implement this procedure in C++. To begin, we write the header file, factor.h.

Program 5.1: Header file factor.h for the first version of the factor procedure.

```
1  #ifndef FACTOR_H
2  #define FACTOR_H
3
4  /**
5   * Factor an integer n. The prime factors are saved in the second
6   * argument, flist. It is the user's responsibility to be sure that
7   * flist is large enough to hold all the primes. If n is negative, we
8   * factor -n instead. If n is zero, we return -1. The case n equal to
9   * 1 causes this procedure to return 0 and no primes are saved in
10  * flist.
```

```
11    *
12    * @param n the integer we wish to factor
13    * @param flist an array to hold the prime factors
14    * @return the number of prime factors
15    */
16   long factor(long n, long* flist);
17
18   #endif
```

Next is the C++ source code.

Program 5.2: Source file `factor.cc` for the first version of the `factor` procedure.

```
1    #include "factor.h"
2
3    long factor(long n, long* flist) {
4
5      // If n is zero, we return -1
6      if (n==0) return -1;
7
8      // If n is negative, we change it to |n|
9      if (n<0) n = -n;
10
11     // If n is one, we simply return 0
12     if (n==1) return 0;
13
14     // At this point we know n>1
15
16     int idx = 0;       // index into the flist array
17     int d = 2;         // current divisor
18
19     while (n>1) {
20       while (n%d == 0) {
21         flist[idx] = d;
22         ++idx;
23         n /= d;
24       }
25       ++d;
26     }
27     return idx;
28   }
```

Lines 5–12 deal with the exceptional cases.

Line 16 sets up the variable `idx`. Throughout the procedure `idx` contains the index of the next location in the `flist` array where we record the prime factors. We set `idx` equal to zero so the first prime we record goes into the first element of `flist`.

Line 17 sets up the variable `d`. This is the current divisor we are testing.

The heart of the procedure lies in lines 19–26. We keep trying to find factors of n as long as there are factors to be found (i.e., as long as n holds a value greater than 1). On line 20 we test n for divisibility by `d`. If successful, we record `d` in the array `flist` at location `idx`, increase `idx` by one so it refers to the next available cell in `flist`, and divide out the factor of `d` from n. We keep doing this until n is no longer divisible by `d`. At that point, we increase `d` by one and continue.

At the end, `idx` has been increased once for every prime factor we found. So we simply return its value at line 27.

Here is a `main` to test the `factor` procedure. This program prints out the prime factorization of all integers from 1 to 100.

Program 5.3: A `main` to test the `factor` procedure.

```
1   #include "factor.h"
2   #include <iostream>
3   using namespace std;
4
5   /**
6    * A program to test the factor procedure.
7    */
8
9   int main() {
10
11      long flist[100];    // place to hold the factors
12
13      for (long n=1; n<=100; n++) {
14          int nfactors = factor(n,flist);
15          cout << n << "\t";
16          for (int k=0; k<nfactors; k++) cout << flist[k] << " ";
17          cout << endl;
18      }
19  }
```

On line 11 we declare `flist` to hold 100 `long` integers. The `for` loop on lines 13–18 requests that we factor n for all values from 1 to 100. The call to `factor` is on line 14. We save the number of factors found in a variable named `nfactors` and the factors themselves populate the elements of the array `flist`.

Line 15 prints the current value of n and then a tab character. The sequence \t stands for a tab; this way the list of factors ends up nicely arranged.

Line 16 prints out the factors of n separated by spaces and then line 17 starts a new line on the screen.

The output looks like this.

```
1
2       2
3       3
4       2 2
5       5
6       2 3
7       7
8       2 2 2
9       3 3
10      2 5
11      11
                        (many lines deleted)
95      5 19
96      2 2 2 2 2 3
97      97
```

```
98        2  7  7
99        3  3  11
100       2  2  5  5
```

5.4   A procedure to calculate Euler's totient

With the `factor` procedure built, our next step is to build a procedure to calculate Euler's totient. The `totient` procedure takes a `long` argument (n) and returns a `long` result ($\varphi(n)$). Here is the header file `totient.h`.

Program 5.4: Header file for the `totient` procedure.

```
1   #ifndef TOTIENT_H
2   #define TOTIENT_H
3
4   /**
5    * Euler's totient function.
6    * @param n the number whose totient we seek
7    * @return the number of elements in {1,2,...,n} that are relatively
8    * prime to n.
9    */
10
11  long totient(long n);
12
13  #endif
```

We use Theorem 5.4 to design the `totient` procedure. For example, let $n = 36{,}750$. Factoring n gives $36{,}750 = 2 \times 3 \times 5 \times 5 \times 7 \times 7$. Then

$$\varphi(n) = (2-1) \times (3-1) \times 5 \times 5 \times (5-1) \times 7 \times (7-1) = 8400.$$

If the prime p appears e times in the factorization of n, then in $\varphi(n)$ it contributes $p^{e-1} \times (p-1)$.

The procedure to calculate $\varphi(n)$ begins by factoring n and saving the result in an array, `flist`. We then step through `flist` one element at a time. If `flist[k]` equals `flist[k+1]`, then we multiply by `flist[k]`; otherwise, we multiply by `flist[k]-1`. To do this, we use an expanded version of the `if` statement. The expanded version is called an `if-else` statement and is structured like this:

```
if (condition) {
   statements1;
}
else {
   statements2;
}
```

When this structure is encountered, the `condition` is evaluated. If it evaluates to TRUE then `statements1` are executed and `statements2` are skipped. However, if condition evaluates to FALSE, then statements1 are skipped and statements2 are executed.

We need to be careful when we reach the end of the array; we do not want to access the element past the end of the array.

Here is the code.

Program 5.5: The code for the totient procedure.

```
1   #include "totient.h"
2   #include "factor.h"
3
4   long totient(long n) {
5       // handle special cases
6       if (n <= 0) return 0;
7       if (n == 1) return 1;
8
9       // factor n
10      long flist[100];
11      long nfactors = factor(n, flist);
12
13      long ans = 1;
14
15      for (long k=0; k<nfactors; k++) {
16
17          if (k < nfactors-1) {
18              if (flist[k] == flist[k+1]) {
19                  ans *= flist[k];
20              }
21              else {
22                  ans *= flist[k]-1;
23              }
24          }
25          else {
26              ans *= flist[k]-1;
27          }
28      }
29      return ans;
30  }
```

After factoring n, we set up a variable named ans that holds the result to be returned (line 13).

Lines 15–28 form the core of the procedure. This is a for loop that steps through the array flist. If we are not yet at the last element of the array (checked on line 17) we compare the current element of the array and the next element of the array for equality. If they are equal, we multiply ans by flist[k] (line 19) but otherwise (see the else on line 21) we multiply by flist[k]-1 (line 22).

If we are at the last element of the array (the else on line 25), then the proper factor is flist[k]-1.

We test the totient procedure with this simple main:

```
#include "totient.h"
#include <iostream>
using namespace std;
```

```
/**
 * A program to test the totient procedure.
 */
int main() {
  for (int k=1; k<=100; k++) {
    cout << k << "\t" << totient(k) << endl;
  }
  return 0;
}
```

The resulting output looks like this:

```
1       1
2       1
3       2
4       2
5       4
6       2
7       6
8       4
9       6
10      4
11      10
            (many lines omitted)
90      24
91      72
92      44
93      60
94      46
95      72
96      32
97      96
98      42
99      60
100     40
```

5.5 The Sieve of Eratosthenes: `new` and `delete[]`

The `totient` procedure we developed in the previous section works well, but relies on an inefficient factoring procedure. Because we expect to be calculating $\varphi(k)$ for millions of different values of k, it is worth the effort to make the procedure as efficient as possible.

One of the inefficiencies in the `factor` procedure arises from the lack of a table of prime numbers. If we had a table of prime numbers we could avoid wasted steps. If we only needed to factor one number or only wanted one value of φ, it might not be worth the effort. However, because we plan to compute φ millions of times, we can greatly increase the speed of our program by first building a table of primes.

An efficient method to build a table of primes is known as the *Sieve of Eratosthenes*. To find all the primes up to some value N, we write down all the integers from 2 up to N (we skip 1). We circle 2 (it's prime) and then cross off all other multiples of 2. The first unmarked number is 3. We circle 3 then cross off all other multiples of 3. Notice that 4 is crossed off, so the next unmarked number is 5. We circle 5, and then cross off all other multiples of 5. We continue in this manner until we reach N. At the end, the circled numbers are exactly the primes; every other entry in the table has been crossed off. The algorithm is illustrated in Figure 5.2.

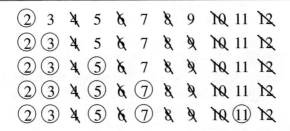

Figure 5.2:   Illustrating the Sieve of Eratosthenes algorithm.

We now create the `sieve` procedure. In a header file, `sieve.h` we declare the procedure as follows.

Program 5.6: The header file for a procedure to build a table of primes via the Sieve of Eratosthenes.

```
 1  #ifndef SIEVE_H
 2  #define SIEVE_H
 3
 4  /**
 5   * The Sieve of Eratosthenes: Generate a table of primes.
 6   *
 7   * @param n upper limit on the primes (i.e., we find all primes
 8   * less than or equal to n).
 9   * @param primes array to hold the table of primes.
10   * @return the number of primes we found.
11   */
12
13  long sieve(long n, long* primes);
14
15  #endif
```

The parameter n is an upper bound on the primes we seek. The array `primes` is a place to hold the primes found by the procedure. It is the responsibility of the procedure that calls `sieve` to make sure that `primes` is big enough to hold the array.

For example, if we wish to generate all primes up to ten million (10^7), how big does `primes` need to be? The prime number theorem gives us an estimate. The number of primes less than or equal to n is approximately $n/\log n$. In the case $n = 10^7$, this gives an estimated 620,421 primes. In fact, the number is a bit higher: 664,579 to be exact. To be safe, the calling procedure should make sure that `primes` is (say) 1.5 times the estimate in the prime number theorem.

The code for the `sieve` procedure introduces a new idea. The procedure requires an array to hold the sieve. If n is 10^7, then this table requires 10^7 entries. For efficiency's sake, we want the individual entries in the table to use as few bytes as possible. We could declare this table to be made up of `short` integers, but the `char` type is only one byte on most systems. Let's call the table `theSieve`. We may be tempted to begin our program as follows.

```
long sieve(long n, long* primes) {
  char theSieve[n];  // array for the sieve
```

Unfortunately, this is illegal. In declaring an array, the size of the array must be a constant; it may not be a number that can't be known until the program is run.

We could declare `theSieve` to be a huge array (say, of size ten million). However, if we wish to use this procedure for larger values of n we would need to rewrite the program.

A better solution is to create an array using C++'s `new` operator. This operator allows us to create an array whose size is determined when the program is running.

The beginning of the `sieve` procedure looks like this:

```
long sieve(long n, long* primes) {

  if (n<2) return 0;  // no primes unless n is at least 2.

  char* theSieve;
  theSieve = new char[n+1];  // hold the marks
```

The initial `if` statement handles the case when a user might call `sieve` with, say, n negative. The next two lines are the interesting part.

The line `char* theSieve;` declares `theSieve` to be a name that can refer to an array of `char`s. However, this name may not be used yet because the array is not created. The star on the type name `char` is important; without it, the name `theSieve` would refer to a single character, not an array.

At this point we have a name to use for the array, but no space in memory to hold the array. The next line allocates the space. The statement

```
theSieve = new char[n+1];
```

does two things. First, it requests a block of $n + 1$ pieces of memory, each big enough to hold a `char`. Second, it causes `theSieve` to refer to the start of that block of memory. In the program, we want to use the entries up to `theSieve[n]`; this is why we requested $n + 1$ array elements.

There is an important difference between an array that is declared with the usual sort of statement (such as `long primes[10];`) and an array that is created with `new` (such as `long* primes = new long[10];`). Arrays created with the standard declaration automatically disappear when the procedure in which they are defined exits; at that point, the memory they use is automatically freed to be used by other parts of the program. However, an array allocated using the `new` statement remains reserved until it is explicitly released.

A block of memory allocated by the use of `new` must later be released with a `delete[]` statement like this:

```
delete[] theSieve;
```

There must be one and only one `delete[]` balancing each use of `new`. If the `delete[]` is missing, the array persists until the program exits. If one has repeated requests to `new` (without matching `delete[]`s), then more and more memory is tied up in arrays until the computer runs out of memory to honor the requests. This situation is known as a *memory leak*.

On the other hand, if one tries to perform a `delete[]` on the same block of memory more than once, the gates of the underworld will open and demons will rule the earth. Or your program might crash.

The `sieve` program follows.

Program 5.7: The `sieve` procedure.

```
1   #include "sieve.h"
2
3   long sieve(long n, long* primes) {
4
5       if (n<2) return 0;   // no primes unless n is at least 2.
6
7       char* theSieve;
8
9       theSieve = new char[n+1];   // hold the marks
10
11      // Names of marks to put in theSieve
12      const char blank   = 0;
13      const char marked  = 1;
14
15      // Make sure theSieve is blank to begin
16      for (long k=2; k<=n; k++) theSieve[k] = blank;
17
18      long idx = 0; // index into the primes array
19
20      for (long k=2; k<=n; k++) {
21        if (theSieve[k]==blank) {   // we found an unmarked entry
22          theSieve[k] = marked;     // mark it as a prime
23          primes[idx] = k;          // record k in the primes array
24          idx++;
25
26          // Now mark off all multiples of k
27          for(long d=2*k; d<=n; d+=k) theSieve[d] = marked;
28        }
29      }
30      delete[] theSieve;
31      return idx;
32  }
```

Line 9 allocates the array `theSieve`. The matching `delete[]` is on line 30.

On lines 12 and 13 we create names for the marks we use in the array `theSieve`. We use two types of mark to distinguish the two types of cells: blank cells and marked cells. We could have simply used the values 0 and 1 in the program, but unnamed constants are to be shunned. By giving these names we make the code more understandable. The `const` qualifier in the declaration tells the compiler that these values, `blank` and `marked`, never change. This does two good things. It prevents us from accidentally writing code that would change these symbols and it enables the compiler to produce more efficient object code.

Line 16 ensures that the array `theSieve` is entirely populated with `blank` (i.e., zero) before we begin. In a perfect world arrays are given to you filled with sensible default values (such as zero). However, it is folly to rely on this. An initial run through the array to make sure it is in the state we hope is quick and easy.

The variable `idx` on line 18 is an index into the array `primes`. It refers to the next available cell in `primes` at all points in the program. At the end, it will have been incremented once for every prime we record, and so it will hold the number of primes found. That is why we use `idx` as the return value on line 31.

The sieving takes place on lines 20–29. When we come to an entry k in `theSieve` that is blank it must be a prime. We mark that location, record the number k in `primes` (and increment `idx`). Then (line 27) we place a mark in every cell that is a multiple of k.

Here is a `main` to test the `sieve` procedure.[2]

Program 5.8: A program to test the `sieve` procedure.

```
1   #include "sieve.h"
2   #include <iostream>
3   using namespace std;
4
5   const long N = 10000000;          // ten million
6   const long TABLE_SIZE = 800000; // prime number theorem overestimate
7
8   /**
9    * A program to test the sieve procedure.
10   */
11
12  int main() {
13    long primes[TABLE_SIZE];
14    long np = sieve(N,primes);
15
16    cout << "We found " << np << " primes" << endl;
17
18    cout << "The first 10 primes we found are these: " << endl;
```

---

[2]Note: On Windows computers this program might crash because of line 13. Some computers place a limit on the maximum size array one can declare. The solution is to allocate large arrays dynamically. That is, replace line 13 with this: `long* primes; primes = new long[TABLE_SIZE];` Remember to `delete[] primes;` before the end of the program.

```
19    for (long k=0; k<10; k++) cout << primes[k] << " ";
20    cout << endl;
21
22    cout << "The largest prime we found is " << primes[np-1] << endl;
23
24    return 0;
25  }
```

Finding all the primes up to ten million is quick. The output from the program appeared on my screen in under four seconds.

```
We found 664579 primes
The first 10 primes we found are these:
2 3 5 7 11 13 17 19 23 29
The largest prime we found is 9999991
```

5.6   A faster totient

With a table of primes at our disposal, we can calculate $\varphi(n)$ without first factoring n; here's how. For each prime p in the table, we check if p divides n. If so, we replace n by $(n/p)(p-1)$. In the end, we have calculated

$$n\left(\frac{p_1 - 1}{p_1}\right)\left(\frac{p_2 - 1}{p_2}\right)\cdots\left(\frac{p_t - 1}{p_t}\right)$$

which, by Theorem 5.4, is $\varphi(n)$.

The procedure we create has two arguments. The declaration of this function (in the file totient.h) looks like this:

```
long totient(long n, const long* primes);
```

Ignore the const keyword for a moment.

The first argument is n: the number for which we wish to calculate φ.

The second argument is a table of primes. The type of this argument is long* which indicates that primes holds the starting position of an array of long integers. The table of primes is not duplicated; what we pass to the totient procedure is the address of the table.

The procedure returns a long: the totient of n.

Now we consider the extra word const in the declaration. When we pass an array to a procedure it is possible for the procedure to change the values held in the array. Indeed, we relied on that fact when we created the sieve procedure. Recall that sieve is declared as long sieve(long n, long* primes);. The sieve procedure receives the array address primes and can then populate that array with the desired values.

In the case of our new `totient` procedure, we use the values housed in the array `primes`, but we do not alter them. The `const` qualifier asserts that the procedure `totient` does not modify any element in the array `primes`.

Although the procedure `totient` would work equally well without the `const` qualifier, it is a good habit to declare arguments as `const` when appropriate. If (by mistake) the code in your procedure is capable of changing elements in the array, the compiler will complain and help you spot the error.

The code for the new `totient` procedure is this:

Program 5.9: A faster `totient` procedure that employs a table of primes.

```
1  #include "totient.h"
2
3  long totient(long n, const long* primes) {
4    if (n<=0) return 0;
5
6    long ans = n;
7    for (long k=0; primes[k] <= n; k++) {
8      if (n%primes[k]==0) {
9        ans /= primes[k];
10       ans *= primes[k]-1;
11     }
12   }
13   return ans;
14 }
```

The program is fairly straightforward. We make a copy of n in a variable named `ans`. (This isn't necessary, but improves clarity.) In lines 7–12 we consider all primes that are less than or equal to n. If such a prime p is a factor of n, we modify `ans` by dividing out p and then multiplying by $p-1$ (lines 9–10). In the end, `ans` holds $\varphi(n)$.

The only caveat is that we must be sure that the array `primes` contains all the prime factors of n. One way to do this is to generate all primes up to n using `sieve`. For example, the following `main` tests the faster `totient` procedure.

```
#include "totient.h"
#include "sieve.h"
#include <iostream>
using namespace std;

/**
 * A main to test the faster version of Euler's totient on
 * the integers from 1 to 100.
 */

int main() {

  const int N = 100;    // testing up to N

  long primes[10*N]; // table of primes
  sieve(10*N, primes);
```

```
for (long k=1; k<=N; k++) {
  cout << k << "\t" << totient(k,primes) << endl;
}
}
```

The output is a two-column table. The first column contains the integers from 1 to 100, and the second column contains Euler's totient of these.

5.7 Computing p_n for large n

Recall from the beginning of this chapter that we can calculate p_n by the following formula,

$$p_n = \frac{1}{n^2}\left[-1 + 2\sum_{k=1}^{n}\varphi(k)\right].$$

In this section we write a program to calculate p_n for n equal to one million.

The main part of the program adds $\varphi(k)$ as k goes from 1 to one million. Because we know p_n is around 0.6, we expect the final sum to be around 0.6×10^{12} which is larger than a long on a system where sizeof(long) is 4 bytes. (A 32-bit integer can store values up to about two billion, but not in the trillions.) So we need to use a long long (or __int64); fortunately on my computer this is an 8-byte quantity and can hold values that are nearly 10^{19}. This is more than adequate to the task.

Calculating this sum takes many minutes (but not many hours). We can request the program to report its progress along the way. In the program we present, we report p_k whenever k is a multiple of 10^5. Here is the program.

Program 5.10: A program to calculate p_n for n equal to one million.

```
1   #include "totient.h"
2   #include "sieve.h"
3   #include <iostream>
4   #include <iomanip>
5   using namespace std;
6
7   /**
8    * A program to calculate the probability that two integers chosen in
9    * {1,2,...,n} are relatively prime. This probability is calculated
10   * for values of n up to ten million.
11   */
12
13  int main() {
14
15    const long N = 1000000;          // one million
16    const long TABLE_SIZE = 200000;  // prime number th'm overestimate
17
18    // set up the table of primes
```

```
19    long* primes;
20    primes = new long[TABLE_SIZE];
21    long np;
22    np = sieve(2*N,primes);
23
24    long long count=0;     // sum of phi(d) from 1 to n
25
26    cout << setprecision(20);
27    for (long k=1; k<=N; k++) {
28      count += totient(k, primes);
29      if (k%100000==0) {
30        cout << k/1000 << " thousand \t";
31        cout << double(2*count-1) / (double(k)*double(k)) << endl;
32      }
33    }
34    return 0;
35  }
```

Notice there is a new header included at line 4. The `iomanip` header provides devices to change the output style. In our case, we want to print out more digits of p_n than usual. This occurs on line 26. The `cout << setprecision(20);` statement modifies `cout` so that it prints up to 20 decimal digits for `double` real numbers (trailing zeros, if any, are not printed). The `iomanip` header is needed to define `setprecision`. (See Section 14.6 for more information on adjusting output format.)

Lines 19–22 are used to generate the table of primes.

The variable `count`, declared on line 24, is used to accumulate the sum of $\varphi(k)$. As discussed, this needs to be type `long long` because the final sum exceeds the maximum value a `long` can hold.

The core of the program is on lines 27–33. This is a `for` loop that increments `count` by $\varphi(k)$ as k goes from one to one million. On line 29 we check if k is divisible by 100 thousand; if so, we report p_k at that time.

Here is the output of the program (which took about 25 minutes to run on my computer).

```
100  thousand    0.6079301507
200  thousand    0.60792994587500004
300  thousand    0.60792774407777783
400  thousand    0.60792759136874996
500  thousand    0.607928317404
600  thousand    0.6079276484527778
700  thousand    0.60792730424285712
800  thousand    0.60792796007343752
900  thousand    0.6079273649074074
1000 thousand    0.60792710478300005
```

5.8 The answer

It certainly appears that p_n is converging and that the limit, to six decimal places, is 0.607927. The next step, necessarily, takes us beyond C++; we need to recognize this number to formulate (and prove!) a conjecture.

Fortunately, there are good tools for this step. Neil Sloane's *On-Line Encyclopedia of Integer Sequences* is a remarkable resource that takes us directly to the answer. Visit http://www.research.att.com/~njas/sequences/ and enter the sequence of digits into the sequence search engine: 6 0 7 9 2 7 and press the SEARCH button. After a brief delay, the answer emerges:

$$\frac{1}{\zeta(2)} = \frac{6}{\pi^2} = 0.6079271018540\ldots$$

is the number we seek. (The site also gives several references to this well-known problem.)

We close this chapter with a sketch of the proof.

Sweeping all worries about convergence under the rug, consider two large integers. What is the probability they are not both even (a necessary condition for the numbers to be relatively prime)? Each has a $\frac{1}{2}$ chance of being even, so the probability neither has a factor of 2 is $\left(1 - \frac{1}{4}\right)$. More generally, the probability neither has a prime p as a common factor is $\left(1 - 1/p^2\right)$. So the limit of p_n is

$$\prod_p \left(1 - \frac{1}{p^2}\right)$$

where the product is over all primes.

Recall that $\zeta(2)$ is given by

$$\zeta(2) = \sum_{n=1}^{\infty} \frac{1}{n^2}.$$

This can be expressed as a product. The idea is to factor n^2 into even powers of its prime divisors. The product representation is

$$\zeta(2) = \prod_p \left(1 + \frac{1}{p^2} + \frac{1}{p^4} + \frac{1}{p^6} + \cdots\right)$$

where the product is over primes p. To see why this works, consider the term (in the sum) $1/120^2$. We factor 120 as $2^3 \times 3 \times 5$. Expanding the product representation, the term $1/60^2$ appears by taking $1/2^6$ from the first factor, $1/3^2$ from the second, $1/5^2$ from the third, and 1 from all the other factors.

Notice that the factors in the product representation are geometric series. Therefore

$$\zeta(2) = \prod_p \left(\frac{1}{1 - \frac{1}{p^2}} \right)$$

and so

$$\frac{1}{\zeta(2)} = \prod_p \left(1 - \frac{1}{p^2} \right)$$

as desired.

5.9 Exercises

5.1 Calculate $\varphi(100)$, $\varphi(2^9)$, and $\varphi(5!)$.

5.2 What is wrong with this program and how can the mistake be repaired?

```
#include <iostream>
using namespace std;
int main() {
    int n;
    cout << "Enter n: ";
    cin >> n;
    int vals[n];
    // do stuff with the vals array
    return 0;
}
```

5.3 Solve the pair of congruences $x \equiv 3 \pmod{20}$ and $x \equiv 5 \pmod 9$.

5.4 Write a procedure to solve problems such as Exercise 5.3. It may be declared like this:

```
long crt(long a1, long n1, long a2, long n2);
```

(The name `crt` stands for Chinese Remainder Theorem.)

The procedure should be designed to solve the pair of recurrences

$$x \equiv a_1 \pmod{n_1} \qquad x \equiv a_2 \pmod{n_2}$$

where n_1 and n_2 are relatively prime positive integers. The return value is the solution mod $n_1 n_2$.

How should the procedure handle a situation in which n_1 and n_2 are not of this form?

5.5 In Program 5.7 we defined two values named `blank` and `marked` and used them to populate an array named `theSieve`. For the marks we used `char` type values (and the array was declared to be `char*`). However, it would be more logical to use `bool` values because each cell in `theSieve` takes only one of two possible values and the Boolean `true` or `false` makes perfect sense in this context.

Why did we use `char` type values instead of `bool`?

5.6 Write a program that fills an array with Fibonacci numbers F_0 through F_{20} and then prints them out in a chart.

5.7 Write a program that fills two arrays a and b with integers according to this recurrence:

$$a_0 = b_0 = 1 \qquad a_n = b_{n-1} \qquad b_n = a_{n-1} + 2b_{n-1}.$$

The program should then print out a table in which each row is of the form

$$k \qquad a_k \qquad b_k \qquad \frac{a_k}{b_k}.$$

Conjecture a value for $\lim_{n \to \infty} a_n/b_n$. And, of course, prove your conjecture.

5.8 Write a procedure to find the maximum value in an array of `long` integers. The inputs to the procedure should be the array and the number of elements in the array. The output should be the maximum value in the array.

5.9 Write a procedure that generates an array of Fibonacci numbers as its return value. The input to the procedure should be an integer $n \geq 2$ that specifies the desired size of the array. Here is how such a procedure would appear in a `main()`:

```
int main() {
    long* fibs;
    fibs = make_fibs(10);
    for (int k=0; k<10; k++) cout << fibs[k] << " ";
    cout << endl;
    return 0;
}
```

This code should produce the output:

```
1 1 2 3 5 8 13 21 34 55
```

In addition, there is a subtle bug in the `main()`. What is it?

5.10 Create a procedure `long fibs(int n)` to return the nth Fibonacci number. The procedure should work as follows. The first time the procedure is called, it creates a table of Fibonacci numbers holding F_n through F_{40}. Then it returns the value held in the table it created. On all subsequent calls, it does not need to recompute any Fibonacci numbers, but simply returns the value in the table it built during its first invocation.

If the input parameter is out of range (either less than 0 or greater than 40) the procedure should return -1.

5.11 What happens when `new` asks for more memory than your computer can provide? Write a program that repeatedly requests `new` for large blocks of memory without ever releasing those blocks with `delete[]`. That is, your program should have a severe, deliberate memory leak.

5.12 A computational experiment yields the following result: `5.8598744820`. Propose a conjecture.

Part II

Objects

Chapter 6

Points in the Plane

The first part of this book introduces the fundamental data types (`long`, `double`, `bool`, etc.), arrays of these types, and procedures.

C++ provides the ability to define new types of data that can be used just as the basic types are. New data types that we create are called *classes*. In this chapter we create a class called `Point` that represents a point in the Euclidean plane. When we want a variable to represent an integer quantity, we declare it to be type `long` (or one of the other integer types). Likewise, once we have the `Point` class set up, we can declare a variable to represent a point in the plane like this:

```
Point X;
```

We call X an *object* of type `Point`.

6.1 Data and methods

A class definition specifies the *data* that describe an object of the class and the *methods* to inspect and manipulate objects.

For the class `Point` we need to hold data that specify a point's location in the plane. There are two natural ways we might do this: rectangular coordinates (x, y) or polar coordinates (r, θ). Later in this chapter, when we write the C++ files to create the `Point` class, we choose the rectangular representation.

A class is more than a way to bundle data together. A class also specifies operations that may be performed on its objects. Here are some things we might want to know about points and actions we might want to perform on points.

- Learn a point's rectangular coordinates (x and y).

- Learn a point's polar coordinates (r and θ).

- Change one (or both) of a point's rectangular coordinates.

- Change one (or both) of a point's polar coordinates.

- Rotate a point about the origin through a given angle.

- Check if two points are equal.

- Find the distance between two points.

- Find the midpoint between two points.

- Print a point's coordinates using a statement of the form `cout<<P<<endl;`.

We perform these various tasks by means of procedures. Many of the procedures are part of the definition of the class itself and are invoked by a different syntax. For example, we create a procedure to change a point's *x* coordinate called `setX`. To set a point *P*'s *x* coordinate to -4.5, we use the following special syntax,

```
P.setX(-4.5);
```

The `setX` procedure is part of the definition of the class `Point`. Computer scientists call such procedures *member functions* but as we reserve the word *function* for its mathematical meaning, in this book we call such procedures *methods*. The term *method* is also used by many computer scientists.

A C++ class is a bundle that combines *data* that describe its objects and *methods* for inspecting and manipulating the objects.

Once a class is defined, we can use it in procedures just as with any other C++ data type. For example, here is a procedure named `dist` that computes the distance between two points.

```
double dist(Point P, Point Q) {
  double dx = P.getX() - Q.getX();
  double dy = P.getY() - Q.getY();
  return sqrt(dx*dx + dy*dy);
}
```

Notice that `dist` is a procedure that works on two arguments (both of type `Point`) and returns a real number answer of type `double`. Inside this procedure we use `Point`'s `getX` and `getY` methods to access the *x* and *y* coordinates of the two points. (The `sqrt` procedure is defined in a C++ header file so this code requires a `#include <cmath>` directive.)

The `dist` procedure is not a method (i.e., not a member of the `Point` class); it is simply a procedure that uses the `Point` data type. It is invoked using the usual syntax; to calculate the distance between two points we call `dist(P,Q)`. However, `getX` is a method of the `Point` class; it is used to reveal the point's *x* coordinate. Because it is a class method, it is invoked using the special syntax `P.getX()`.

Let's see how this all works.

6.2 Declaring the `Point` class

When we create a new procedure we break the definition into two files: a header file (whose name ends with `.h`) that declares the procedure and a code file (whose name ends with `.cc`) that specifies the algorithm.

Likewise, a class is defined in two files: a `.h` file that declares the class and a `.cc` file that gives the algorithms for its various methods.

The declaration for a class looks like this:

```
class ClassName {
    declarations for data and methods;
};
```

Notice the required semicolon after the class declaration's close brace—it is easy to forget.

We now present the declaration for the `Point` class. There are several new ideas present in this declaration and we examine each. As an aid to readability, we omit all the documentation from the file `Point.h`. It is extremely important to include comments in the header file that explain what the class represents and what each of its various methods does. The time you take to write these comments will be repaid tenfold when you subsequently write programs that use your classes.

Here is the header file.

Program 6.1: Header file `Point.h` for the `Point` class (condensed version).

```
 1  #ifndef POINT_H
 2  #define POINT_H
 3  #include <iostream>
 4  using namespace std;
 5
 6  class Point {
 7
 8  private:
 9    double x;
10    double y;
11
12  public:
13    Point();
14    Point(double xx, double yy);
15    double getX() const;
16    double getY() const;
17    void setX(double xx);
18    void setY(double yy);
19    double getR() const;
20    void setR(double r);
21    double getA() const;
22    void setA(double theta);
23    void rotate(double theta);
24    bool operator==(const Point& Q) const;
25    bool operator!=(const Point& Q) const;
26
27  };
28
29  double dist(Point P, Point Q);
30  Point midpoint(Point P, Point Q);
31  ostream& operator<<(ostream& os, const Point& P);
32
33  #endif
```

Let's examine this file in detail.

- To begin, lines 1, 2, and 33 are the usual mechanism to prevent double inclusion of the header file.

- The declaration of the `Point` class spans lines 6 through 27. Line 6 announces the declaration. The keyword `class` tells us that we are defining a new class that we have chosen to name `Point`. The open brace on line 6 is matched by the close brace on line 27, and the declaration is between these. The semicolon on line 27 ends the declaration.

- Ignore for now the keywords `private`, `public`, and `const`. We return to them subsequently.

- Lines 9–10 specify the data for objects of this class. The data are simply two real numbers giving the *x* and *y* coordinates; these are named (quite sensibly) x and y.

- Lines 13 and 14 declare the *constructors* for the class `Point`. A constructor is a method that is invoked when an object of type `Point` is declared.

  When a procedure contains a variable of type `long`, the variable must be declared before it can be used. Likewise, programs that use variables of type `Point` must declare those variables as such; a statement such as the following is required,

  ```
  Point P;
  ```

  This statement creates a variable named P and then invokes a method (named `Point`) that initializes the data held in P. In our case, this is extremely simple; the member variables x and y are set equal to zero. This basic constructor is declared on line 13; the code that sets x and y equal to zero is in another file (named `Point.cc`) that we examine in detail later.

  Line 14 declares a second constructor. This constructor takes two real arguments (named xx and yy). This constructor enables us to declare a point using the following syntax,

  ```
  Point Q(-3.2, 4.7);
  ```

  This declaration creates a new point at location $(-3.2, 4.7)$.

- Lines 17–23 declare the methods for the class `Point`.

  The methods `getX` and `getY` are used to learn the values held by x and y, and the methods `setX` and `setY` are used to modify the coordinates.

  Similarly, `getR` and `getA` are used to learn the point's polar coordinates and `setR` and `setA` are used to change them.

  Methods to inspect and to modify the data held in an object are extremely common. We need methods such as these because a well-designed class forbids direct access to its data. This is called *data hiding*.

Line 23 declares a procedure named `rotate`. Invoking `rotate` causes the point to be relocated at a new location by rotating the point about the origin through a specified angle. For example, if the point P is situated at coordinates $(1,2)$, then after the statement `P.rotate(M_PI/2);` the point will be located at $(-2,1)$. (The symbol `M_PI` is defined[1] in the `cmath` header; it stands for π.)

- Lines 24–25 are *operator* declarations. C++ does not know how to compare two objects of type `Point`. If we want a statement of the form `if(P==Q)` ... in our program, we need to specify how points are to be compared. When C++ encounters an expression of the form `P==Q` or `P!=Q`, it needs to know exactly what to do.

  In the case of the class `Point`, the procedures are quite simple. We study these operators in detail later in this chapter. For now, please observe the ampersand (`&`) in the argument list; operators are invoked using call by reference.

- Lines 29–31 are not part of the `Point` class declaration. These lines declare three procedures that are relevant to dealing with `Points` and could have been declared in a separate header file. However, it is more convenient to have everything pertinent to the `Point` class declared in the same file.

  The procedures on lines 29–30 are used to find the distance and midpoint between two points, respectively.

  The procedure on line 31 is used in statements of the form `cout << P;`. This is explained later.

6.3 Data hiding

We now examine the keywords `private` and `public` in the declaration of the `Point` class.

The declaration of the `Point` class is divided into two sections. The first section is labeled `private:` and under that label are the two data members of the class: `x` and `y`. There are no methods in `Point`'s private section.

Data (and methods) in the private section are accessible only to the class's methods. For example, the `getX` and `setY` methods have access to the data `x` and `y`. Any other procedure (such as `midpoint`) cannot access anything that is in the private section.

The reason for putting data in the private section is to protect those data from tampering. Tampering by whom? You, of course! For a simple class such as `Point`, there is not much that you can harm if you were able to manipulate the data directly.

---

[1] The symbol `M_PI` might not be defined by all C++ compilers.

However, later when we build more complicated classes (such as `Permutation`), if you access the data directly you could easily put the object into an invalid state.

Another reason for hiding your data (from yourself) is the ability to change the implementation. For the `Point` example, you may decide that it was a bad idea to save data in rectangular coordinates because the vast majority of your work is in polar. If your other programs were able to access directly the data in the `Point` class, you would need to rewrite all your programs were you to decide to change the internal representation of `Point` to polar.

However, by hiding the data, your other programs cannot rely on how you represent points. All they use are the various *get* and *set* methods. You can rewrite your class's methods and then all programs that use the `Point` class will work. In a sense, other procedures that use `Point` won't "know" that anything is different.

The second part of the declaration follows the `public:` label. All of the methods in the public section are accessible to any procedure that uses objects of the `Point` class. It is possible to have a public data member of a class, but this is a bad idea. Because some classes (not yours!) do use public data, I will tell you how to use public data, but you must promise me that you will *never* make use of this ability in your own classes.

Suppose we had placed x and y in the public section of the class declaration. If P is an object of type `Point`, then we could refer to the coordinates of the point using the notation `P.x` and `P.y`. For example, in lieu of

```
Point P;
P.setX(-4.2);
P.setY(5.5);
```

we could write

```
Point P;
P.x = -4.2;
P.y = 5.5;
```

Here is a mathematical analogy to data hiding. In analysis, all we need to know about the real number system is that it is a complete ordered field. Two analysts—let's call them Annie and Andy—may have different preferences on how to define the reals. Annie prefers Dedekind cuts because these make the proof of the least upper bound property easy. Andy, however, prefers equivalence classes of Cauchy sequences because it is easier to define the field operations. In both cases, the analysts' first job is to verify that their system satisfies the complete ordered field axioms. From that point on, they can "forget" how they defined real numbers; the rest of their work is identical. The "private" data of real numbers is either a Dedekind cut or a Cauchy sequence; we need not worry about which. The "public" part of the real numbers are the complete ordered field axioms.

6.4 Constructors

A *constructor* is a method that is invoked when an object of a class is declared. For example, the following code

```
Point P;
```

invokes the `Point()` constructor method for P. Recall (lines 13 and 14 in Program 6.1) that we declared two constructors for the `Point` class in the `public` section. The declarations look like this:

```
Point();
Point(double xx, double yy);
```

The first thing to notice is that the name of the method exactly matches the name of the class; this is required for constructors.

The second thing to notice is that no return type is specified. It is not unusual for procedures not to return values, but such procedures are declared type `void` to designate this fact. However, constructors are not declared type `void` and it is understood that they do not return values.

We are ready to write the actual program for the `Point` constructors. The first version of the constructor takes no arguments. In a separate file (that we call `Point.h`) we have the following.

```
#include "Point.h"

Point::Point() {
    x = y = 0.;
}
```

Please see Program 6.2, lines 4–6 on pages 109–110. The `#include` directive is necessary so the C++ compiler is aware of the `Point` class declaration. In a sense, the header file is an announcement of the class and the `.cc` file fills in the details.

The name of the constructor is `Point::Point()`. The first `Point` means "this is a member of the `Point` class" and the second `Point` means this is a method named `Point`. Because there is no return type and the name of the method matches the name of the class, it must be a constructor.

The method's action is simple; it sets the member variables x and y equal to zero. (The single line `x = y = 0.;` is equivalent to writing `x = 0.;` and `y = 0.;` as two separate statements.)

Notice that x and y are not declared here because they are already declared in the `class Point` declaration. These are member variables of the class `Point`. Even though they are marked `private`, any method that is a member of the `Point` class may use these variables.

There is a second constructor that accepts two arguments. The purpose of this constructor is so we may declare a variable this way:

```
Point Q(-5.1, 0.3);
```

This statement declares a `Point` variable named Q in which x equals −5.1 and y equals 0.3. The code that accomplishes this is:

```
Point(double xx, double yy) {
  x = xx;
  y = yy;
}
```

Notice that the arguments to this constructor are named xx and yy. Although C++ might allow it, we must not name these arguments x and y. We choose names that are similar to the names of the member variables x and y. Some people like the names x_ and y_ for such arguments, but these are more difficult to type.

6.5   Assignment and conversion

Constructors serve a second purpose in C++; they can be used to convert from one type to another. Recall that `cout << 7/2;` writes 3 to the screen because 7 and 2 are integers. To convert 3, or an `int` variable x, to type `double`, we wrap the quantity inside a call to `double` as a procedure like this: `double(7)` or `double(x)`.

In the case of the `Point` class, we did not provide a single-argument constructor. However, we can convert a *pair* of `double` values to a `Point` like this:

```
Point P;
....
P = Point(-5.2, 4.2);
```

The last statement causes an unnamed `Point` object to be created (with $x = -5.2$ and $y = 4.2$) and then be copied into P.

Here is another example in which one `Point` is assigned to another:

```
Point P;
Point Q(-2., 4.);
P = Q;
```

When the first line executes, the point P is created and is located at $(0,0)$ (this is the standard, zero-argument constructor). When the second line executes, the point Q is created as it is located at $(-2,4)$. Finally, line three executes and the value of Q is copied to P. By default, C++ simply copies the x and y values held in Q to the corresponding data in P. Effectively, the assignment `P = Q;` is equivalent to the pair of statements `P.x = Q.x; P.y = Q.y;`. (Such statements cannot appear outside a method of the class `Point` because the data are private.)

This is the default assignment behavior. Sometimes more sophisticated action must be taken in order to process an assignment; we explain when this is necessary and how it is accomplished in Chapter 11.

There is no natural way to convert a single `double` to a `Point`, so we don't define one. We could, however, decide that the real value *x* should be converted to the point

$(x,0)$. In that case, we would add the line `Point(double xx);` to the declaration of the `Point` class (in `Point.h`) and add the following code to `Point.cc`.

```
Point::Point(double xx) {
   x = xx;
   y = 0.;
}
```

Had we done this, a `main` program could contain the following code.

```
Point P;
double x;
....
P = Point(x);
```

6.6   Methods

We now examine the methods `getX`, `setX`, and so on. (See lines 15–23 of Program 6.1.) The first of these is declared `double getX() const;`. The word `double` means that this method returns a real value of type `double`. Next, `getX()` gives the name of the method. The empty pair of parentheses means that this method takes no arguments. (We deal with `const` in a moment.)

We specify the code for this method in the separate file `Point.cc`. Here are the relevant lines.

```
double Point::getX() const {
   return x;
}
```

The beginning exactly matches the corresponding line in `Point.h` except for the prefix `Point::` in front of `getX`. This prefix tells the C++ compiler that the code that follows specifies the `getX` method of the class `Point`. (If the `Point::` prefix were forgotten, the compiler would have no way of knowing that this procedure is part of the `Point` class and would complain that x is undeclared.)

The code couldn't be simpler. The value in the member variable x is returned. This method is able to access x (even though x is `private`) because `getX` is a member of the class `Point`.

Consider the following code.

```
Point P;
Point Q(4.5, 6.0);
cout << P.getX() << endl;
cout << Q.getX() << endl;
```

The first call to `getX` is for the object P. In P, the variable x holds 0, and so 0 is printed when the third line executes. The second call to `getX` is for a different object, Q. When it is invoked, the value 4.5 is returned because that is what is held in Q's member variable x.

Next we consider the code for the setX method. In the header file, this was declared like this: void setX(double xx);. In Point.cc, the code for this method
looks like this:

```
void Point::setX(double xx) {
  x = xx;
}
```

The return type is void because this method does not return any values. The full
name of the method is Point::setX; the prefix identifies this as a method of the
class Point. The method takes one double argument (named xx). Between the
braces is the code for this method: we simply assign the value in the argument xx to
the member variable x.

There is an important difference between getX and setX. The getX method does
not alter the object for which it was called. The statement cout << P.getX();
cannot affect the object P in any way. We certify this by adding the keyword const
in the declaration and definition of the getX method. When the word const appears
after the argument list of a method, it is a certification that the method does not alter
the object to which it belongs. If, in writing the Point::getX method we had a
statement that could change the object (such as x = M_PI;), the compiler would
complain and refuse to compile the code.

Whenever you create a method that is not intended to modify the state of an object,
include the keyword const after the argument list.

Notice that the setX method does not include const after its argument list. This
is because setX is designed to modify the state of the object on which it is invoked.

The definitions of the other get and set methods are similar; see lines 13–58 in
Program 6.2. One special feature is the use of the atan2 procedure in getA. The
atan2 procedure is defined in the cmath header file. It is an extension of the usual
arctan function. Calling atan2(y,x) returns the angle to the point (x, y) regardless
of which quadrant of the plane contains the point. (See Appendix C.6.1 for a list of
mathematical functions available in C++.)

Finally, we examine the rotate method; it is declared like this:

```
void rotate(double theta);
```

In polar coordinates, this method moves a point from (r, α) to $(r, \alpha + \theta)$.

To implement this method, we take advantage of the fact that we have already
created the getA and setA procedures. We use the first to determine the current
angle and the second to change that angle.

Here is the C++ code in Point.cc.

```
void Point::rotate(double theta) {
  double A = getA();
  A += theta;
  setA(A);
}
```

The return type is `void` because this method returns no value. The full name of the method is `Point::rotate` which identifies this as a method in the `Point` class whose name is `rotate`. The method takes a single argument of type `double` named `theta`. The word `const` does not follow the argument list because this method may alter the object on which it is invoked.

The first step is to figure out the current angle of this point; we do this by calling the `getA()` method. In, say, a `main` program, if we want to know the polar angle of a point `P` we would invoke the procedure like this: `P.getA()`. Here, we see the call `getA()` not appended to any object. Why? To what object does `getA()` apply?

Remember that this code is defining a method for the class `Point` and is invoked with a call such as `P.rotate(M_PI);`. Once inside the `rotate` procedure, a disembodied call to `getA` means, apply `getA` to the object for which this method was called. So, if elsewhere we have the call `P.rotate(M_PI);`, once we enter the `rotate` procedure, unadorned calls to `getA` refer to `P`. Likewise, the call to `setA(A)` on the penultimate line is applied to the object on which `rotate` was called.

In other words, when we invoke `P.rotate(t);` (where `t` is a `double` and `P` is a `Point`), the following steps are taken. First a `double` variable named `A` is created and is assigned to hold the polar angle of `P` (which was calculated via a call to `getA`). Next, `A` is increased by `t`. Finally, the polar angle of `P` is changed via a call to `setA`. At the end (the close brace) the variable `A` disappears because it is local to the `rotate` procedure.

6.7 Procedures using arguments of type `Point`

There is nothing special about writing procedures that involve arguments of type `Point`. We declare two such procedures, `dist` and `midpoint`, in `Point.h` (lines 30–31). They are declared outside the class declaration for `Point` because they are not members of the `Point` class. Recall that they are defined as follows.

```
double dist(Point P, Point Q);
Point midpoint(Point P, Point Q);
```

The code for these procedures resides in `Point.cc`. Here we examine the code for `midpoint`.

```
Point midpoint(Point P, Point Q) {
   double xx = ( P.getX() + Q.getX() ) / 2;
   double yy = ( P.getY() + Q.getY() ) / 2;
   return Point(xx,yy);
}
```

Notice that we do not include the prefix `Point::` in the name of this procedure; `midpoint` is not a member of the class `Point`. It is simply a procedure that is no different from, say, the `gcd` procedure we created in Chapter 3.

The procedure takes two arguments and has a return value, all of type `Point`. The code in the procedure is easy to understand. We average the two x and the two y coordinates of the points to find the coordinates of the midpoint. We must use the `getX` procedure and we cannot use `P.x`. The latter is forbidden because this procedure is not a member of the class `Point` and so the `private` protection blocks direct access to the two data elements, `x` and `y`.

The `return` statement involves a call to the two-argument version of the constructor. An unnamed new `Point` is created by calling `Point(xx,yy)` and that value is sent back to the calling procedure. For example, suppose the `main` contains the following code.

```
Point P(5,8);
Point Q(6,2);
Point R;

R = midpoint(P,Q);
```

When `midpoint` is invoked, copies of `P` and `Q` are passed to `midpoint`. Then `midpoint` performs the relevant computations using these copies and creates a point at coordinates $(5.5, 5)$. This return value is then copied to `R`.

The repeated copying is a consequence of call by value. Because a `Point` does not contain a lot of data, we need not be concerned. (On my computer, call by value requires the passage of 16 bytes of data for each `Point` whereas call by reference only passes 4 bytes. This difference is not significant.)

However, for larger, more complicated objects call by reference is preferable. In some cases, call by reference is mandatory; such is the case for the procedures we consider next.

6.8   Operators

C++ gives programmers the ability to extend the definitions of various operations (such as +, *, ==, etc.) to new data types. This is known as *operator overloading*.

For the `Point` class, we overload the following operators: == for equality comparison, != for inequality comparison, and << for output. The first two are implemented in a different way than the third.

We want to be able to compare two `Point`s for equality and inequality. The mechanism for doing this is to define methods for both the == and != operators. (In C++, these are called operators, although as mathematicians we might prefer to call them relations. In this book, we use the C++ terminology to stress the fact that when == is encountered, it triggers the execution of instructions.)

The == operator takes two arguments (the `Point` variables to its left and right) and returns an answer (either `true` or `false`, i.e., a `bool` value). The declaration of == is inside the `Point` class declaration; see line 24 of Program 6.1. We repeat it here.

```
bool operator==(const Point& Q) const;
```

Let's examine this piece by piece.

- The `bool` means that this method returns a value of type `bool` (i.e., either TRUE or FALSE).

- The name of this method is `operator==`. The keyword `operator` means we are ascribing a meaning to one of C++'s standard operations. (You cannot make up your own operator symbols such as `**`.)

- The method takes only one argument, named Q. This is surprising because `==` seems to require two arguments (on the left and right). Because this is a member of the `Point` class, the left argument is the object on which it is invoked and the right argument is the object Q. That is to say, if the expression A==B appears in a program, this method will be invoked with Q equal to B. Later, inside the code for `==`, an unadorned use of x stands for A's x. To use B's data (which is the same as Q's data), we use the syntax Q.x.

- The type of Q is `const Point& Q`. The `Point` means that Q is a variable of type `Point`. The ampersand signals that this is a call by reference. Call by reference is required for operators. This means that we do not copy the argument into a temporary variable named Q. Rather, Q refers to the variable that was handed to the method.

  The `const` inside the parentheses is a promise that this method does not modify Q.

- The keyword `const` appears a second time after the argument list. This `const` is a promise that the procedure does not modify the object on which it is invoked. For the expression A==B, the trailing `const` is a promise that this procedure does not modify A. (The `const` inside the parentheses is a promise that B is unaltered.)

  (There is an alternative way to declare operators that use two arguments. We could have chosen to declare `operator==` as a procedure that is *not* a member of the `Point` class. In that case, we would declare it after the closing brace of the `class` declaration. The declaration would look like this.

  ```
  bool operator==(const Point& P, const Point& Q);
  ```

  The decision to include `==` as a member of the class is somewhat arbitrary; it is mildly easier to write the code when it is a member of the class.)

The declaration of the `!=` operator is analogous.

Now we need to write the code that is invoked when we encounter an expression of the form A==B. This code resides in the file `Point.cc`. Here it is:

```
bool Point::operator==(const Point& Q) const {
  return ( (x==Q.x) && (y==Q.y) );
}
```

The beginning of this code matches the declaration in `Point.h`. The only difference is that we prepend `Point::` to the procedure name (`operator==`) to signal that this is a member of the `Point` class. This also implies that the left argument of `==` must be a `Point`.

The program has only one line. It checks if the x and y coordinates of the invoking object match those of `Q`. That is, the statement `A==B` causes this code to compare `A.x` to `B.x` and `A.y` to `B.y`. The variable `Q` is a reference to `B` and the unadorned `x` and `y` refer to the data in `A`.

The declaration for the `!=` operator is nearly identical. Within the `class` declaration we have this:

```
bool operator!=(const Point& Q) const;
```

We could write the procedure (in `Point.cc`) this way:

```
bool Point::operator!=(const Point &Q) const {
  return ( (x != Q.x) || (y != Q.y) );
}
```

This works just as does the `==` operator; we compare the x and y coordinates of the two points to see if they are different.

However, we present a different method for creating the `!=` method. It is conceivable that the steps taken to check if two objects are equal are complicated. It would be useful if we could use the `==` method in defining the `!=` method. We would simply see if the two objects are equal, and then compute the "not" of the result (using the `!` operator).

The problem is this: Only one of the two arguments is named. In the declaration only the right hand argument to `!=` has a name. The left argument is nameless. This does not create a problem if all we need to do is access the left argument's data; we simply refer to the data elements by name. In `x != Q.x` the first `x` is the left argument's `x`. How can we refer to the left argument in its entirety? The solution is this: Use `*this`. The expression `*this` is a reference to the object whose method is being defined. This enables us to build on the `==` procedure and use it in writing the `!=` procedure. Here is the code (from `Point.cc`).

```
bool Point::operator!=(const Point &Q) const {
  return ! ( (*this) == Q );
}
```

The `return` statement first invokes the `==` method on two objects. The left-hand argument is `*this` referring to the object on which `!=` is called and the right-hand argument is `Q`. For example, if this were invoked by the expression `A!=B` in some other procedure, then `*this` is `A` and `Q` is `B`. By *is* we mean that `*this` is not a copy of `A`, but `A` itself. Likewise, `Q` is not a copy of `B`, but is `B` itself.

(Extra for experts: The expression `*this` consists of two parts: the operator `*` and the pointer `this`. The `this` pointer always points to the location that holds the object whose method is being defined. The ∗ operator *dereferences* the pointer; this means that ∗this is the object housed at location this. In other words, ∗this is the object for whom the current method is being applied.)

The final operator we consider in this chapter is the << operator. We want to be able to write points to the computer screen with a statement of the form cout<<P;. The name of the procedure that does this is, not surprisingly, operator<<. However, we cannot declare this within the boundary of the class declaration because the object on the left is not of type Point. What is the type of cout? This has been hidden from you because cout is defined in the header file iostream.

The object cout is of type ostream (which stands for *output stream*) and is declared in the header iostream. Therefore, the expression cout << P contains two arguments: the left argument is of type ostream and the right is of type Point. Furthermore, the result of this expression is also of type ostream. A statement of the form

```
cout << P << " is in quadrant I" << endl;
```

contains the operator << three times. The order of operations rules for C++ (which you do not need to know) insert implicit parentheses into this expression to dictate the order in which the three different << procedures are called. With the hidden parentheses revealed, the expression looks like this:

```
( (cout << P) << " is in quadrant I" ) << endl;
```

The leftmost << is executed first. The effect of this operation is to print P on the screen (we see how that is done in a moment) and the result of the operation is to return cout. (We do not return a copy of cout, but the object itself—we explain how to do that in a moment.)

After the first << executes and returns cout what remains is this:

```
( cout << " is in quadrant I" ) << endl;
```

Now the second << is called. The effect is to send the character array to the screen, and again cout is returned. Finally we are left with this:

```
cout << endl;
```

This causes the a new line to be started on the screen.

With this background we are ready to declare and define the << operator for Point. In the file Point.h (but outside the class declaration) we have the following (line 31 of Point.h).

```
ostream& operator<<(ostream& os, const Point& P);
```

Let's examine this declaration piece by piece.

- The return type of this procedure is ostream&. When we invoke the << operator via the expression cout << P the result of the procedure is cout, an object of type ostream. The ampersand indicates that this procedure returns a reference to (not a copy of) the result. We explain how this is accomplished when we examine the code.

- The name of this procedure is `operator<<`.

- The procedure takes two arguments. The first is the left-hand argument to `<<` and the second is taken from the right.

  The first (left) argument is of type `ostream`. We call this argument `os` which stands for "output stream". Recall that `<<` can be used with either `cout` or `cerr`; these are both objects of type `ostream`. The call is by reference, not by value. Objects of type `ostream` are large and complicated; we do not want to make a copy of them. Furthermore, operators should be invoked using call by reference. This argument is not declared `const` because the act of printing changes the data held in the `ostream` object.

  The second (right) argument is of type `Point`. Again, we use call by reference because that is what C++ requires for operators. Printing a `Point` does not affect its data; we certify this by flagging this argument as `const`.

When we execute the statement `cout << P`, in effect we create a call that could be thought of like this: `operator<<(cout,P)`.

Next we write the code for the `<<` operator. A nice format for the output is to print the two coordinates separated by a comma and enclosed in parentheses. Here is the code that makes this work.

```
ostream& operator<<(ostream& os, const Point& P) {
  os << "(" << P.getX() << "," << P.getY() << ")";
  return os;
}
```

Watch what happens when we encounter the statement `cout << A << endl;`. The left `<<` is executed first. We pass `cout` and `A` to the procedure. The first argument `os` becomes `cout` and the second argument becomes `A`. (Remember, in call by reference, the arguments are not copies of the calling arguments, but the arguments themselves.)

Effectively, the first line inside the procedure is this:

```
cout << "(" << A.getX() << "," << A.getY() << ")";
```

(The argument `os` becomes `cout` and the argument `P` becomes `A`.) If `A` is the point $(2,5)$, this line causes $(2,5)$ to be written on the computer's screen.

The second line in the procedure returns `os`. Because the return type of this procedure is `ostream&` (as opposed to plain `ostream`), this return does not send a copy of `os` back to the invoking procedure; it sends `os` itself. In this example, `os` is precisely `cout`, so the result of this procedure is `cout` (and not a copy of `cout`).

Therefore, after the first `<<` in `cout << A << endl;` finishes, the statement is reduced to this: `cout << endl;`.

To summarize, here is the file `Point.cc` in its entirety.

Program 6.2: Code for the `Point` class methods and procedures.

```
1   #include "Point.h"
2   #include <cmath>
3
4   Point::Point() {
5     x = y = 0.;
6   }
7
8   Point::Point(double xx, double yy) {
9     x = xx;
10    y = yy;
11  }
12
13  double Point::getX() const {
14    return x;
15  }
16
17  double Point::getY() const {
18    return y;
19  }
20
21  void Point::setX(double xx) {
22    x = xx;
23  }
24
25  void Point::setY(double yy) {
26    y = yy;
27  }
28
29  double Point::getR() const {
30    return sqrt(x*x + y*y);
31  }
32
33  void Point::setR(double r) {
34    // If this point is at the origin, set location to (r,0)
35    if ( (x==0.) && (y==0.) ) {
36      x = r;
37      return;
38    }
39
40    // Otherwise, set position as (r cos A, r sin A)
41    double A = getA();
42    x = r * cos(A);
43    y = r * sin(A);
44  }
45
46  double Point::getA() const {
47    if ( (x==0.) && (y==0.) ) return 0.;
48
49    double A = atan2(y,x);
50    if (A<0) A += 2*M_PI;
51    return A;
52  }
53
54  void Point::setA(double theta) {
```

```
55      double r = getR();
56      x = r * cos(theta);
57      y = r * sin(theta);
58   }
59
60   void Point::rotate(double theta) {
61      double A = getA();
62      A += theta;
63      setA(A);
64   }
65
66   bool Point::operator==(const Point& Q) const {
67      return ( (x==Q.x) && (y==Q.y) );
68   }
69
70   bool Point::operator!=(const Point &Q) const {
71      return ! ( (*this) == Q );
72   }
73
74   double dist(Point P, Point Q) {
75      double dx = P.getX() - Q.getX();
76      double dy = P.getY() - Q.getY();
77      return sqrt(dx*dx + dy*dy);
78   }
79
80   Point midpoint(Point P, Point Q) {
81      double xx = ( P.getX() + Q.getX() ) / 2;
82      double yy = ( P.getY() + Q.getY() ) / 2;
83      return Point(xx,yy);
84   }
85
86   ostream& operator<<(ostream& os, const Point& P) {
87      os << "(" << P.getX() << "," << P.getY() << ")";
88      return os;
89   }
```

Once a class has been created (or even during its development) it's important to test its features. Here is a program to check the various aspects of the `Point` class.

Program 6.3: A program to check the `Point` class.

```
1    #include "Point.h"
2    #include <iostream>
3    using namespace std;
4
5    /**
6     * A main to test the Point class.
7     */
8
9    int main() {
10     Point X;        // Test constructor version 1
11     Point Y(3,4);   // Test constructor version 2
12
13     cout << "The point X is " << X << " and the point Y is "
14          << Y << endl;
15     cout << "Point Y in polar coordinates is ("
```

```
16              << Y.getR() << "," << Y.getA() << ")" << endl;
17
18      cout << "The distance between these points is "
19              << dist(X,Y) << endl;
20      cout << "The midpoint between these points is "
21              << midpoint(X,Y) << endl;
22
23      Y.rotate(M_PI/2);
24      cout << "After 90-degree rotation, Y = " << Y << endl;
25
26      Y.setR(100);
27      cout << "After rescaling, Y = " << Y << endl;
28
29      Y.setA(M_PI/4);
30      cout << "After setting Y's angle to 45 degrees, Y = " << Y << endl;
31
32      Point Z;
33      Z = Y;   // Assign one point to another
34      cout << "After setting Z = Y, we find Z = " << Z << endl;
35
36      X = Point(5,3);
37      Y = Point(5,-3);
38
39      cout << "Now point X is " << X << " and point Y is " << Y << endl;
40      if (X==Y) {
41         cout << "They are equal." << endl;
42      }
43
44      if (X != Y) {
45         cout << "They are not equal." << endl;
46      }
47
48      return 0;
49  }
```

The output of this `main` follows.

```
The point X is (0,0) and the point Y is (3,4)
Point Y in polar coordinates is (5,0.927295)
The distance between these points is 5
The midpoint between these points is (1.5,2)
After 90-degree rotation, Y = (-4,3)
After rescaling, Y = (-80,60)
After setting Y's angle to 45 degrees, Y = (70.7107,70.7107)
After setting Z = Y, we find Z = (70.7107,70.7107)
Now point X is (5,3) and point Y is (5,-3)
They are not equal.
```

6.9   Exercises

6.1 To complement the `Point` class created in this chapter, create your own `Line` class to represent a line in the Euclidean plane. The class should have the following features.

- Because every line in the Euclidean plane can be represented by an equation of the form $ax + by + c = 0$ (where a and b are not both zero), the `Line` class should hold three private data elements: a, b, and c.

- The class should include the following constructors.
  - A zero-argument default constructor. Choose a sensible behavior for this constructor (e.g., construct the line $y = 0$).
  - A two-argument constructor whose input arguments are both type `Point`. Of course, this should create the line through these points. Hint: In order for the `Line` class to use `Point` objects, you need to have `#include "Point.h"` at the top of your `Line.h` file.
  - A three-argument constructor whose `double` arguments are simply assigned to a, b, and c.

  In the latter two cases, give sensible behaviors in the event that the user gives bad inputs (either the two points are the same or $a = b = 0$).

- Include get methods to inspect the values held in a, b, and c.

- Include methods named `reflectX()` and `reflectY()` that reflect this `Line` through the *x*- and *y*-axis, respectively.

- Include a method to check if a given `Point` is on the `Line`.

- Include a method that generates a `Point` on the `Line`.

- Include an `==` operator to check if two `Line` objects are the same. Note: This is not as simple as checking that a, b, and c are the same for the two lines.

- Include an operator `<<` for printing a `Line` to the screen (e.g., `cout<<L;` where L is type `Line`). Pick a sensible format for the output.

- Include a procedure (not a method in the class) that calculates the distance between a `Point` and a `Line`. The procedure should accept the two arguments in either order: `dist(P,L)` or `dist(L,P)`.
  Hint: The standard C++ procedures `sqrt(x)` (for $\sqrt{x}$) and `fabs(x)` (for $|x|$) may be of assistance here. Some systems may require the directive `#include <cmath>` to use these. See Appendix C.6.1.

6.2 To test the `Line` class from Exercise 6.1, a programmer wrote the following code.

```
#include "Line.h"
#include <iostream>
using namespace std;

int main() {
   Point X(5,3);
   Point Y(-2,8);
   Line L(X,Y);
   cout << "X = " << X << endl;
   cout << "Y = " << Y << endl;
   cout << "The line L through X and Y is  " << L << endl;

   Point Q;
   Q = L.find_Point();
   cout << "Q = " << Q << " is a point on L" << endl;
   Line M(X,Q);
   cout << "The line M through X and Q is " << M << endl;
   cout << "Are lines L and M the same?\t" << (L==M) << endl;
   cout << "Is Y incident with M?\t" << M.incident(Y) << endl;
   cout << "Distance from Y to M is zero?\t"
        << (dist(Y,M)==0) << endl;
   return 0;
}
```

In this program, we establish two points $X = (5,3)$ and $Y = (-2,8)$, and the line L through them.

Next we construct a point Q on L and then we construct another line M through X and Q. Because the points X, Y, and Q are collinear, it must be the case that L and M are the same line. However, when the code is run, we see the following output.

```
X = (5,3)
Y = (-2,8)
The line L through X and Y is  [5,7,-46]
Q = (0,6.57143) is a point on L
The line M through X and Q is [3.57143,5,-32.8571]
Are lines L and M the same?    0
Is Y incident with M?   0
Distance from Y to M is zero?   0
```

The program reports that $L \neq M$, that Y is not on the line M, and that Y is a nonzero distance away from M. All of these are mathematically incorrect. What's wrong? How might these problems be addressed?

6.3 In the `Line` class developed in these exercises, we represent a line as a triple (a,b,c) standing for the equation $ax + by + c = 0$. Alternatively, we could represent a `Line` as a pair of `Point`s. How would the header and code files for the `Line` class need to be modified were we to decide to switch to this alternative? How would programs that use the `Line` class need to be modified?

6.4 Create a procedure to test if two `Line` objects represent intersecting lines and, if not, to find their `Point` of intersection.

6.5 Create a `LineSegment` class. The data for the class should be two `Point` objects representing the end points of the segment. You may consider making these data elements `public` even though this practice is usually discouraged. Why might this be acceptable in this case?

6.6 Suppose we wish to add a `translate` method to the `Point` class developed in this chapter. The effect of invoking `P.translate(dx,dy)` would be to move the point from its current location (x,y) to the new location $(x+dx,y+dy)$. In addition (and this is the point of this exercise), this method should return the new value of `P`. For example, consider this code:

```
Point P(4,5);
cout << P.translate(1,2) << endl;
```

This should print `(5,7)` on the computer's screen.

This procedure is declared by adding the following line to the `public` section of the `Point` declaration in `Point.h`.

```
Point translate(double dx, double dy);
```

Explain how to code this procedure in the `Point.cc` file.

6.7 In Exercise 6.6 we considered how to add a `translate` procedure to the `Point` class. With this addition in place, consider the following code in a `main()`.

```
Point P(3,5);
(P.translate(1,2)).translate(10,10);
cout << P << endl;
```

This causes `(4,7)` to be printed on the screen.

We might have expected `(14,17)` to have been printed. Explain the behavior of the code.

How can you modify the code to achieve the desired behavior.

Hint: Normally a `return X;` statement returns a copy of `X`. Examine how we coded `operator<<` for the `Point` class to implement a different behavior for `return`.

Chapter 7

Pythagorean Triples

A *Pythagorean triple* is a list of three integers (a,b,c) such that $a^2 + b^2 = c^2$. Such a triple is called *primitive* if the integers are nonnegative and $\gcd(a,b,c) = 1$. For example, $(3,4,5)$ is a Pythagorean triple because $3^2 + 4^2 = 5^2$. Although $(30,40,50)$ and $(-3,4,-5)$ are also Pythagorean triples, they are not primitive.

In this chapter and the next we develop the C++ machinery necessary to find all primitive Pythagorean triples with $a, b, c \leq 1000$ (or any other given value).

7.1 Generating Pythagorean triples

There is a simple method to generate Pythagorean triples. Let z be a Gaussian integer; that is, $z = m + ni$ where $m, n \in \mathbb{Z}$. Consider $|z^4|$. On the one hand,

$$|z^4| = \left(|z|^2\right)^2 = \left(m^2 + n^2\right)^2$$

but

$$|z^4| = \left|z^2\right|^2 = \left|(m^2 - n^2) + (2mn)i\right|^2 = (m^2 - n^2)^2 + (2mn)^2.$$

Therefore

$$(m^2 - n^2)^2 + (2mn)^2 = (m^2 + n^2)^2. \tag{$*$}$$

Because $m, n \in \mathbb{Z}$, it follows that $(m^2 - n^2, 2mn, m^2 + n^2)$ is a Pythagorean triple. This can also be checked directly by expanding both sides $(*)$.

Although every triple of the form $(m^2 - n^2, 2mn, m^2 + n^2)$ is a Pythagorean triple, not all Pythagorean triples arise in this manner. For example, $(9,12,15)$ cannot be so expressed. [*Proof.* If $(9,12,15)$ were of this form, then $2mn = 12$, so $mn = 6$. This implies that $\{m,n\} = \{\pm 2, \pm 3\}$ or $\{\pm 1, \pm 6\}$ none of which yields $(9,12,15)$.]

However, it can be shown that every *primitive* Pythagorean triple is of the form $(m^2 - n^2, 2mn, m^2 + n^2)$. We use this as fact to create a class that represents primitive Pythagorean triples.

7.2   Designing a primitive Pythagorean triple class

We want to design a C++ class to represent primitive Pythagorean triples. The first question is: What shall we call this class? The most descriptive name is the verbose `PrimitivePythagoreanTriple` but it would be painful to have to type that repeatedly. On the other extreme, we could call the class `PPT`, but that would be too cryptic. We settle on a compromise solution: `PTriple`.

How shall a `PTriple` store its data? The easiest solution is to save the triple (a,b,c). Although we could omit c because it can be calculated from a and b, the small bit of extra memory used by keeping c also is not a problem. We don't want to consider $(3,4,5)$ different from any permutation of $(3,4,5)$, so let us agree that a `PTriple` should hold the three integers in nondecreasing order. The three integers are held in `long` variables named `a`, `b`, and `c`.

Every class needs a constructor. Based on our discussion, the constructor should process a pair of integers (m,n) to produce the triple $(m^2 - n^2, 2mn, m^2 + n^2)$. Unfortunately, life is not as simple as assigning

$$a \leftarrow m^2 - n^2, \quad b \leftarrow 2mn, \quad \text{and} \quad c \leftarrow m^2 + n^2.$$

To begin, this may result in a or b negative. Even if both are positive, they may be in the wrong order (we want $0 \le a \le b$). Finally, the three values might not be relatively prime. The only thing of which we can be confident is that $m^2 + n^2$ is nonnegative and the largest of the three.

To handle these issues the constructor needs to correct the signs of a and b, put them in their proper order, and divide through by $\gcd(a,b)$.

One last worry. In case $m = n = 0$, the triple constructed would be $(0,0,0)$. Technically, this is not primitive because the entries are not relatively prime. We have two choices: we can allow $(0,0,0)$ to be a valid `PTriple` or we can forbid it. In either case, the constructor needs to treat this possibility as a special case. In this book, we decide to allow $(0,0,0)$ as a valid `PTriple`.

We also need to provide a constructor with no arguments. This is what is invoked by a declaration of the form `PTriple P;`. What Pythagorean triple should this create? In other words, what should be the default primitive Pythagorean triple? There are three natural choices: $(0,0,0)$, $(0,1,1)$, or $(3,4,5)$. The last is the smallest triple in which none of the elements is zero. For better or worse, we decide that $(0,0,0)$ is the default. If you are unhappy with any of these decisions, you can certainly choose to proceed differently.

What methods do we need? We need to know the values held in `a`, `b`, and `c`, so we plan for methods `getA()`, `getB()`, and `getC()` to report those values. We do not want methods that can alter these values because we do not want to allow a user to put a `PTriple` into an invalid state. If we had a `setA` method, then `P.setA(2);` would result in a `PTriple` that cannot possibly be a Pythagorean triple.

We also want operators to compare `PTriples`. In addition to the `==` and `!=` operators, we define a `<` operator. We need a less-than operator so we can sort `PTriples` in an array and remove duplicates. (Remember, our goal is to find all primitive Pythagorean triples with $0 \leq a \leq b \leq c \leq 1000$.) We have some choice as to how to define $<$ for PTriples; we just need to ensure that our definition of $<$ results in a total order. One simple solution is to order the triples lexicographically. We put $(a,b,c) < (a',b',c')$ provided $c < c'$, or else $c = c'$ and $b < b'$. Of course, if $c = c'$ and $b = b'$, then $a = a'$ and the triples are identical.

Finally, it is useful to define the $<<$ operator so we can print PTriples to the computer's screen.

We are now ready to implement the PTriple class.

7.3 Implementation of the `PTriple` class

As planned, the PTriple class has three private data elements of type long named a, b, and c. The constructors ensure that the following conditions are always met.

$$0 \leq a \leq b \leq c, \qquad \gcd(a,b) = 1, \qquad a^2 + b^2 = c^2. \qquad (*)$$

The only exception is that we allow $(a,b,c) = (0,0,0)$.

We create the class with two constructors: one that takes no arguments and one that takes two long arguments. The first constructs $(0,0,0)$ and the latter uses the method described at the beginning of the chapter. It is the constructor's responsibility to make sure that the conditions in $(*)$ are met. We relegate that to a private helper method named reduce. (Strictly speaking, we do not need a separate procedure for these steps, but we want to illustrate an instance of a private method for a class.)

The public methods for PTriple are getA, getB, getC, and the operators ==, !=, and <. Finally, we have an operator<< procedure that is not a member of the PTriple class.

We now present the header file PTriple.h. In this case, we have not removed the comments. This header file introduces some new ideas that we discuss after we present the file.

Program 7.1: Header file for the PTriple class.

```
1   #ifndef PTRIPLE_H
2   #define PTRIPLE_H
3   #include <iostream>
4   using namespace std;
5
6   /**
7    * A PTriple represents a reduced Pythagorean triple. That is, a
8    * sequence of three nonnegative integers (a,b,c) in nondecreasing
9    * order for which a^2+b^2=c^2 and (a,b,c) are relatively prime. The
10   * relatively prime requirement means that we only deal with primitive
11   * Pythagorean triples. We allow (0,0,0).
12   */
```

```
13  class PTriple {
14  private:
15    /// Shorter leg of the triple
16    long a;
17    /// Longer leg of the triple
18    long b;
19    /// Hypotenuse
20    long c;
21
22    /**
23     * This private method makes sure the triple elements are
24     * nonnegative, in nondecreasing order, and relatively prime.
25     */
26    void reduce();
27
28  public:
29    /**
30     * Default constructor. Makes the triple (0,0,0).
31     */
32    PTriple() {
33      a = b = c = 0;
34    }
35
36    /**
37     * Construct from a pair of integers. Given integers m and n, we
38     * make a Pythagorean triple by taking the legs to be 2mn and
39     * m^2-n^2 and the hypotenuse to be m^2+n^2. We then make sure the
40     * three numbers are nonnegative, in nondecreasing order, and then
41     * divide out by their gcd. For example PTriple(2,1) creates the
42     * famous (3,4,5) triple.
43     */
44    PTriple(long m, long n);
45
46    /// What is the shorter leg of this triple?
47    long getA() const {
48      return a;
49    }
50
51    /// What is the longer leg of this triple?
52    long getB() const {
53      return b;
54    }
55
56    /// What is the hypotenuse of this triple?
57    long getC() const {
58      return c;
59    }
60
61    /**
62     * Check if this PTriple is less than another.  The ordering is
63     * lexicographic starting with c, then b. That is, we first compare
64     * hypotenuses. If these are equal, then we compare the longer leg.
65     * @param that Another PTriple
66     * @return true if this PTriple is lexicographically less than that.
67     */
68    bool operator<(const PTriple& that) const;
```

```
69
70      /**
71       * Check if this PTriple is equal to another.
72       */
73      bool operator==(const PTriple& that) const {
74        return ( (a==that.a) && (b==that.b) && (c==that.c) );
75      }
76
77      /**
78       * Check if this PTriple is not equal to another.
79       */
80      bool operator!=(const PTriple& that) const {
81        return !( (*this) == that );
82      }
83    };
84
85    /// Send a PTriple to an output stream
86    ostream& operator<<(ostream& os, const PTriple& PT);
87
88    #endif
```

The class declaration begins on line 13. Within the `private` section we see the three `long` variables that hold the triple's data. Thinking of a right triangle with side lengths a, b, c, we call a the shorter leg, b the longer leg, and c the hypotenuse.

Also within the `private` section is the `reduce()` method. This is a method invoked by the two-argument constructor `PTriple(m,n)`; we discuss this in more detail below.

The `public` section begins on line 28 beginning with the two constructors. The zero-argument constructor `PTriple()` is not declared in the way we expect. Normally, a procedure declaration does not include the procedure's code. All we expect is this:

```
PTriple();
```

Instead, we see what we would normally expect to find in the `.cc` file.

```
PTriple() {
  a = b = c = 0;
}
```

This is a special feature of C++. Procedures may be defined in header files and are known as *inline* procedures. An inline procedure combines the declaration and definition of a procedure at a single location in the header file.

It is possible to make any procedure into an inline procedure. The keyword `inline` is used to announce this fact. However, for procedures that are members of classes (i.e., methods), the keyword is optional. For all other procedures (i.e., ordinary procedures that are not class members), the `inline` keyword is mandatory.

When should you use inline procedures as opposed to the usual technique of a declaration in the `.h` file and definition in the `.cc` file? If the method's code is only a line or two, then it is a good idea to write it as an inline procedure. Several of `PTriple`'s methods are extremely short and we write those as inline procedures as well. Others (such as the two-argument constructor or the `operator<`) are more involved, so we use the usual technique for those.

There are advantages and disadvantages to writing inline code. It is convenient to write the code for a method in the same file where it is defined. The C++ compiler produces more efficient code from procedures that are written inline. (However, smart C++ compilers can figure out when it is worthwhile to convert an ordinary procedure into an inline procedure.) However, the object code for programs that use inline procedures takes up a bit more disk space. Also, it takes a bit more time to compile programs with inline procedures.

Sometimes it is mandatory to write methods as inline procedures. We explain that when the time comes (Chapter 12).

The procedures that are not specified by inline code in the header file are defined in the file `PTriple.cc` that we present next.

Program 7.2: Program file for the `PTriple` class.

```
1   #include "PTriple.h"
2   #include "gcd.h"
3
4   PTriple::PTriple(long m, long n) {
5     a = 2*m*n;
6     b = m*m - n*n;
7     c = m*m + n*n;
8
9     reduce();
10  }
11
12  void PTriple::reduce() {
13    // Nothing to do if a=b=c=0
14    if ((a==0) && (b==0) && (c==0)) return;
15
16    // Make sure a,b are nonnegative (c must be)
17    if (a<0) a = -a;
18    if (b<0) b = -b;
19
20    // Make sure a <= b
21    if (a>b) {
22      long tmp = a;
23      a = b;
24      b = tmp;
25    }
26
27    // Make sure a,b,c are relatively prime
28    long d = gcd(a,b);
29    a /= d;
30    b /= d;
31    c /= d;
32  }
33
34  bool PTriple::operator<(const PTriple& that) const {
35    if (c < that.c) return true;
36    if (c > that.c) return false;
```

```
37
38    if (b < that.b) return true;
39    return false;
40 }
41
42 ostream& operator<<(ostream& os, const PTriple& PT) {
43    os << "(" << PT.getA() << "," << PT.getB() << ","
44       << PT.getC() << ")";
45    return os;
46 }
```

The code for the constructor `PTriple(m,n)` is broken into two parts. Given the input integers m,n we set

$$a \leftarrow 2mn, \quad b \leftarrow m^2 - n^2, \quad \text{and} \quad c \leftarrow m^2 + n^2.$$

Then we need to fix a few things up. We need to make sure that a and b are non-negative, that $a \leq b$, and that $\gcd(a,b,c) = 1$. We relegate these chores to the private `reduce` method (which is invoked by the `PTriple` constructor on line 9).

The code for the < operator follows. This operator is used to compare a given `PTriple` with another. If R and P are variables of type `PTriple`, the expression R<P causes the `operator<` procedure belonging to R to execute.

We first compare the c values of R and P; this happens on lines 35–36. The first comparison is `c < that.c`. The first c is the hypotenuse of the left-hand argument of < (i.e., R). The unadorned c refers to the c for the object on which the method was invoked. The c of the right-hand argument needs to be specified, so we must refer to it as `that.c` because the right-hand argument is named `that`. Because P is passed by reference (required for operators) to <, `that.c` is exactly the same variable as P.c.

In short: When R<P is encountered, the unadorned c on line 35 is the c data member of R and the `that.c` on line 35 is the c data member of P.

7.4 Finding and sorting the triples

Our goal is to find all primitive Pythagorean triples (a,b,c) with $0 \leq a \leq b \leq c \leq 1000$. Our strategy is this.

- First, we create an array of `PTriples` by calling `PTriple(m,n)` over a suitably large range of m and n. Because every primitive Pythagorean triple can be obtained from $(m^2 - n^2, 2mn, m^2 + n^2)$ we do not need to consider values of m or n greater than $\sqrt{1000}$.

  The upper bound of 1000 is, of course, arbitrary. Our design permits an arbitrary upper bound of N.

- As the Pythagorean triples are created, we save them in an array. Because the upper bound on a, b, c is arbitrary (N is specified by user input), the array we create needs to be dynamically allocated.

  How big should this array be? In the worst-case scenario, all of the Pythagorean triples we create might be primitive and different from one another. (This is actually a gross overestimate, but serves our purposes.) Because we iterate over all m and n with $1 \le m, n \le \sqrt{N}$, an array that can accommodate N values is large enough.

- Once the array is populated, we sort it. It was in anticipation of the need to sort the array that we defined the < operator.

- Finally, we read through the array printing out unique elements.

Here is the program that does all those steps; we explain the key features after we present the code.

Program 7.3: A program to find Pythagorean triples.

```
1   #include "PTriple.h"
2   #include <iostream>
3   #include <cmath>
4   using namespace std;
5   /**
6    * Find all primitive Pythagorean triples (a,b,c) with 0 <= a <= b <=
7    * c <= N where N is specified by the user.
8    */
9
10  int main() {
11    PTriple* table;       // table to hold the triples
12    long N;               // upper bound on triples.
13
14    // Ask the user for N
15    cout << "Please enter N (upper bound on triples) --> ";
16    cin >> N;
17    if (N <= 0) return 0;    // nothing to do when N isn't positive
18
19    // Allocate space for the table
20    table = new PTriple[N];
21
22    // Populate the table with all possible PTriples
23    long idx = 0;            // index into the table
24    long rootN = long( sqrt(double(N)) );
25
26    for (long m=1; m<=rootN; m++) {
27      for (long n=1; n<=rootN; n++) {
28        PTriple P = PTriple(m,n);
29        if (P.getC() <= N) {
30          table[idx] = P;
31          idx++;
32        }
33      }
34    }
```

```
35
36    // Sort the table
37    sort(table, table+idx);
38
39    // Print out nonduplicate elements of the table
40    cout << table[0] << endl;
41    for (int k=1; k<idx; k++) {
42      if (table[k] != table[k-1]) {
43        cout << table[k] << endl;
44      }
45    }
46
47    // Release memory held by the table
48    delete[] table;
49    return 0;
50  }
```

Note: Some compilers may require #include <algorithm> in order to use the sort procedure (line 37).

Now for the analysis of the code.

- We need a table to hold the PTriples and so we declare a variable table of type PTriple* (pointer to a PTriple, i.e., the head of an array). Because we do not know how large this array needs to be, we cannot declare it with a statement of the form PTriple table[1000];.

- Lines 15–17 ask the user to give us an integer N. If the user gives us an integer that is less than 1, we end the program.

- Line 20 allocates space for table.

- The variable idx, declared on line 23, is used to access elements in the array table.

- The variable rootN is set to $\lfloor \sqrt{N} \rfloor$. The syntax, unfortunately, is awkward. The C++ sqrt procedure (which may require #include <cmath>) takes and returns values of type double. So we first need to cast the value of N into a double, calculate its square root, and then cast back to type long. When a double is converted into a long, the digits to the right of the decimal place are discarded. Therefore, line 24 computes the desired $\lfloor \sqrt{N} \rfloor$.

- Lines 26–34 run through all possible values of m,n with $1 \leq m,n \leq \sqrt{N}$. For each pair, we generate a primitive Pythagorean triple. If its c-value is no larger than N (line 29) we insert its value in the table and increment idx. Therefore, once we finish this double for-loop, the variable idx holds the number of entries we made into the array table.

- Sorting of the table takes place on line 37. The C++ sort procedure can be used on any array of elements provided the type of those elements can be compared with a < operator. Thus, `sort` can be used to sort an array of `long`s or `double`s. In order to sort an array of `PTriples`, we just need to provide a < operator.

The `sort` procedure may require the `algorithm` header.

The sorting is invoked with an unusual syntax: `sort(table, table+idx);`. The first argument, `table`, makes sense. The `sort` procedure needs to know what array to sort. It's the second argument that is difficult to understand.

The second argument to `sort` needs to be a pointer to the location of the first element *beyond the end of the array*. In other words, if the array holds five elements (indexed 0 through 4), then the second argument to sort needs to be a pointer to the nonexistent *sixth* element of the array (index 5).

More generally, `sort` can be used to sort a contiguous block of elements of an array. The first argument to `sort` should be a pointer to the start of the block and the second argument should be a pointer to the first element *after* the end of the block.

We know that `table` is a pointer to the first element of the array `table`. That is, `table` holds the memory address of `table[0]`. In C++, `table+1` evaluates to the address in memory that holds `table[1]`. So `table+idx` is the memory location holding `table[idx]`. However, at this point in our program, `idx` holds a number one greater than the location where we last placed a `PTriple`. This is what `sort` wants and so this is what we do.

Bottom line: To sort an array named, say, `table` that contains, say, 25 elements, the statement we use is this: `sort(table, table+25);`. It's OK to forget all this discussion about pointers and just remember this:

```
sort( table_name, table_name + number_of_elements);
```

- Lines 40–45 step through the elements of `table`. If the current element is different from the previous one, we print it.

- Finally, we release the memory allocated to `table` on line 48. Strictly speaking, this isn't necessary because we are at the end of the program; the memory would be reclaimed automatically. Nevertheless, it is a good habit to be sure every `new` is matched with a `delete[]`.

Here is what we see when the program is run (for $N = 100$).

```
Please enter N (upper bound on triples) --> 100
(0,1,1)
(3,4,5)
(5,12,13)
(8,15,17)
(7,24,25)
(20,21,29)
(12,35,37)
(9,40,41)
```

```
(28,45,53)
(11,60,61)
(33,56,65)
(16,63,65)
(48,55,73)
(36,77,85)
(13,84,85)
(39,80,89)
(65,72,97)
```

We see that there are 17 primitive Pythagorean triples with $0 \le a \le b \le c \le 100$. This implies that the array we created, `table`, used more memory than we really needed.

Let's complain about this program.

- It wastes memory. This is not a terrible problem because holding hundreds, or even hundreds of thousands of Pythagorean triples is well within the capacity of even the most modest computers. Still, we may encounter other situations in which we need to be careful about the amount of memory we use.

- It was annoying that we needed to figure out how much memory to set aside in `table`. It would be much easier if the table could adjust its size to suit our needs, rather than requiring us to figure out how big to make it.

- The call to `sort` is still bugging me. Adding an object of type `PTriple*` and an object of type `long` just seems wrong. (It isn't wrong, but it is confusing.)

We deal with all these complaints in the next chapter.

7.5 Exercises

7.1 Create an `Interval` class to represent closed intervals on the real line: $[a,b] = \{x \in \mathbb{R} : a \le x \le b\}$. Implement this class entirely within an `Interval.h` file and without any code file. To do this, you will need to make all methods and procedures inline.

The class should include the following features.

(a) Two constructors: a zero-argument constructor that creates a default interval (say, $[0,1]$) and a two-argument interval that creates the interval with the specified end points. In response to either `Interval(3,4)` or `Interval(4,3)`, the interval $[3,4]$ should be constructed.

(b) Get methods to reveal the end points of the interval.

(c) Comparison operators for equality (`==`) and inequality (`!=`).

(d) An `operator<` for lexicographic sorting of the intervals. That is, $[a,b] < [c,d]$ provided $a < c$ or ($a = c$ and $b < d$).

(e) An `operator<<` for output to the screen.

7.2 In Exercise 7.1 you created an `Interval` class. Use that class to investigate the following problem. Let $I_1, I_2, \ldots, I_n$ be a collection of random intervals. What is the probability that one of these intervals intersects all of the others?

By *random interval* we mean an interval whose end points are chosen independently and uniformly from $[0,1]$ (with the understanding that $[a,b] = [b,a]$).

Write a program that generates n random intervals and then checks to see if one of those intervals meets all of the others. It should repeat this experiment many times and report the frequency with which it meets success.

7.3 Write a procedure that finds the median in a list of real numbers. Declare the procedure as

```
double median (const double* array, long nels);
```

where `array` is the list of values and `nels` is the number of elements in the array.

Warning: Your procedure should not modify the list (hence the use of the keyword `const`), but should use its values to find the median. Still, a good way to find the middle element(s) is to work with a sorted array. If you wish to use the C++ `sort` procedure, include these lines in your code:

```
#include <algorithm>
using namespace std;
```

Note: If the length of the list is even, there is no single middle value, but rather two "middle" values. In this case, take the median to be the average of those two middle values.

7.4 Write a program to count the number of primitive Pythagorean triples in which (a) one of the legs has length k and (b) in which the hypotenuse has length k, for all k with $1 \le k \le 100$.

Chapter 8

Containers

In this chapter we explore a variety of container classes available in C++. The C++ arrays provide little functionality and can be difficult to use. It is easy to make mistakes using arrays: the arrays might not be big enough to hold the data, we might access elements beyond the end of the array, there is no convenient way to enlarge an array after it has been created, and it is easy to forget a `delete[]` statement for an array that has been dynamically allocated.

In lieu of holding values in arrays, C++ provides the means to create arraylike structures that can change size on demand, unordered collections (sets and multisets), and other useful containers (stacks, queues, double-ended queues, and simple lists). These containers are easy to use.

8.1 Sets

In Chapter 7 we generated primitive Pythagorean triples and collected our results in an array. We then needed to process the array to remove duplicates. Alternatively, with each new triple generated, we could have scanned the array to see if the new triple was already recorded. This, however, increases the processing time and does not solve the problem that we do not know in advance how large to make the array.

The C++ set class behaves much as a finite mathematical set. In a C++ `set`, all the elements of the set must be the same type. That is, the elements of the `set` may be all `long`s or all `PTriple`s; the `set` cannot contain a mixture of types.

To use `set`s in your program, you need to include the `set` header file with the directive `#include <set>`.

To declare a `set` one needs to specify the type of element stored in the set. This is done using the syntax `set<type> set_name;`. For example, to declare a set of integers, use the following statement,

```
set<long> s;
```

The variable `s` is now a set that can hold `long` integer elements. Any C++ type (either innate or a class you define) can be held in a `set` with only one proviso: The type must have `<` and `==` operators defined.

There are fundamental operations we need to be able to perform on sets such as adding an element to a set, deleting an element from the set, and so on. Here are the important methods you need to know to use C++ sets. In each case, S and T stand for sets of elements of some type, and e stands for variable of that type. (For example, S and T are sets of longs and e is a long.)

Add an element Use the statement S.insert(e);. This causes a copy of e to be inserted into the set unless, of course, the set already contains this value.

Delete an element Use the statement S.erase(e);. If the set contains the value held by e, that value is removed from the set. If the value is not in the set, nothing changes.

Delete all elements in a set Use the statement S.clear();.

Is an element in the set? Use the method S.count(e). This returns the number of times that e's value appears in the set. This is either 0 or 1.

How many elements are in this set? The method S.size() returns the cardinality[1] of S.

Is a set empty? The method S.empty() returns a bool value: true when S is empty and false otherwise.

Check if two sets are the same The operators == and != work as expected. The expression S==T returns true if S and T contain the same elements.

Copy one set to another The statement S = T; overwrites S with a copy of T.

There are other fundamental operations that one would like to perform on sets including finding the smallest (or largest) element in a set, printing all the elements in a set, and so on. The C++ set type can handle these tasks, but they are more complicated. Before we explain how to do these, we return to Pythagorean triples. We employ a strategy that is similar to the one we used in Program 7.2. We ask the user to input N. We then generate all pairs of integers (m,n) with $1 \le m, n \le \sqrt{N}$. We use these to generate primitive Pythagorean triples. We examine each triple to see if $c \le N$; if so, we add it to a set S. In the end, we print all the elements of the set. The difficult part of the program is printing the set! Here is the code; the explanation follows.

---

[1]The container classes in the Standard Template Library have size() methods that report the number of elements held in the container. The value returned by size() method has type size_t which is often an unsigned value. Consequently, if you compare the return value from size() to, say, a long integer value, the compiler may issue a warning or an error. To fix this problem, you may need to convert the result of the size() method to a long value. Fortunately, this is simple. Replace X.size() with long(X.size()) and all should be well.

Program 8.1: A program to find Pythagorean triples using sets.

```
1   #include "PTriple.h"
2   #include <set>
3
4   int main() {
5     set<PTriple> S;              // Set to hold the PTriples we find
6
7     // Ask the user for N
8     cout << "Please enter N (upper bound on triples) --> ";
9     long N;
10    cin >> N;
11    if (N <= 0) return 0;  // Nothing to do when N isn't positive
12
13    // We only need to run the constructor arguments up to the square
14    // root of N.
15    long rootN = long (sqrt(double(N)) );
16
17    // Run through possibilities and if appropriate, add to the set S.
18    for (int k=1; k<=rootN; k++) {
19      for (int j=1; j<=rootN; j++) {
20        PTriple P(j,k);
21        if (P.getC() <= N)
22          S.insert(P);
23      }
24    }
25
26    // Print out the elements of the set
27    set<PTriple>::iterator si;             // Iterator for S
28    for (si = S.begin(); si != S.end() ; si++) {
29      cout << *si << endl;
30    }
31
32    return 0;
33  }
```

We begin by declaring a set S (line 5) whose job is to hold the triples we find. Lines 8–11 ask the user for N and line 15 calculates $\lfloor \sqrt{N} \rfloor$.

Lines 18–24 step through all values of m, n and use these to generate primitive Pythagorean triples. If the triple's hypotenuse is no greater than N (line 21) then we add it to the set (line 22).

The last part of the program (lines 27–30) prints the set S to the computer's screen. To do this, we need an object called an *iterator* (explained next).

At the end of the program, we do not need to delete the set S. The set is designed to release the memory it consumes when the procedure in which it was declared terminates.

8.2   Set iterators

To access the 8th element of an array a is easy; we just type a[7]. (Remember that the first element of the array is a[0].) However, there is no method to access the 8th element of a set. When programmers built the C++ set class, they chose a design that would make the insertion, deletion, and element-query operations as efficient as possible. These requirements are incompatible with rapid access to the *n*th element for arbitrary *n*.

Nonetheless, it is necessary for users to be able to access the elements held in a set. To do so, they can fetch the elements sequentially using an object called an *iterator*.

A set iterator is an object that refers to elements of a set. The iterator specifies a "current" element of the set. There are three important operations set iterators can perform:

(a) They can report the value at the "current" location,

(b) They can advance to the next element in the set, and

(c) They can step back to the previous element of the set.

An iterator for a set is declared with a statement like this:

```
set<long>::iterator si;
```

This declares si to be an iterator for sets that contain long integers. The declaration does not give si any particular value; it is not ready to report on elements of a set.

C++ set objects include a method named begin(). This method returns an iterator to the first (i.e., smallest) element of the set. So, if S is a set of long integers, we can initialize si with a statement like this:

```
si = S.begin();
```

Note that si is *not* a long integer; it is an iterator. Hence a statement of the form cout << si does not print the smallest element of the set. To access the value that the iterator considers to be current, we use this syntax: *si. (The * operator is designed specifically to mimic the dereferencing notation used by pointers.) Therefore, the following code does what it says.

```
set<long> S;
set<long>::iterator si;
....
if (!S.empty()) {
  si = S.begin();
  cout << "The smallest element in S is " << *si << endl;
}
else {
  cout << "The set S is empty" << endl;
}
```

We push an iterator forward and backward using the increment ++ and decrement -- operators, respectively. That is, the statement si++ (or ++si) advances the iterator to the next element of a set and si-- (or --si) moves the operator back one step.

We need to take care that we don't run past the end of the set or rewind before its beginning. The set class provides the methods begin() and end() for this purpose.

As we mentioned, S.begin() returns an iterator for the first element of a set, S. Unfortunately, S.end() does *not* return an iterator to the last element of the set! Instead, it returns an iterator one step past the end. (This design is consistent with the idea that if an array a contains n elements, then a[n] is one step past the end of the array.) If the set s is empty, S.begin() and S.end() give identical results.

(In addition to begin and end, there is a pair of reverse iterators named rbegin and rend. Invoking S.rbegin() returns an iterator to the last element of the set and S.rend() returns an iterator to a position one step before the first element. These are useful if we want to step through the elements of a set from largest to smallest.)

It is possible to compare iterators with the operators == and !=.

With these ideas in place, here is how to print all the elements of a set to the computer's screen.

```
set<long> S;
set<long>::iterator si;
....
for (si = S.begin(); si != S.end(); si++) {
  cout << *si << " ";
}
cout << endl;
```

This code prints the elements of s on a single line with spaces separating the elements.

Suppose we wish to sum the elements of s. Here is how we can do it.

```
long sum = 0;
for (si = S.begin(); si != S.end(); si++) {
  sum += *si;
}
```

You may not use set iterators to change a value in a set. The expression *si may not appear to the left of an = sign (assignment operator).

Please refer to lines 27–30 of Program 8.1. There we declare an iterator si for sets of Pythagorean triples. We then use it to print (one per line) all the elements of the set s.

Here is how to use iterators to create procedures to check if one set is a subset of another and to compute the union and intersection of sets.[2]

---

[2]The code given here compiles and runs as expected when using the g++ compiler, but not with Microsoft Visual Studio. Here's why and how to fix the problem.
Explanation: Some iterators are capable of modifying the containers to which they refer (not set, but

```
bool subset(const set<long>& A, const set<long>& B) {
  if (A.empty()) return true;
  set<long>::iterator si;
  for (si=A.begin(); si!=A.end(); si++) {
    if (B.count(*si) == 0) return false;
  }
  return true;
}

void Union(const set<long>& A, const set<long>& B, set<long>& C) {
  C = A;
  set<long>::iterator si;
  for (si = B.begin(); si != B.end(); si++) {
    C.insert(*si);
  }
}

void Intersection(const set<long>& A, const set<long>&B, set<long>& C){
  C.clear();
  set<long>::iterator si;
  for (si = A.begin(); si != A.end(); si++) {
    if (B.count(*si)>0) {
      C.insert(*si);
    }
  }
}
```

The first procedure checks if $A \subseteq B$. The next two procedures set C equal to $A \cup B$ and $A \cap B$, respectively.

In all three procedures arguments A and B are declared const set<long>&. This means that A and B are sets of long integers and are passed by reference. Call by reference is important in this instance. The sets might contain tens of thousands of elements. If we used call by value (i.e., if we omitted the &), then the set would be copied and this takes time. By passing a reference, the procedure receives its data instantaneously. The const keyword is an assurance that the code that follows does not alter the sets A or B.

Argument C in the second and third procedures holds the result of the computation. In order for the procedures to modify argument C it must be a reference variable (hence the & is required).

The union and intersection procedures do not return a result; they are both type void. It might be tempting to declare the union procedure like this:

others such as vector and list discussed later in this chapter). If an argument to a procedure is a container and declared to be a const reference, then using an iterator referring to that container could violate the const guarantee (because the iterator might modify the object). The g++ compiler allows us to use set iterators for const objects as there is no danger that the set can be modified through the use of the iterator. Visual Studio, however, takes a stricter approach. It does not allow us to use an iterator that refers to a const container. The solution is use a restricted form of an iterator called a const_iterator; we discuss these on pages 145–147.
Solution: To make this code usable in the stricter context of Visual Studio, change all instances of iterator to const_iterator.

```
set<long> Union(const set<long>& A, const set<long>& B);
```

Inside the procedure, we could build the union in a set named C and conclude with `return C;`.

Such a procedure would work, but it would be inefficient. The problem is that the return variable would be copied to the calling procedure.

We named the union procedure with a capital letter: `Union`. Why not name this simply `union`? The problem we are avoiding is that `union` is a C++ keyword (not one that is of interest to mathematicians). We cannot name a procedure `union`, just as we cannot name a procedure `if` or `for`. (See Section 15.5.3 if you are curious.)

These procedures work perfectly for sets of `long` integers, but cannot be used for sets of `double` numbers. In C++ there is a mechanism to create a procedure schema (called a *template*) so that one procedure can be used with arguments of many different types. This is explained later (in Chapter 12).

It is possible to initialize an iterator with methods other than `begin` and `end`. Another useful method is `find`. If A is a set and e is an element, then `A.find(e)` returns an iterator for A that is focused on e. However, if e is not in A, then `A.find(e)` signals its inability to find e in the set by returning `A.end()`, that is, an iterator that is one step beyond the end of the set.

8.3 Multisets

The directive `#include <set>` also provides the `multiset` type. A `multiset` may contain an element repeatedly. For a set, the `count` method returns only 0 or 1; for multisets, it returns the multiplicity of the element, and this may be any nonnegative integer.

A `multiset` iterator is declared in the expected way:

```
multiset<type>::iterator mi;
```

If an element appears several times in a set, `*mi` gives that value repeatedly as mi is increased. Here is an example.

Program 8.2: A program to demonstrate the use of `multiset`.

```
1  #include <iostream>
2  #include <set>
3  using namespace std;
4
5  int main() {
6    multiset<long> A;
7
8    for (int k=1; k<=5; k++) A.insert(k);  // A <- {1,2,3,4,5}
9    A.insert(3);  // we put an extra 3 into A
10
```

```
11    cout << "The size of A is " << A.size() << endl;
12
13    // print the set
14    multiset<long>::iterator ai;
15    cout << "The elements of A are: ";
16    for (ai = A.begin(); ai != A.end(); ai++) {
17      cout << *ai << " ";
18    }
19    cout << endl;
20    return 0;
21  }
```

The output from this program is this:

```
The size of A is 6
The elements of A are: 1 2 3 3 4 5
```

8.4   Adjustable arrays via the `vector` class

Arrays in C++ can be created in two ways. First, we can specify in our program how large the array should be:

```
long primes[100];
```

Every time the program is run, the array `primes` has the same size. Alternatively, the array can be created while the program is running with a size that is determined at that time:

```
long *primes;
long n;
...
primes = new long[n];
...
delete[] primes;
```

Although this allows the size of an array to be set by the program while it is running, it does not provide all the functionality we might like. For example, if while the program is running we discover that the array is too small, there is no simple way to increase its size. Alternatively, once the array is populated with values, we may discover that we have overestimated its size. There is no simple way to shrink the array down to just the size we need (and thereby release the extra memory). Finally, dynamically allocated arrays should be released with a `delete[]` statement when they are no longer needed. However, it is easy to forget to do this. For example, a procedure might look like this:

```
void procedure() {
  long *list;
  ...
  list = new long[n];
```

```
...
if (something_bad) {
  cout << "Trouble happens" << endl;
  return;
}
...
cout << "All's well that ends well" << endl;
delete[] list;
}
```

If something_bad evaluates to true the procedure ends early. In that case, the procedure never reaches the delete[ ] statement and the memory allocated to list is never released. We have a memory leak. When an array is passed to another procedure, the procedure that receives the array cannot determine the number of elements the array contains; that information would need to be passed as a separate argument.

C++ provides a remedy to the many woes that plague ordinary arrays: "smart" arrays are called vectors. A vector holds elements of a given type in a list. The elements of the vector are accessed exactly as elements in an array. Before declaring any vector variables, the directive #include <vector> is required. Then, to create a vector of, say, long integers, we can use one of the following,

```
vector<long> alpha;
vector<long> beta(100);  // round parentheses, not square brackets!
```

The first creates a vector of longs of size zero and the latter creates a vector of longs that (initially) can hold 100 values.

To learn the size of an array, use the size method: for example, beta.size() returns 100 (based on the declaration above).

Because beta has size 100, its 100 elements can be accessed just as an array is: from beta[0] to beta[99]. It is an error to access beta[100] because that would be beyond the capacity of the vector. Unfortunately, this error may happen silently because vectors do not check if the index they are given is in the proper range.[3]

If the array we declared is not large enough, we can ask it to grow using the resize method.

```
vector<long> beta(100);
beta[99] = 23;
beta.resize(110);
beta[100] = -4;   // OK; we have space up to beta[109]
```

The amount of memory a vector uses is not always its size. Often, when a vector is resized, it grabs more memory than you requested. The resizing operation is time consuming if the vector does not have spare capacity. For example, suppose

---

[3]The vector class provides a method named at( ) that may be used in place of the square brackets notation; that is, we can write vec.at(k) in place of vec[k]. The at( ) procedure checks that its argument is within range. If the index is illegal, then at( ) signals the error by *throwing an exception*; this is a concept discussed in Section 15.3.

the vector beta initially has size 100, and then (as in the example) you resize it to 110 (i.e., with the statement beta.resize(110);). Here is what happens behind the scenes.

- The vector can only hold 100 elements and is out of space. So it requests a new block of memory of size 200 (even though you only asked for a resize to 110).

- The elements held in the old block of memory are copied to the new block of memory.

- The old block of memory is released.

At this point, the size of the vector is 110, but behind the scenes extra space has been grabbed for future expansion. If at this point you request beta.resize(120), no recopying of data is necessary. The vector simply grows into the memory it has already set aside. However, if you request beta.resize(250), the reallocation procedure happens again.

You can inspect and control the amount of extra space a vector holds if you wish. The method capacity() returns the maximum size to which the vector can grow without going through the time-consuming reallocation process. A statement such as beta.reserve(1000); causes the vector to set aside room for 1000 elements. This statement does not affect the *size* of beta; it simply sets aside room for future expansion of the size.

It is not necessary to use capacity and reserve. You may use the default behavior and that should serve you well nearly all of the time.

Invoking the clear() method on a vector erases all its elements and sets it size to zero. The method empty() returns true in case the size of the vector is zero.

One more method: invoking beta.max_size() returns the maximum possible capacity a vector may have on your computer. (On my computer, the result is just over one billion.)

A special type of vector is one that holds bool values. If we declare a Boolean array, such as bool flags[1024]; the array holds each true/false value in at least one byte (or worse). This is an inefficient use of the computer's memory. Instead, we can declare flags like this:

```
vector<bool> flags(1024);
```

With this, we can access individual elements exactly as if flags were an array (type bool*) but the memory use is much more efficient. Each byte of memory can hold eight bits. This is not significant when dealing with only an array whose size is in the thousands, but it becomes an issue when the array has millions or billions of entries. On the other hand, accessing individual entries in a vector<bool> is slower than accessing elements of an array. The vector<bool> needs to do extra work to access individual bits held in its memory.

To illustrate the use of `vector` objects, let us revisit the Sieve of Eratosthenes. We use a `vector<long>` to hold the table of primes. The sieving part of the procedure uses a `vector<bool>`.

The header file `vector-sieve.h` declares the new version like this:

```
long sieve(long n, vector<long>& primes);
```

The first argument gives an upper bound on the primes to be generated. The second is a place to hold the primes. In this instance, we do not need to worry if the object `primes` is large enough to hold the values. It is resized as needed as the algorithm runs. The return value is the number of primes found. Here is the code.

Program 8.3: The Sieve of Eratosthenes revisiting using `vector` classes.

```
1   #include "vector-sieve.h"
2
3   long sieve(long n, vector<long>& primes) {
4     primes.clear();  // erase the sieve
5
6     if (n < 2) return 0;  // no primes < 2
7
8     // Make a table of boolean values. true = prime and false =
9     // composite. We initialize the table to all true.
10    vector<bool> theSieve;
11    theSieve.resize(n+1);
12    for (long k=2; k<=n; k++) theSieve[k] = true;
13
14    long idx = 0;
15    for (long k=2; k<=n; k++) {
16      if (theSieve[k]) {
17        primes.resize(idx+1);
18        primes[idx] = k;
19        idx++;
20        for (long d = 2*k ; d<=n; d+=k) theSieve[d] = false;
21      }
22    }
23    return primes.size();
24  }
```

We know that the sieve table, `theSieve`, needs to run up to `theSieve[n]` so we immediately resize it to hold n+1 values (line 11). The rest of the program does the usual sieving procedure. Each time we add an element to the table (when the condition on line 16 is satisfied), we increase the size of `primes` by one and insert the newly found prime into the last position.[4] Occasionally, the computer needs to reserve larger and larger chunks of memory to hold the growing `vector`. In this program, we trust the default behavior. However, we could have monitored the capacity of the vector and increased it (say, by 100,000 cells) each time it was exhausted.

---

[4]There is an alternative way to add one element to the end of a `vector`: use the push_back method. That is, if `vec` is a `vector`, then `vec.push_back(x)` increases the size of `vec` by one and puts a copy of `x` into the newly created last position.

Here is a main to illustrate the use of the new `sieve` procedure.

```
#include "vector-sieve.h"
#include <iostream>
using namespace std;
/// Test the vector version of sieve.
int main() {
  vector<long> primes;
  long N;

  cout << "Find primes up to what value? ";
  cin >> N;

  // Generate the primes
  sieve(N, primes);

  cout << "We generated " << primes.size() << " primes" << endl;
  cout << "The largest of which is " << primes[primes.size()-1]
       << endl;
  return 0;
}
```

Running this program with N equal to one billion gives the following in a matter of minutes.

```
Find primes up to what value? 1000000000
We generated 50847534 primes
The largest of which is 999999937
```

This would not have been possible on my computer using the old version of `sieve` because the sieve table would have exhausted all available memory; the space efficiency of `vector<bool>` made this possible.

8.5   Ordered pairs

Ordered pairs occur frequently in mathematics and C++ has a convenient mechanism for handling them. The two entries in an ordered pair need not be the same type. To use the C++ type `pair`, you first need the directive `#include <utility>`. Then, the following declaration creates an ordered pair named `couple` whose first entry is a `long` integer and whose second entry is a `double` real number:

```
pair<long, double> couple;
```

The C++ `pair` does not hide its data in a `private` section, so it is easy to extract and modify the entries. The two data members are named `first` and `second`. To set the pair `couple` we defined above to $(6, \pi)$, we use these statements:

```
couple.first = 6;
couple.second = M_PI;
```

Ordered pairs are convenient for procedures that return two values. Rather than modify call-by-reference arguments (as we did in the extended gcd procedure), a procedure can return a pair containing the two quantities of interest.

For example, here is a procedure to simulate the roll of a pair of dice. It returns an ordered pair of random integers (x, y) with $1 \leq x, y \leq 6$.

```
#include <utility>
#include "uniform.h"
using namespace std;

pair<long, long> dice() {
  long a = unif(6);
  long b = unif(6);
  return make_pair(a,b);
}
```

The return statement uses the make_pair procedure; make_pair is a convenient mechanism for the creation of ordered pairs. The two arguments to make_pair can be any C++ type; the compiler knows the types of the arguments and creates a pair in which first and second have the correct type.

Ordered pairs can be compared for equality using the == and != operators. If < is defined for the two types held in the pair, then < can be used to compare the pairs; the comparison is lexicographic comparing first first and then second.

8.6 Maps

A C++ vector can be thought of as a function defined on the finite domain $\{0, 1, 2, \ldots, n-1\}$. The values this function takes may be any C++ type.

C++ provides a generalization of vector called map. A map behaves much as does a (mathematical) function $f : A \to B$ where A is a finite set. Recall that a mathematical function is a set f of ordered pairs (a, b) with the property that if $(a, b), (a, c) \in f$, then $b = c$. Similarly, a C++ map is a container that holds key/value pairs (k, v) with the property that for each key k there can be at most one pair (k, v) held in the map.

A map object is declared like this:

```
#include <map>
...
map<key_type, value_type> m;
```

The #include <map> directive is necessary to define the map type. The types key_type and value_type can be any C++ types as long as key_type can be ordered using <. For example, we can declare a map as map<long,double> f; and then f acts as a function from a finite subset of $\mathbb{Z}$ to $\mathbb{R}$.

Once declared, there are natural operations we can perform with `maps`. In the examples that follow, `f` is a map declared as `map<long,double>`. The variable `k` is a key (hence of type `long`) and the variable `v` is a value (hence of type `double`).

Set $f(k) = v$ for a given key k and value v To insert the key-value pair (k, v) into a map f, the simplest thing to do is to use the following statement,

```
f[k] = v;
```

Alternatively, one can use the `insert` method to add the pair (k, v) to f. The statement looks like this: `f.insert(make_pair(k,v));`. Clearly, the syntax `f[k]=v;` is simpler and clearer.

Note that if a function value is already defined for $f(k)$, then the statement `f[k]=v;` overwrites the old value for $f(k)$.

Determine if $f(k)$ is defined for a given k For this we use the `count` method. The expression `f.count(k)` returns the number of key/value pairs in `f` that contain the key `k`; this is either 0 (`f[k]` is undefined) or 1 (`f[k]` is defined).

Determine the value v associated with a key k In other words, given k, find $f(k)$. The easy way to do this is to use `f[k]`. Provided `f[k]` is defined, this returns the value associated with the key `k`. Thus `f[k]` may appear on either the left or right side of an assignment statement, or in any expression we like.

This leads to the question, what happens when we have a statement such as `cout << f[5] << endl;` but we have not yet defined `f[5]`? First, you should be careful in writing your programs so that this situation does not arise. What happens is that seeing that a value for `f[5]` is needed, the computer assigns a value to `f[5]`. This value is some sort of default value provided by the value's type. In this example, because the values are of type `double`, a tacit `f[5] = 0.;` takes place.

It is risky to rely on default behaviors and much better to be careful in your programming so that you check if `f[5]` is defined (with a statement such as `if(f.count(5)>0){...}`).

Undefine $f(k)$ Given a key k, we might wish to reset `f[k]` to undefined by erasing the key-value pair with key k. The `erase` method does this: `f.erase(5);` deletes $(5, v)$ is there is such a pair in `f`.

Determine the number of key-value pairs in a map The `size` method does just this task. The expression `f.size()` returns the number of key-value pairs held in `f`.

Reset a map to its empty state The statement `f.clear();` clears the map. This results in `f.size()` evaluating to zero.

Check if a map is empty The statement `f.empty()` yields `true` if `f.size()` is zero, and `false` otherwise.

If we wish to examine, one by one, all the pairs held in a map we need to use a map *iterator*. The method for doing this is similar to the one we examine for sets. The declaration for a map iterator looks like this:

```
map<key_type,value_type>::iterator mi;
```

At this point, the map iterator mi is declared, but does not refer to any part of any map. The map class provides the methods begin and end that are analogous to the same-named methods for set. The expression f.begin() returns an iterator that refers to the first ordered pair held in f (assuming f is not empty). The expression f.end() gives an iterator that is positioned one place past the end of the map.

The variable mi is not an element of the map f, but rather is a device for extracting the members of the map. Because a map is a collection of ordered pairs, the expression *mi returns an object of type pair<key_type,value_type>. Let's see how this works with an example.

Program 8.4: A program to illustrate the use of maps.

```
1   #include <iostream>
2   #include <map>
3   using namespace std;
4
5   /**
6    * A program to illustrate the use of maps.
7    */
8
9   int main() {
10    map<long, double> f;   // f is a function from integers to reals
11
12    f[-3] = 0.5;
13    f[2]  = M_PI;
14    f[6]  = 11;
15    f[0]  = -1.2;
16    f[6] = exp(1.);    // notice we are overwriting f[6]
17
18    for (long k=0; k<10; k++) {
19      cout << "The value of f[" << k << "] is ";
20      if (f.count(k) > 0) {
21        cout << f[k] << endl;
22      }
23      else {
24        cout << "undefined" << endl;
25      }
26    }
27
28    cout << "There are " << f.size() << " ordered pairs held in f"
29         << endl <<  "They are: ";
30
31    map<long,double>::iterator mi;
32    for (mi = f.begin(); mi != f.end(); mi++) {
33      long k   = (*mi).first;
34      double v = (*mi).second;
35      cout << "(" << k << "," << v << ") ";
36    }
```

```
37      cout << endl;
38
39      return 0;
40  }
```

The map f is declared on line 10. Lines 12–16 set various values for f[k]. Note that we define f[6] on line 14, but then it is overwritten (with value *e*) on line 16.

Line 18 steps through key values from 0 to 9. If the function is defined for that key, we print out the corresponding value; otherwise, we announce that it is undefined.

Line 31 declares a map iterator mi that we use in lines 32–37 to print out all the ordered pairs held in the map. As mi steps through the map, we extract the data to which it refers. Remember that *mi is a pair, and so to access its two entries, we use first and second. The expression (*mi).first gives the key and (*mi).second gives the corresponding value.[5]

The output of this program follows.

```
The value of f[0] is -1.2
The value of f[1] is undefined
The value of f[2] is 3.14159
The value of f[3] is undefined
The value of f[4] is undefined
The value of f[5] is undefined
The value of f[6] is 2.71828
The value of f[7] is undefined
The value of f[8] is undefined
The value of f[9] is undefined
There are 4 ordered pairs held in f
They are: (-3,0.5) (0,-1.2) (2,3.14159) (6,2.71828)
```

An interesting use for maps is the implementation of look-up tables. Suppose we want to create a procedure for a function f and calculating $f(x)$ is time consuming. It would be useful if the procedure could remember past values that it calculated. That way, we would never have to calculate $f(x)$ twice for the same value of x.

For example, consider the following recursively defined sequence of numbers a_n for $n \in \mathbb{Z}^+$. Let $a_1 = 1$. For $n > 1$, let

$$a_n = \sum_{d|n, d<n} a_d.$$

For example,

$$a_{12} = a_1 + a_2 + a_3 + a_4 + a_6 = 1 + 1 + 1 + 2 + 3 = 8.$$

If we program a procedure to calculate a_n recursively, then the computation of a_{12} requests the values $a_1, a_2, a_3, a_4,$ and a_6. Except for a_1 (the base case) each of these spawns additional calls to the procedure. To prevent this inefficiency, we program the procedure to remember the values it already calculated.

---

[5]There is an alternative to the notation (*mi).first. Instead, we can write mi->first. The C++ expression a->b is defined to mean (*a).b.

To do this, we include a map as a *static* variable in the procedure. A static variable retains its state even after the procedure exits. Here is the code for the procedure that we call `recur`.

Program 8.5: A procedure that remembers values it has already calculated.

```
1   #include "recur.h"
2   #include <map>
3   using namespace std;
4
5   long recur(long n) {
6
7     static map<long, long> lookup;
8
9     if (n <= 0) return 0;
10    if (n == 1) return 1;
11
12    if (lookup.count(n) > 0) return lookup[n];
13
14    long ans = 0;
15    for (long k=1; k<=n/2; k++) {
16      if (n%k == 0) ans += recur(k);
17    }
18
19    lookup[n] = ans;
20
21    return ans;
22  }
```

The look-up table is declared on line 7. It is a map in which keys and values are both of type `long`. It is declared `static` so that its state is preserved between procedure calls.

Lines 8 and 9 handle the base cases. If the user gives an illegal (i.e., nonpositive) value for n, we return 0. For the case $n = 1$, we return $a_1 = 1$.

Next (line 12) we check if we have already calculated a_n for this value of n; if so, we simply return the value we previously computed.

Otherwise (we have not previously computed a_n) we calculate a_n by summing a_d over proper divisors of n (lines 15–17). (Please note that this is done in a rather inefficient manner; we sacrificed efficiency here for the sake of pedagogic clarity.)

Finally, before returning the answer (held in `ans`) we record the value for a_n in the look-up table (line 19).

This procedure is more than twice as fast as a conventional recursive procedure. There is, however, a price to be paid. As more and more a_n values are calculated, they occupy memory. This is a classic time/memory tradeoff. One flaw with this code is that there is no way to clear the memory consumed by the look-up table. We could design the code so that if the user sends a negative value to the procedure, then the memory is released. That is, we could add the following statements to the program.

```
if (n<0) {
  lookup.clear();
```

```
    return 0;
}
```

Just as the `set` type can be extended to the `multiset` type, there is also a class named `multimap` that allows multiple values to be associated with a given key.

8.7   Lists, stacks, and assorted queues

There are additional object container classes available in C++ and we discuss some of them here. Each has its strengths and weaknesses. We give a brief overview of each and then delve into a few details. All of these containers support the following operations.

Erase all elements The statement `C.clear();` erases all the elements in the container `C`.

Determine the size of the container Use `C.size()`.

Check if the container is empty Use `C.empty()`.

Make a copy of the container Use `new_C = C;`.

8.7.1   Lists

A `list` is a container that holds values in a linear structure. One can rapidly insert new elements at the beginning, end, or anywhere in the middle of a list. Deletion of elements at any point in the list is also efficient. However, to access, say, the 17th element of a list, one has to go to the beginning of the list and step forward repeatedly until we arrive at the desired element. There is no way to check if a given element is in the list except by stepping through the list element by element.

To use a `list` in your program, start with the directive `#include <list>`. To declare a variable to be a `list` of elements of type, say, `long`, use a statement such as `list<long> L;`.

Elements of a list can be accessed through iterators. To declare an iterator for a list, use a statement such as this:

```
list<long>::iterator Li;
```

If `L` is a list, `L.begin()` is an iterator for the first element of the list (assuming the list is not empty) and `L.end()` is an iterator that is one step past the end of the list.

Here are some common tasks that one can perform on a list.

Insertion To insert an element `e` at the start of a list `L`, use `L.push_front(e)`. Now `e` is the first value on the list and all the previously held values follow. To insert at the end of the list, use `L.push_back(e);` and now e is the last value held.

More generally, if `Li` is an iterator into a list `L`, then `L.insert(Li,e)` inserts e into the list in front of the element pointed to by `Li`. For example, if the list is $(1,3,5,6,5,-7,2)$ and `Li` points at the -7, then the statement `L.insert(Li,17);` modifies the list so it now holds $(1,3,5,6,5,17,-7,2)$.

The statements `L.push_front(e);` and `L.insert(L.begin(),e);` are equivalent. Similarly, `L.push_back(e);` and `L.insert(L.end(),e);` are equivalent.

Deletion To delete the first element of a list, use `L.pop_front();` and to delete the last element of the list use `L.pop_back();`.

To delete an element referred to by an iterator `Li`, use `L.erase(Li);`.

To delete all elements equal to e, use `L.remove(e);`.

It is also possible to remove all elements that satisfy a given condition. To do this, create a procedure that returns a `bool` value. For example, here is a procedure that checks if an integer is even.

```
bool is_even(long n) {
   return (n%2 == 0);
}
```

Now, to remove all even elements from a list of `long` integers, use the statement `L.remove_if(is_even);`.

Sorting If the list holds elements for which < is defined, the statement `L.sort();` sorts the list into ascending order.

Once a list is sorted, the statement `L.unique();` removes duplicate values.

Modification of a value To change a value held in a list, you need an iterator `Li` focused on its spot in the list. Then you simply assign to `*Li`. For example, here is some code that changes the value held at the second position in a list (assuming the list has at least two elements).

```
list<long> L;
...
list<long>::iterator Li;
Li = L.begin();
Li++;    // now refers to 2nd element of list
*Li = -51;
```

Here is a sample program to illustrate these ideas. There are two new ideas in this program. First, we included all the procedures in one file. Each procedure is defined fully before it is used; therefore the `main()` procedure comes last. In general it is better to put procedures into separate files and their declarations into a header (.h) file; it is just easier to present these all as one file for pedagogic purposes.

Second, we introduce a variation on the iterator concept: a `const_iterator`. We explain that after we present the program.

Program 8.6: A program to demonstrate the use of `list`s.

```cpp
1   #include <iostream>
2   #include <list>
3   using namespace std;
4
5   /// A procedure to print a list to cout
6   void print_list(const list<long>& L) {
7     list<long>::const_iterator Li;
8     for (Li = L.begin(); Li != L.end(); Li++) {
9       cout << *Li << " ";
10    }
11    cout << endl;
12  }
13
14  /// Check if an integer is even (to illustrate remove_if)
15  bool is_even(long n) {
16    return (n%2 == 0);
17  }
18
19  /// A main to illustrate various list operations
20  int main() {
21    list<long> L;
22    L.insert(L.begin(),4);
23    L.insert(L.end(),15);
24    L.insert(L.begin(),7);
25    L.push_front(24);
26    L.push_back(5);
27    L.push_front(99);
28
29    list<long>::iterator Li;
30    Li = L.begin();
31    Li++;              // Focus on 2nd element in the list
32    L.insert(Li,0); // Inserts in front of 2nd element
33
34    cout << "Here is the list: ";
35    print_list(L);
36    cout << "The first element of the list is: " << L.front() << endl;
37    cout << "The last element of the list is: " << L.back() << endl;
38    cout << "The list contains " << L.size() << " elements" << endl;
39
40    L.sort();
41    cout << "And now sorted:    ";
42    print_list(L);
43
44    L.pop_back();
45    L.pop_front();
46    cout << "First and last deleted: ";
47    print_list(L);
48
49    L.remove_if(is_even);
50    cout << "Even values deleted: ";
51    print_list(L);
52    return 0;
53  }
```

The first procedure in this program is used to print a `list` to the computer's screen. We pass the list as a reference variable because this is more efficient (hence the `&` on line 6). Had we passed the list by value, a new copy of the list would have been created and that's a waste of time and memory.

Because this procedure does not modify the list, we certify that with the keyword `const` on line 6.

The next step is to declare an iterator for the list and use that to report each value held in the list. Here's the problem. An iterator can be used to modify list values. If the C++ compiler sees you working with an iterator for the argument `L` it worries that you might change the elements held in the list. Suppose we had declared `Li` in the usual way:

```
list<long>::iterator Li;
```

If we then focus `Li` on an element of `L` (e.g., with `Li = L.begin();`) then the compiler gets upset because we now have the ability to modify `L`.

So, instead of using a "full power" iterator, we instead declare `Li` to be a "read only" iterator—an iterator that can learn the values held in the list, but cannot modify them. Such an iterator is known as a `const_iterator` and it is declared like this (see also line 7):

```
list<long>::const_iterator Li;
```

Now the compiler can stop worrying that we might modify `L` in this procedure. The rest of the `print_list` procedure is straightforward.

(Aside: We could have defined an `operator<<` to send lists to the computer's screen. Then we could write statements like this: `cout << L;`.)

The `is_even` procedure on lines 15–17 is used to illustrate `remove_if` later in the program (line 50).

The `main` procedure appears on line 20 and begins by declaring `L` to be a list of integers. We first insert 4 at the beginning of the list, then 15 at the end, and then 7 at the beginning. The list now stands at $(7, 4, 15)$.

Next we insert 24 at the beginning, 5 at the end, and 99 at the beginning. Now the list is $(99, 24, 7, 4, 15, 5)$.

On lines 30–31 we focus `Li` on the second element of the list (which holds the value 24). The statement `L.insert(Li,0);` inserts the value 0 in front of 24; now the list holds $(99, 0, 24, 7, 4, 15, 5)$.

Lines 34–38 are self-explanatory.

Line 40 sorts the list; it now holds $(0, 4, 5, 7, 15, 24, 99)$.

On lines 44–45 we delete the first and last elements of the list. Now the list holds $(4, 5, 7, 15, 24)$.

On line 49 we delete all elements that evaluate to `true` when processed by the `is_even` procedure. This reduces the list to $(5, 7, 15)$.

Here is the output from the program.

```
Here is the list: 99 0 24 7 4 15 5
The first element of the list is: 99
The last element of the list is: 5
The list contains 7 elements
And now sorted:    0 4 5 7 15 24 99
First and last deleted: 4 5 7 15 24
Even values deleted: 5 7 15
```

8.7.2 Stacks

A `stack` is a container that holds elements in a linear structure. New elements can be added only at one end of the `stack` (called the *top* of the stack). The only element that can be deleted is the one at the top of the stack, and it is also the only one that is visible.

Before declaring a stack variable, use the directive `#include <stack>`. Now to declare a stack of, say, `double` values, use a declaration like this:

`stack<double> S;`

There are four methods you need to know to work with stacks.

Check if the stack is empty The `empty()` method returns `true` if the stack contains no elements.

Add an element to the top of a stack Use the `push` method: `S.push(x);`.

Learn the value held at the top of the stack Use `S.top()`. You may also use the `top` method to modify the topmost value, as in `S.top()=M_PI;`.

Remove the top value held in the stack The statement `S.pop();` removes the top element from the stack.

8.7.3 Queues

A `queue` is a container that holds elements in a linear structure. New elements are added only at one end of the queue (called the *back*). The only element that can be removed from the queue (and the only one whose value is visible) is at the opposite end of the queue (called the *front*).

To use queues in your program, use the directive `#include <queue>`. Declare your queue like this:

`queue<long> Q;`

Here is what you need to know to use queues.

To check if the queue is empty `Q.empty()` returns `true` if the queue is empty and `false` otherwise.

To add an element to the back of the queue `Q.push(e)` inserts the value in `e` to the end of the queue.

To see the value at the front of the queue `Q.front()` gives the value held at the head of the queue.

To delete the value at the front of the queue Use the statement `Q.pop();`.

It is also possible to look (and modify) the last element in the queue with the `back()` method, but this is not in the spirit of queues. The usual mode of operation with queues is to insert values at the back of the queue and not deal with them again until they emerge at the front.

8.7.4 Deques

A `deque` is a double-ended queue (hence its name). It is a container that holds elements in a linear structure. New elements can be inserted or deleted at either end. It is possible to access (and modify) any element of a `deque` quickly using array notation (e.g., `D[4]`). These containers incorporate features of stacks, queues, and vectors.

To use a `deque` in your program, use the directive `#include <deque>` and declare your variables like this:

```
deque<long> D;
```

Here are the most important things you can do with a `deque`.

Add elements To add a value to the back use `D.push_back(e);` and to add a value to the front use `D.push_front(e);`. The "front" of the deque is considered its beginning and the "back" is its end.

Delete elements The methods `pop_front()` and `pop_back()` delete the first and last elements of the deque, respectively.

Inspecting and modifying elements The methods `front()` and `back()` can be used to get the first/last values in the deque, and to modify those values. A statement such as `D.front()=34;` changes the first element's value to 34. Of course, these methods require that the deque be nonempty.

Alternatively, one can use square brackets to get or to modify any value held in a deque. Index 0 always refers to the first (front) element of the deque. Index `D.size()-1` is used for the last. To change the second entry in a deque, one would use a statement like this: `D[1]=86;`. As with arrays and vectors, it is important not to give an out-of-range index.

Deques support iterators (declared like this: `deque<long>::iterator Di;`), but it is easier to use square brackets to work with individual entries.

Here is a short program to illustrate the use of deques; its output follows.

Program 8.7: A program to illustrate the **deque** container.

```
1   #include <iostream>
2   #include <deque>
3   using namespace std;
4
5   int main() {
6     deque<long> D;
7
8     D.push_back(17);
9     D.push_back(23);
10    D.push_front(-9);
11    D.push_front(5);
12
13    D[1] = 0;
14
15    cout << "The first element of the deque is " << D.front() << endl;
16    cout << "The last element of the deque is " << D.back() << endl;
17    cout << "Here is the entire structure: " ;
18    for (long k=0; k<D.size(); k++) {
19      cout << D[k] << " " ;
20    }
21    cout << endl;
22
23    return 0;
24  }
```

```
The first element of the deque is 5
The last element of the deque is 23
Here is the entire structure: 5 0 17 23
```

8.7.5 Priority queues

A `priority_queue` is a container that holds its elements in a treelike structure. The elements must be comparable via the < operation. Elements can be inserted at any time but only the largest is visible and deletable.

To use a `priority_queue` include the <queue> header and declare your variable like this:

```
priority_queue<long> PQ;
```

The relevant operations for a `priority_queue` are these.

Adding a value Use `PQ.push(e);`.

Getting the largest value held Use `PQ.top()`.

Deleting the largest value held Use `PQ.pop();`.

A short program to illustrate these concepts, and its output, follow.

Program 8.8: Demonstrating the use of the `priority_queue` container.

```
1   #include <iostream>
2   #include <queue>
3   using namespace std;
4
5   int main() {
6     priority_queue<long> PQ;
7
8     PQ.push(5);
9     PQ.push(12);
10    PQ.push(0);
11    PQ.push(-7);
12
13    cout << "The elements we pushed into the priority_queue are ";
14    while (!PQ.empty()) {
15      cout << PQ.top() << " ";
16      PQ.pop();
17    }
18    cout << endl;
19    return 0;
20  }
```

```
The elements we pushed into the priority_queue are 12 5 0 -7
```

8.8 Exercises

8.1 In mathematics, the elements of a set can themselves be sets. Show how to declare a set of sets of integers in C++ and write a short program that creates the set $\big\{\{1,2,3\},\{4,5\},\{6\}\big\}$.

8.2 Write a procedure to print a `set` of `long` integers to the screen. The elements of the set should be enclosed between curly braces and separated by commas. (Be sure that the last element of the set is not followed by a comma.) The output should resemble these examples:

```
{1,2,3}      {-1}      {0,1}      {}
```

8.3 Suppose we wish to use sets of complex numbers in a program. As a test, we create this simple program:

```
#include <complex>
#include <set>
using namespace std;

int main() {
```

```
    complex<double> z(2.,3.);
    set< complex<double> > A;
    A.insert(z);
    return 0;
}
```

This program creates a complex number $z = 2 + 3i$ and a set of complex numbers A into which we insert z. Unfortunately, the computer fails to compile this code and, instead, prints out a long stream of error messages that includes a complaint that there is

```
no match for const std::complex<double>& <
const std::complex<double>&' operator
```

What is wrong and how can we fix this?

8.4 Let $a_1, a_2, a_3, \ldots$ be the sequence of numbers defined recursively by

$$a_1 = 1 \qquad \text{and} \qquad a_n = \sum_{d \mid n, d < n} a_d.$$

These are the values calculated by Program 8.5 (page 143).

Prove that a_n is the number of ordered factorizations of n. That is, the number of ways to write $n = f_1 f_2 \cdots f_t$ where the f_i are integers with $f_i > 1$. For example, $a_6 = 3$ because the ordered factorizations of 6 are

$$6 \quad 2 \times 3 \quad 3 \times 2.$$

We have $a_1 = 1$ because the empty product evaluates to 1.

Find another combinatorial description of this sequence.

8.5 Create a class to represent integer partitions. Given a nonnegative integer n, a *partition* of n is a multiset of positive integers whose sum is n; the elements of the multiset are called the *parts* of the partition.

Name your class `Partition` and give it the following capabilities:

- A constructor that creates the empty partition of 0.
- An `add_part` method for adding a part.
- A `get_sum` method for learning the sum of the parts in the partition (i.e., the number partitioned by this `Partition`).
- An `nparts` method that reports the number of parts in the partition.
- A `get_parts` that returns the parts of the partition in a `vector<int>` container.
- An `operator<` to give a total order on the set of integer partitions.
- An `operator<<` for writing `Partition` objects to the screen. Choose an attractive output format. For example, the partition $\{1,1,3,4\}$ of 9 can be written to the screen as `9 = 4+3+1+1`.

8.6 Use the `Partition` class that you developed in Exercise 8.5 to create a program that prints out all the partitions of a positive integer n, where n is specified by the user.

8.7 Create a procedure to calculate binomial coefficients $\binom{n}{k}$ by the following algorithm. When $k = 0$ or $k = n$, set $\binom{n}{k} = 1$. Otherwise, use the Pascal's triangle identity: $\binom{n}{k} = \binom{n-1}{k-1} + \binom{n-1}{k}$. This can be done recursively, but if the recursion is done naively, the same binomial coefficients are recalculated many times. Instead, devise a procedure that never calculates any binomial coefficient more than once.

8.8 Write a procedure to convert an array of `long` integers into a `vector<long>`.

8.9 *Sorting vectors.* Suppose `values` is a `vector<long>` of size 10 and we wish to sort the elements held in `values` into ascending order. We might consider the statement `sort(values, values+10)`; but that is incorrect (and the compiler generates an error message).

Next we guess that (like a `list`) `vector` defines a `sort` method. So we try `values.sort()`; but this is also incorrect.

The difficulty with the first approach is that `values` is an object of type `vector<long>` and, unlike a C++ array, is *not* a pointer to the first element. So the expression `values+10` is illegal. The difficulty with the second approach is that `vector` objects do not define a `sort` method.

How do we sort elements held in a `vector`?

Hint: In place of pointers to the first and one-past-the-last elements usually required by `sort`, we can use iterators.

8.10 Suppose we have an iterator that refers to an element of a set, and then we delete that element from the set. What can you say about the iterator now?

8.11 In light of Exercise 8.10, write a procedure that deletes all odd elements from a set of integers. Such a procedure would be declared such as this:

```
void delete_odds(set<long>& A);
```

8.12 In many instances we want to perform an action on all the elements of a container. To do this, we typically use an iterator like this:

```
set<long> A;
...
set<long>::iterator sp;
for (sp = A.begin(); sp != A.end(); ++sp) {
  // do something with the value *sp
}
```

Because this structure is so frequent, the Standard Template Library provides a procedure named `for_each` that acts just as the code above does. The `for_each` procedure is defined in the `algorithm` header.

A call to `for_each` looks like this:

```
for_each(cont.begin(), cont.end(), proc);
```

Here `cont` is a container (such as a `set` or `vector`) and `proc` is a procedure that takes a single argument of the same type as `cont` houses. Furthermore, `proc` does not have a return value. For example, if `cont` is a
`vector<double>`, then `proc` would be declared `void proc(double x);`.

The statement `for_each(cont.begin(),cont.end(),proc);` is equivalent to this code:

```
vector<double>::iterator vi;
for (vi=cont.begin(); vi!=cont.end(); ++vi) {
  proc(*vi);
}
```

Use the `for_each` procedure to create a `print_set` procedure that prints out
a set of `long` integers to `cout`. The format of the output should look like this:
`{ 1 3 9 }.`

8.13 A *simplicial complex* can be defined combinatorially as a finite set $\mathscr{S}$ of finite
nonempty sets with the property that if $A \in \mathscr{S}$ and $\emptyset \neq B \subseteq A$, then $B \in \mathscr{S}$.
The sets in $\mathscr{S}$ are called the *simplices* of the simplicial complex.

For example, this topological simplicial complex

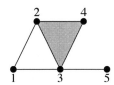

can be written combinatorially as

$$\Big\{ \{1\},\{2\},\{3\},\{4\},\{5\},\{1,2\},\{1,3\},\{2,3\},\{2,4\},\{3,4\},\{3,5\},\{2,3,4\} \Big\}.$$

Create a class called `SimplicialComplex` that holds such structures. For the
sake of efficiency, store only the maximal simplices in the complex. For the
example given above, the maximal simplices are $\{1,2\}$, $\{1,3\}$, $\{2,3,4\}$, and
$\{3,5\}$.

The class should contain a basic constructor (that creates an empty simplicial
complex) and a method to add new simplices. The `add` method should be
careful to deal with the following two situations.

- If the new simplex X is already in the simplicial complex do not add it
  again; we need to check that X is not a subset of one of the maximal
  simplices in $\mathscr{S}$.

- If the new simplex X is not in $\mathcal{S}$, but contains some of the simplices already in $\mathcal{S}$, then we need to update the set of maximal simplices to include X and to delete those that are proper subsets of X.

The class should also contain a method to check if a given simplex X is a member of the complex.

If you are feeling adventurous, create an `erase` method to remove simplices from the complex. Of course, when a simplex X is deleted from $\mathcal{S}$, we must also delete all simplices that contain X. We then need to be careful to update our collection of maximal simplices. If we delete the simplex $\{3,4\}$ from the example in the figure, then we must also delete $\{2,3,4\}$. After the deletion, the simplex $\{2,3\}$ is maximal.

Chapter 9

Modular Arithmetic

Let n be a positive integer. The set $\mathbb{Z}_n$ is $\{0, 1, \ldots, n-1\}$. The set $\mathbb{Z}_n$ is a ring with the following operations.

$$x + y = (x + y) \bmod n \qquad \text{and} \qquad x \cdot y = (xy) \bmod n$$

The goal for this chapter is to develop a C++ class for working in $\mathbb{Z}_n$. Most of the C++ techniques used in this chapter have already been developed in previous chapters, but a few new ideas are presented here as well. The creation of a C++ class for representing elements of $\mathbb{Z}_n$ is necessary for our later work in general finite fields.

9.1 Designing the Mod type

In order to design a C++ class to emulate $\mathbb{Z}_n$ we need to decide what sort of data is stored in each object as well as the methods and operations to be performed on these objects. We also need a name for the new class and we choose the name Mod.

Each object of type Mod represents an element of $\mathbb{Z}_n$ for some positive integer n. The element 8 in $\mathbb{Z}_{10}$ is different from the element 8 in $\mathbb{Z}_{11}$. (Consider the result of the operation $8 + 8$.) Thus, each Mod object needs to hold two integers: the value in $\mathbb{Z}_n$ and the modulus, n. We call these val and mod and we declare these as private members of the Mod class. To represent 8 in $\mathbb{Z}_{11}$, we set val equal to 8 and mod equal to 11.

We need a constructor to create Mod objects, and the natural choice is to have a constructor with two arguments: one that specifies the value and one that specifies the modulus. However, all classes should provide a default constructor that takes no arguments. What sort of object should Mod() construct? A natural choice is to set val to zero, but what of the modulus? One idea is to choose a default modulus that is used when a user does not specify a modulus. We are then faced with a decision: What should that default modulus be? Rather than impose a solution, we let the user decide. So we need a mechanism to set the default modulus. The implementation of this leads us to some new C++ concepts (static class variables and methods) and we explain these later in this chapter.

Now that we have the concept of a default modulus we may also create a single-argument constructor. A call to Mod(x) should create a new Mod object with value x in the default modulus.

So far, the `Mod` class looks like this:

```
class Mod {
private:
   long val;
   long mod;
public:
   Mod();
   Mod(long x);
   Mod(long x, long m);
};
```

The operations and methods we want to perform with `Mod` objects are these.

- Given a `Mod` object, we need to inspect and to modify both its value and its modulus. Changing either the value or the modulus may require us to change the other because a value $x \in \mathbb{Z}_n$ must satisfy $0 \le x \le n-1$.

- Given two `Mod` objects, we should be able to check whether they are equal. In addition, we define a < operator to compare `Mod` objects; this enables us to store `Mod` objects in containers such as a `set` that require a < operator.

- We want to be able to perform the usual arithmetic operations:

  ```
  x+y; x-y; x*y; x/y;
  x += y; x -= y; x *= y; x /= y;
  -x;
  ```

  For these operations, we need to be concerned about two situations that may arise: combining objects of different moduli and division by a noninvertible element of $\mathbb{Z}_n$. We handle these by returning an *invalid* `Mod` object. This invalid value is represented internally with value and modulus equal to zero. (Valid `Mod` objects have a positive modulus.)

  Furthermore, it is convenient to be able to combine `Mod` and `long` objects with a single operation. For example, suppose `x` and `y` are `Mod` objects, then a statement such as `y = x+1;` should assign to `y` a value one greater than `x`'s and the same modulus as `x`.

- In addition to the operations listed above, we want to perform exponentiation a^k where $a \in \mathbb{Z}_n$ and $k \in \mathbb{Z}$. Negative exponentiation gives a valid result provided a is invertible.

9.2   The code

We now present the `Mod.h` and `Mod.cc` files. The header file is long; to make it more manageable for you to read we have removed most of the comments.

In the sections that follow, we work our way through the various features of the `Mod` class that these files implement.

Program 9.1: Header file for the Mod class, Mod.h.

```
1   #ifndef MOD_H
2   #define MOD_H
3   #include <iostream>
4   using namespace std;
5
6   const long INITIAL_DEFAULT_MODULUS = 2;
7
8   class Mod {
9   private:
10    long mod;
11    long val;
12    static long default_modulus;
13    void adjust_val() {
14      val = val%mod;
15      if (val<0) val += mod;
16    }
17
18  public:
19    Mod() {
20      mod = get_default_modulus();
21      val = 0;
22    }
23
24    Mod(long x) {
25      mod = get_default_modulus();
26      val = x;
27      adjust_val();
28    }
29
30    Mod(long x, long m) {
31      if (m <= 0) {
32        val = 0;
33        mod = 0;
34      }
35      else {
36        mod = m;
37        val = x;
38        adjust_val();
39      }
40    }
41
42    long get_val() const { return val; }
43
44    void set_val(long x) {
45      if (mod == 0) return; // no change for an invalid object
46      val = x;
47      adjust_val();
48    }
49
50    long get_mod() const { return mod; }
51
52    void set_mod(long m) {
53      if (m <= 0) {
54        mod = 0;
```

```
55        val = 0;
56      }
57      else {
58        mod = m;
59        adjust_val();
60      }
61    }
62
63    static void set_default_modulus(long m) {
64      if (m <= 0) {
65        default_modulus = INITIAL_DEFAULT_MODULUS;
66      }
67      else {
68        default_modulus = m;
69      }
70    }
71
72    static long get_default_modulus() {
73      if (default_modulus <= 0)
74        set_default_modulus(INITIAL_DEFAULT_MODULUS);
75      return default_modulus;
76    }
77    bool is_invalid() const { return mod==0; }
78
79    bool operator==(const Mod& that) const {
80      return ( (val==that.val) && (mod==that.mod) );
81    }
82
83    bool operator==(long that) const {
84      return (*this) == Mod(that,mod);
85    }
86
87    bool operator!=(const Mod& that) const {
88      return ( (val != that.val) || (mod != that.mod) );
89    }
90
91    bool operator !=(long that) const {
92      return (*this) != Mod(that,mod);
93    }
94
95    bool operator<(const Mod& that) const {
96      if (mod < that.mod) return true;
97      if (mod > that.mod) return false;
98      if (val < that.val) return true;
99      return false;
100   }
101
102   Mod add(Mod that) const;
103
104   Mod operator+(const Mod& x) const { return add(x); }
105
106   Mod operator+(long x) const { return add(Mod(x,mod)); }
107
108   Mod operator+=(const Mod& x) {
109     *this = add(x);
110     return *this;
```

```
111    }
112
113    Mod operator+=(long x) {
114      *this = add(Mod(x,mod));
115      return *this;
116    }
117
118    Mod operator-() const { return Mod(-val,mod); }
119
120    Mod operator-(const Mod& x) const {
121      return (*this) + (-x);
122    }
123
124    Mod operator-(long x) const {
125      return (*this) + (-x);
126    }
127
128    Mod operator-=(const Mod& x) {
129      *this = add(-x);
130      return *this;
131    }
132
133    Mod operator-=(long x) {
134      *this = *this + (-x);
135      return *this;
136    }
137
138    Mod multiply(Mod that) const;
139
140    Mod operator*(const Mod& x) const { return multiply(x); }
141
142    Mod operator*(long x) const { return multiply(Mod(x,mod)); }
143
144    Mod operator*=(const Mod& x) {
145      *this = multiply(x);
146      return *this;
147    }
148
149    Mod operator*=(long x) {
150      *this = multiply(Mod(x,val));
151      return *this;
152    }
153
154    Mod inverse() const;
155
156    Mod operator/(const Mod& x) const { return multiply(x.inverse()); }
157
158    Mod operator/(long x) const {
159      return multiply(Mod(x,mod).inverse());
160    }
161
162    Mod operator/=(const Mod& x) {
163      *this = multiply(x.inverse());
164      return *this;
165    }
166
```

```
167    Mod operator/=(long x) {
168       *this = multiply(Mod(x,mod).inverse());
169       return *this;
170    }
171
172    Mod pow(long k) const;
173
174 };
175
176 ostream& operator<<(ostream& os, const Mod& M);
177
178 inline bool operator==(long x, const Mod& y) {
179    return (y==x);
180 }
181
182 inline bool operator!=(long x, const Mod& y) {
183    return (y!=x);
184 }
185
186 inline Mod operator+(long x, Mod y) {
187    return y+x;
188 }
189
190 inline Mod operator-(long x, Mod y) {
191    return (-y) + x;
192 }
193
194 inline Mod operator*(long x, Mod y) {
195    return y*x;
196 }
197
198 inline Mod operator/(long x, Mod y) {
199    return y.inverse() * x;
200 }
201
202 #endif
```

Program 9.2: Source file for the Mod class, Mod.cc.

```
 1 #include "Mod.h"
 2 #include "gcdx.h"
 3
 4 long Mod::default_modulus = INITIAL_DEFAULT_MODULUS;
 5
 6 ostream& operator<<(ostream& os, const Mod& M) {
 7    if (!M.is_invalid()) {
 8       os << "Mod(" << M.get_val() << "," << M.get_mod() << ")";
 9    }
10    else {
11       os << "INVALID";
12    }
13    return os;
14 }
15
16 Mod Mod::add(Mod that) const  {
```

```
17    if (is_invalid() || that.is_invalid()) return Mod(0,0);
18    if (mod != that.mod) return Mod(0,0);
19    return Mod(val + that.val, mod);
20  }
21
22  Mod Mod::multiply(Mod that) const {
23    if (is_invalid() || that.is_invalid()) return Mod(0,0);
24    if (mod != that.mod) return Mod(0,0);
25    return Mod(val * that.val, mod);
26  }
27
28  Mod Mod::inverse() const {
29    long d,a,b;
30    if (is_invalid()) return Mod(0,0);
31
32    d = gcd(val, mod, a, b);
33
34    if (d>1) return Mod(0,0); // no reciprocal if gcd(v,x)!= 1
35    return Mod(a,mod);
36  }
37
38  Mod Mod::pow(long k) const {
39    if (is_invalid()) return Mod(0,0);   // invalid is forever
40
41    // negative exponent: reciprocal and try again
42    if (k<0) {
43      return (inverse()).pow(-k);
44    }
45
46    // zero exponent: return 1
47    if (k==0) return Mod(1,mod);
48
49    // exponent equal to 1: return self
50    if (k==1) return *this;
51
52    // even exponent: return ( m^(k/2) )^2
53    if (k%2 == 0) {
54      Mod tmp = pow(k/2);
55      return tmp*tmp;
56    }
57
58    // odd exponent: return ( m^((k-1)/2) )^2 * m
59    Mod tmp = pow((k-1)/2);
60    return tmp*tmp*(*this);
61  }
```

9.3   The default modulus: Static class variables and methods

The design of the Mod type calls for the notion of a *default modulus*. The default modulus is employed when the user does not specify a modulus. This occurs with the single-argument constructor `Mod(long)` (described in detail in Section 9.4). Consider the following code.

```
Mod x(6);
Mod y;
y = 5;
Mod z;
z = Mod(1);
```

For x, we explicitly invoke the single-argument constructor to set x equal to the element 6 in $\mathbb{Z}_n$ where n is the current default modulus. The variable y is first built using the default (no-argument) constructor with value 0 in $\mathbb{Z}_n$ where, as before, n is the default modulus. Then the assignment y = 5; implicitly calls the single-argument constructor to give y the value 5 in $\mathbb{Z}_n$. Finally, the variable z is assigned the value $1 \in \mathbb{Z}_n$; in this case, we explicitly invoke the single-argument constructor.

To create and to use a default modulus value, we need to accomplish the following tasks.

- We need a variable to hold the current value of the default modulus.

- We need an initial value for the default modulus.

- We need the ability to inspect and to change the value of the default modulus.

Let's tackle these one at a time.

Where should the value of the default modulus be held? The simplistic answer is to save it in a variable named `default_modulus` of type `long`. However, this does not completely answer the question; we need to know *where* this variable is declared.

We might make `default_modulus` a variable in the `main` procedure, but there are many drawbacks to this: we need to remember to include the declaration and then we would need to pass it to every `Mod` method that might need it. This makes life too difficult for the programmer.

Could we make `default_modulus` an ordinary member variable for the class `Mod`? No. The problem is that each `Mod` object would hold its own "personal" `default_modulus`. We want one and only one value for `default_modulus` that is common to all `Mod` objects.

Could we make `default_modulus` a global variable? A global variable is one that is accessible to all procedures of a program. This is possible, but undesirable. If `default_modulus` were global, any procedure would be able to modify `default_modulus` and set it to a nonsensical value (such as 0 or -1). We want to hide the variable `default_modulus` and limit access by get and set procedures. The set method would ensure that `default_modulus` always holds a sensible value. Furthermore, global variables are a dangerous programming trick; different parts of the program can access and modify these values in unpredictable ways.

What we need is a private variable that "belongs" to the entire class `Mod` and not to any one particular object of type `Mod`. Such a variable is called a *static* class variable.

Inside the private section of the `Mod` class declaration, we have this (line 12 of Program 9.1, `Mod.h`):

```
static long default_modulus;
```

This line announces that the class `Mod` contains a variable named `default_modulus` (of type `long`) and this value is common to all `Mod` objects. (By contrast, the private class values `mod` and `val` vary from one `Mod` object to another.)

Unfortunately, this one line does not quite finish the job of setting up this variable. Somewhere in our program, but outside the class definition, we need to declare this variable. (The long `class { ... };` just describes the class `Mod`. The actual declaration of a variable takes place in a `.cc` file.)

The declaration of `default_modulus` takes place on line 4 of the file `Mod.cc` (Program 9.2) and looks like this:

```
long Mod::default_modulus = INITIAL_DEFAULT_MODULUS;
```

Let's examine this line one step at a time. First, we are declaring a variable of type `long`, so the keyword `long` begins this line. The name of the variable is `default_modulus`, but it is a member of the `Mod` class; hence, the full name of this variable is `Mod::default_modulus`. Finally, we give this variable a value (rather than letting C++ give it one). We could have typed a specific number, such as 10. Thus `long Mod::default_modulus= 10;` would be acceptable. However, we should avoid nameless constants in our programs. So, the header file (see line 6 of `Mod.h`, Program 9.1) declares a global constant equal to 2 like this:

```
const long INITIAL_DEFAULT_MODULUS = 2;
```

(Global constants are good; global variables are bad.) Thus, whenever the program starts up, the variable `default_modulus` has a known value.

The variable `default_modulus` is hidden from view because it is sequestered in the private section of the class declaration. It is not possible to change this value directly in, say, a `main()` procedure. Only methods belonging to the class `Mod` can access and modify this value. To do this, we create two procedures named `get_default_modulus` and `set_default_modulus`.

If we so chose, we could define such methods inside (i.e., inline) the class declaration like this:

```
class Mod {
  ...
public:
  ...
  void set_default_modulus(long m) { default_modulus = m; }
  long get_default_modulus() { return default_modulus; }
  ...
};
```

(Because these are class methods, the keyword `inline` is not required.)

The problem with this approach is that to change the default modulus, we would need to have a variable of type `Mod` (let's call it x) and use a statement of the form `x.get_default_modulus(17);`. Although this would work, the necessity to connect these methods to a particular `Mod` object doesn't make sense. There's nothing about x that is relevant here. (Note: The proposed code for `set_default_modulus` is too simplistic; it fails to handle improper values for m such as zero or negative integers.)

The better solution is to declare these methods as *static* methods; see lines 63–75 of Mod.h, Program 9.1.

```
static void set_default_modulus(long m) {
  if (m <= 0) {
    default_modulus = INITIAL_DEFAULT_MODULUS;
  }
  else {
    default_modulus = m;
  }
}

static long get_default_modulus() {
  if (default_modulus <= 0)
    set_default_modulus(INITIAL_DEFAULT_MODULUS);
  return default_modulus;
```

The modifier static for a method means that this method is not associated with an object of the given class, but with the class as a whole. So, for example, it would not make sense for either of these procedures to access the member variables val or mod because they are object specific.

Use of these methods requires a special syntax. Remember that these methods are members of the Mod class, but are not tied to any particular object. If we want to use these methods *inside* another Mod method, we can just use their name (either get_default_modulus or set_default_modulus). However, in a procedure such as main() another syntax is required. To use a static method from the class Mod in a procedure, we need to prepend Mod:: to the name of the procedure. Here is an example:

```
Mod::set_default_modulus(10);
cout << "The default modulus is now "
    << Mod::get_default_modulus() << endl;
```

We have seen three usages of the keyword static. Its meaning depends on the context in which it is used.

- A procedure's variable may be declared static. This means that the value held in the variable is preserved between invocations of the procedure.

- A variable in a class declaration may be declared static. This means that this variable is common to all objects of that type.

- A class method may be declared static. This means that this method is not to be associated with any particular object of the class, but with the class as a whole. Consequently, it cannot access any nonstatic member variables of the class.

9.4 Constructors and get/set methods

The Mod class has three constructors. A default constructor that takes no arguments, a one-argument constructor that uses the default modulus, and a two-argument constructor that specifies both value and modulus. The code for three constructors is written inline on lines 19–40 of Mod.h (Program 9.1).

The constructors must ensure that the modulus is positive (or else the resulting object is invalid). When the modulus is provided by the default modulus, we can be assured that the modulus is nonnegative. However, in the two-argument case, we need to check that the modulus specified is positive. If not, we create an invalid Mod object.

The constructors also need to ensure that the value is in the proper range. For example, the user may call Mod(-1, 10). In this case, we should create $9 \in \mathbb{Z}_{10}$. This need to adjust the value so that it lies in the proper range is a recurring issue, so we create a private method to adjust val so that it lies between 0 and mod-1. The adjust_val() method is given inline on lines 13–15 of Mod.h (Program 9.1).

Because the val and mod member variables are private, we need methods to inspect and to modify these. To that end, we specify the inline methods get_val, set_val, get_mod, and set_mod on lines 42–61 of Mod.h (Program 9.1). The get methods are flagged as const because they do not modify the Mod object. The set methods are designed to modify objects. These are designed so that the modulus is nonnegative and the value lies in the proper range.

We allow mod to equal zero to signal an invalid Mod object. We provide a handy is_invalid method (line 77) to check if this is the case.

9.5 Comparison operators

Given Mod objects x and y, we want to be able to check whether they are equal, and to sort them with <. We define the following relational operators:

$$x == y \qquad x \mathrel{!}= y \qquad x < y$$

The == operator is defined inline on lines 79–80 and the != operator on lines 91–93 of Mod.h (Program 9.1).

The < operator sorts Mod objects first on the basis of the modulus, and then on the basis of the value. See lines 95–100 of Mod.h.

In addition to comparing Mod objects to other Mod objects, it is convenient to be able to compare a Mod object to a long integer. For example, consider this code:

```
Mod x(9,10);
if (x == -1) cout << "It worked!" << endl;
```

In this case, we want the -1 interpreted as an element of $\mathbb{Z}_{10}$ and then the comparison ought to evaluate to true.

In C++, we need to define the operators Mod == long and long == Mod separately. The first of these is given on lines 83–85; we repeat that code here.

```
bool operator==(long that) const {
   return (*this) == Mod(that,mod);
}
```

The method is type bool because it returns a true/false result. The single argument is of type long because it is the right-hand argument; the left-hand argument is the Mod object. That is, in the statement x == -1 (where x is a Mod), the left argument is (implicitly) x and the right argument (that) equals -1.

The procedure works by converting that into a Mod object with the same modulus as the invoking object: Mod(that,mod). Then it uses the already defined Mod==Mod operator to compare. In order for an object to use itself we use *this. Therefore, when we encounter x == -1, the expression

```
(*this) == Mod(that,mod)
```

compares x (embodied by *this) with Mod(that,mod) where mod is the modulus of x.

Now we need to write a procedure for expressions of the form long == Mod. This cannot be written as a method inside the Mod class because the left-hand argument is not of type Mod. We therefore need to define this as a procedure outside the curly braces enclosing the Mod class declaration.

The code for long == Mod is on lines 178–180 of Mod.h (Program 9.1) and we repeat that code here.

```
inline bool operator==(long x, const Mod& y) {
   return (y==x);
}
```

The keyword inline is mandatory here (it is optional for inline methods inside the class declaration). Again, the procedure is of type bool as it returns a true/false value. This version of operator== has two arguments; because this is not a method belonging to a class, but rather a free-standing procedure, we need to specify the left and right arguments of == explicitly. The left argument is of type long and the right is of type Mod. The easiest way to see if x==y is true (where x is a long and y is a Mod) is to take advantage of the fact that we already have Mod==long defined; we just invoke y==x.

All three manifestations of != (Mod!=Mod, Mod!=long, and long!=Mod) are defined in the same manner.

9.6   Arithmetic operators

Now we implement the various arithmetic operations for $\mathbb{Z}_n$: addition, subtraction, multiplication, division, and exponentiation.

We begin with addition. Of course, we want to define the + operator when both arguments are type `Mod`. To do that, we create an `operator+` method in the `Mod` class. In addition, it is useful to define `Mod+long` and `long+Mod` operations. We also want the add/assign operation `x += y` where `x` is a `Mod` and `y` is either `Mod` or `long`.

All of these various forms of addition require the same basic underlying operations, we define an `add` method first and all the various + operations can use that to do the work. The `add` method of the `Mod` class is declared on line 102 of `Mod.h` (Program 9.1) as follows: `Mod add(Mod that) const;`. The actual code is found on lines 16–20 of `Mod.cc` (Program 9.2):

```
Mod Mod::add(Mod that) const  {
   if (is_invalid() || that.is_invalid()) return Mod(0,0);
   if (mod != that.mod) return Mod(0,0);
   return Mod(val + that.val, mod);
}
```

Notice that we first check if either the invoking `Mod` object or the argument `that` is invalid; if so, we return an invalid `Mod` object. Also, if the moduli of the two addends are different, we return an invalid `Mod` object. Finally, we add the values of the two objects and return a new `Mod` object containing the sum.

With the `add` method in place, we define the various + operators. The `Mod+Mod` method is on line 104 of `Mod.cc` and the `Mod+long` is on line 106. The `long+Mod` procedure cannot be a member of the `Mod` class (because the `Mod` is not on the left), so it is defined as a procedure on lines 186–188 of `Mod.h`. In each case, the `Mod` object passed is declared `const Mod&` because addition does not modify the addends (hence `const`) and call by reference is required for operators.

Next we define the += operators. The `Mod+=Mod` operator's definition (lines 108–111 of `Mod.h`) is repeated here:

```
Mod operator+=(const Mod& x) {
   *this = add(x);
   return *this;
}
```

Recall that the effect of the statement `a+=x;` is tantamount to `a=a+x;`. C++ allows us to program the += operator to do anything we want, but it makes most sense to adhere to the intended meaning.

The statement `a=a+x;` has the effect of replacing the value held in `a` with the result of the computation `a+x`. The result of `a+=x;` is the new value of `a`. Thus, a compound statement of the form `z=a+=x;` is interpreted as `z = (a+=x)` and is equivalent to `a=a+x; z=a;`. The procedure we write for `Mod+=Mod` should adhere to this behavior.

Thus, the `operator+=` method returns a `Mod` value. The argument, x, is declared as `const Mod&` x because (a) the code does not modify x's value and (b) pass by reference is required for `operator+=`.

To add x to the invoking object we appeal to the `add` method already defined. The statement `*this = add(x);` does the required work. The invoking object calculates the sum of its own value and that of x, and then assigns that value to itself.

Finally, we need to return the value in the invoking object as the result of this method. This is accomplished with the statement `return *this;`.

Next we examine the subtraction methods. Because $a - b$ is equivalent to $a + (-b)$, we begin by defining the unary minus (i.e., negate) operator. The unary operator $-a$ is a zero-argument method in the class `Mod`. The reason we do not need any arguments is because the operator applies to the invoking object. The full definition of unary minus is given inline in `Mod.h` on line 118 (Program 9.1) and repeated here:

```
Mod operator-() const { return Mod(-val,mod); }
```

Now the binary minus can be built using unary minus and addition. The `Mod-Mod` and `Mod-long` operations are implemented as methods within the class `Mod` (see lines 120–126). The `long-Mod` operator cannot be a member of the `Mod` class (because the left-hand argument is not a `Mod`), and so it needs to be a stand-alone procedure outside the class definition. See lines 190–192 of `Mod.h`. Finally, the `Mod-=Mod` and `Mod-=long` methods are given on lines 128–136.

The multiplication operators are created in a manner similar to that of addition. We declare a `multiply` method on line 138 of `Mod.h` and then give the code in `Mod.cc` on lines 22–26. The `Mod*Mod`, `Mod*long`, `Mod*=Mod`, and `Mod*=long` operators are inline members of the `Mod` class (lines 140–152 of `Mod.h`, and `long*Mod` is an ordinary inline procedure outside the class definition (lines 194–196).

We reduce division $a \div b$ to multiplication by a reciprocal; that is, $a \times b^{-1}$. Therefore, we begin by building an `inverse` method for the `Mod` class. This method is declared on line 154 of `Mod.h` as follows: `Mod inverse() const;`. Invoking `b.inverse()` should return the multiplicative inverse of b (in the appropriate $\mathbb{Z}_n$) if possible; otherwise (i.e., b is not invertible) we return an invalid result. This is implemented in `Mod.cc` on lines 28–36; we repeat the code here:

```
Mod Mod::inverse() const {
  long d,a,b;
  if (is_invalid()) return Mod(0,0);

  d = gcd(val, mod, a, b);

  if (d>1) return Mod(0,0); // no reciprocal if gcd(v,x)!= 1
  return Mod(a,mod);
}
```

The code first checks if the invoking `Mod` object is valid; if not, no inverse is possible and we return an invalid result. We then invoke the extended `gcd` procedure to find integers a and b so that a*val+b*mod equals d=gcd(val,mod). If d equals 1, then a holds the value of the inverse.

With the inverse method established, we use it to define Mod/Mod, Mod/long, long/Mod, Mod/=Mod, and Mod/=long.

There are a large number of arithmetic operators, but they all trace their actions back to four methods: add, multiply, unary minus, and inverse. There is an interesting benefit to this approach. Suppose we think of a better way to perform these operations; rather than needing to rewrite myriad operator methods, we just need to update these few to implement the new methods. For example, on a computer for which a long is four bytes, the largest integer value is around two billion. If we are working in a modulus near that limit and we multiply two Mod objects, the intermediate result may overflow the long data type. See line 25 of Mod.cc in which we calculate val * that.val. If val and that.val are both greater than, say, 10^5, then the multiply procedure gives an incorrect result. To fix this problem, we could rewrite the procedure to save val and that.val in long long variables, and then perform the multiplication. We would then reduce modulo mod which brings the values back to within the proper range.

The final operation we implement is exponentiation. Given $a \in \mathbb{Z}_n$ and $k \in \mathbb{Z}$, we want to calculate a^k. If k is negative, we raise a^{-1} to a positive power. Rather than multiply a by itself repeatedly, it is more efficient to use repeated squaring. Note that

$$a^k = \begin{cases} \left(a^{k/2}\right)^2 & \text{if } k \text{ is even, and} \\ \left(a^{(k-1)/2}\right)^2 \cdot a & \text{if } k \text{ is odd.} \end{cases}$$

So, calculating a^k by this strategy uses $O(\log_2 k)$ multiplications and not the far greater $k - 1$ required by the naive method.

We call the method pow and it is declared on line 172 of Mod.h and the code is on lines 38–61 of Mod.cc.

We elected not to define any operator symbol to represent exponentiation. Two natural operator symbols would make sense: ** and ^. Unfortunately, there are problems with both. The first is simply illegal. Because C++ does not have a ** operator for its fundamental types, we are not permitted to use ** as an operator symbol for any other types.

The latter, ^, is permissible because ^ is a valid C++ operator (exclusive or). Within the Mod class we may define this operator such as this:

```
Mod operator^(long k) { return pow(k); }
```

Then, in a procedure (such as main) we could have statements such as a = b^5; in lieu of a = b.pow(5);.

The problem is C++'s order of operation rules. C++ knows that multiplication takes priority over addition. An expression such as a*b+c*d is parsed as expected: (a*b)+(c*d). However, in the C++ hierarchy of operations, ^ has lower priority than addition. Thus a statement of the form `a+b^2` would be parsed as `(a+b)^2`. If we defined `^` for the `Mod` class, we would need to be careful to add unnatural parentheses in our expressions. Fortunately, the `.` in `a.pow(k)` has priority over addition and multiplication. So the expressions `a+b*c.pow(k)` and `a+c.pow(k)` have the desired behavior. There is no way to change C++'s order of operation rules, so we elect not to use `operator^` for exponentiation.

9.7   Writing `Mod` objects to output streams

One last task awaits us: writing `Mod` objects to an output stream such as `cout`. The technique for doing this is the same as for `Point` and `PTriple` objects. In `Mod.h` (line 176) we declare `operator<<` as a two-argument procedure:

```
ostream& operator<<(ostream& os, const Mod& M);
```

Then, in `Mod.cc` (lines 6–14) we give the code. If the `Mod` object is invalid, we write the characters `INVALID`. Otherwise, we write a `Mod` object in a form such as this: `Mod(31,100)`.

9.8   A `main` to demonstrate the `Mod` class

We end this chapter with a simple `main` to demonstrate the use of the `Mod` class.

Program 9.3: A program to illustrate the use of the `Mod` class.

```
1   #include "Mod.h"
2
3   /// A main to test the Mod class
4
5   int main() {
6     Mod x,y,z;
7
8     x.set_default_modulus(11);
9
10    x = Mod(17,10);
11    y = Mod(24);
12    z = -3;
13
14    cout << "x = " << x << endl;
15    cout << "y = " << y << endl;
16    cout << "z = " << z << endl << endl;
17
18    cout << "y+z = " << y+z << endl;
19    cout << "y-z = " << y-z << endl;
```

```
20    cout << "y*z = " << y*z << endl;
21    cout << "y/z = " << y/z << endl << endl;
22
23    cout << "x+3 = " << x+3 << endl;
24    cout << "x-3 = " << x-3 << endl;
25    cout << "x*3 = " << x*3 << endl;
26    cout << "x/3 = " << x/3 << endl << endl;
27
28    cout << "4+x = " << 4+x << endl;
29    cout << "4-x = " << 4-x << endl;
30    cout << "4*x = " << 4*x << endl;
31    cout << "4/x = " << 4/x << endl << endl;
32
33    cout << "-x = " << -x << endl << endl;
34
35    cout << "x^9 = " << x.pow(9) << endl;
36    cout << "x^(-9) = " << x.pow(-9) << endl;
37
38    cout << "-1+y^10 = " << -1+y.pow(10) << endl;
39    cout << "y^2  = " << y.pow(2) << endl;
40    cout << "y^(-2)+1 = " << y.pow(-2)+1 << endl << endl;
41
42    cout << "x == 17\t" << (x == 17) << endl;
43    cout << "x != 17\t" << (x != 17) << endl;
44
45    cout << "17 == x\t" << (17 == x) << endl;
46    cout << "17 != x\t" << (17 != x) << endl << endl;
47
48    return 0;
49  }
```

Here is the output of this program.

```
x = Mod(7,10)
y = Mod(2,11)
z = Mod(8,11)

y+z = Mod(10,11)
y-z = Mod(5,11)
y*z = Mod(5,11)
y/z = Mod(3,11)

x+3 = Mod(0,10)
x-3 = Mod(4,10)
x*3 = Mod(1,10)
x/3 = Mod(9,10)

4+x = Mod(1,10)
4-x = Mod(7,10)
4*x = Mod(8,10)
4/x = Mod(2,10)

-x = Mod(3,10)

x^9 = Mod(7,10)
x^(-9) = Mod(3,10)
```

```
-1+y^10 = Mod(0,11)
y^2   = Mod(4,11)
y^(-2)+1 = Mod(4,11)

x == 17 1
x != 17 0
17 == x 1
17 != x 0
```

9.9   Exercises

9.1 Write a procedure to solve a pair of congruences of the form

$$x \equiv a \pmod{m}$$
$$x \equiv b \pmod{n}$$

where m and n are relatively prime. The existence and uniqueness (in $\mathbb{Z}_{mn}$) of the solution to such a problem is guaranteed by the Chinese Remainder Theorem. Therefore, call your procedure crt. It should take two Mod objects as arguments and produce a Mod value in return. Your procedure should be declared like this:

```
Mod crt(const Mod a, const Mod b);
```

Of course, you may use the procedure you developed in Exercise 5.4.

9.2 Create a class to represent the time of day. Call your class Time and give it the following attributes.

- The data should be held in three private variables representing the hour, minute, and second.

- The default (no-argument) constructor should create a value equal to midnight and a three-argument constructor should create the time specified (using hours from 0 to 23).

- Define addition of a Time object and a (long) integer. If T is type Time and n is type long, then T+n and n+T should be the time n seconds after T. (Of course, n might be negative.)

- Define subtraction, but only in the form T-n but not n-T.

- Define ++ and --; these should increase (decrease) T by one second, respectively.

- Define get_hour(), get_minute(), and get_second() methods.

- Define `ampm()` and `military()` methods to control how the time is printed (see the next bullet). These methods should affect how all `Time` objects are printed.

  Also provide a `is_ampm()` method that returns `true` if the current output style is to use AM/PM and `false` if the current style is military (24 hour).

- Define `<<` for printing `Time` objects to the screen. The style of the output should either be `5:03:24 pm` or `17:03:24` as specified by the user with the methods `ampm()` and `military()`, respectively.

  Note the zero in front of the 3 but not in front of the 5. Midnight should be reported either as `12:00:00 am` or `0:00:00` and noon as `12:00:00 pm` or `12:00:00`, as appropriate.

If you are feeling especially brave, you can create a procedure called `now` that returns the current time of day. You can use `time(0)`; this returns the number of seconds since midnight on a specific date but not necessarily in your time zone (unless your local clock is GMT).

9.3 Create a class named `EuclideanVector` to represent vectors in a Euclidean space $\mathbb{R}^n$, and give it the following attributes.

- There should be a default dimension (as a static class variable). Give static methods for inspecting and adjusting this default dimension.

- There should be a zero-argument constructor that gives the zero vector in the default dimension. There should also be a single-argument constructor `EuclideanVector(int n)` that creates the zero vector in $\mathbb{R}^n$.

- There should be methods to get and set the individual coordinates of the vector.

- There should be a method to learn the dimension of the vector.

- There should be an `operator+` for adding vectors. Decide what the behavior should be in case the two vectors are of different dimension.

- There should be an `operator*` method for scalar multiplication. Be sure to allow both scalar–vector and vector–scalar multiplication.

- There should be an `operator==` and an `operator!=` for comparing vectors for equality.

- There should be an `operator<<` for writing vectors to the computer screen.

9.4 Let S denote the set $\{\sqrt{n} : n \in \mathbb{Z}, n \geq 0\}$. Define an operation $*$ on S by $x * y = \sqrt{x^2 + y^2}$. Create a C++ class to represent elements of the set S that includes an `operator*` that implements S's operation.

Include methods to get the value n, to convert an element of S into a decimal approximation, and an `operator<<` to write elements of S to the screen.

9.5 Create a class to represent Hamilton's quaternions.

The *quaternions* are an extension to the complex numbers. Each quaternion can be written as $a + bi + cj + dk$ where $a, b, c, d \in \mathbb{R}$ and i, j, k are special symbols with the following algebraic properties:

$$i^2 = j^2 = k^2 = -1$$
$$ij = k \qquad ji = -k$$
$$jk = i \qquad kj = -i$$
$$ki = j \qquad ik = -j$$

Addition is defined as expected:

$$(a + bi + cj + dk) + (a' + b'i + c'j + d'k)$$
$$= (a + a') + (b + b')i + (c + c')j + (d + d')k.$$

Multiplication is not commutative (as evidenced by the fact that $ij \neq ji$), but otherwise follows the usual rules of algebra. For example,

$$(1 + 2i + 3j - 4k)(-2 + i + 2j + 5k) = 10 + 20i - 18j + 14k.$$

The set of quaternions is denoted by $\mathbb{H}$.

Include the standard operators +, - (unary and binary forms), *, and /, and the combined assignment forms +=, -=, *=, and /=.

Also include the standard comparison operators == and !=, plus an << operator for writing quaternions to the screen.

Chapter 10

The Projective Plane

In this chapter we develop C++ classes to represent points and lines in the projective plane. Because such points and lines share many properties the classes and methods for their classes are extremely similar. Rather than write two nearly identical classes, we first build a base "projective object" class that implements the common functionality. We then use the idea of *inheritance* to establish classes representing points and lines.

10.1 Introduction to the projective plane, $\mathbb{RP}^2$

The points of the Euclidean plane $\mathbb{R}^2$ are ordered pairs of real numbers (x,y). Lines are point sets of the form $\{(x,y) : ax+by = c\}$ where $a,b,c \in \mathbb{R}$ and $ab \neq 0$.

The projective plane $\mathbb{RP}^2$ is an extension of the Euclidean plane. To the points in $\mathbb{R}^2$ we add additional points "at infinity." These points are in one-to-one correspondence with the slopes of lines in the Euclidean plane. More formally, we say that two lines in the Euclidean plane are equivalent if they are parallel. The points at infinity of the projective plane are in one-to-one correspondence with the equivalence classes of parallel lines. In $\mathbb{RP}^2$, each line from the Euclidean plane is given one additional point corresponding to its slope. Finally, all the points at infinity are deemed a line as well, and this line is called the line at infinity.

Alternatively, the points in $\mathbb{RP}^2$ are in one-to-one correspondence with (ordinary) lines through the origin in $\mathbb{R}^3$. The lines of $\mathbb{RP}^2$ correspond to (ordinary) planes through the origin in $\mathbb{R}^3$.

This leads to a natural way to assign coordinates to the points of $\mathbb{RP}^2$. Every point in $\mathbb{RP}^2$ is assigned a triple of real numbers (x,y,z). Two triples name the same point provided they are nonzero scalar multiples of each other. For example, $(2,4,-10)$ and $(-1,-2,5)$ name the same point. The triple $(0,0,0)$ is disallowed.

The original points of the Euclidean plane correspond to triples in which z is nonzero. The point $(x,y,1) \in \mathbb{RP}^2$ is identified with the point (x,y) in the Euclidean plane. Points at infinity are identified with triples in which $z = 0$.

Lines in the projective plane are sets of the form $\{(x,y,z) \in \mathbb{RP}^2 : ax+by+cz = 0\}$ where $a,b,c \in \mathbb{R}$ and are not all zero. For example, $(1,2,3)$, $(1,1,3)$, and $(1,-2,0)$ are collinear points as they all lie on the line $\{(x,y,z) : 2x+y-z = 0\}$. It is convenient to name lines as a triple $[a,b,c]$:

$$[a,b,c] = \{(x,y,z) : ax + by + cz = 0\}.$$

We use square brackets to name lines so we don't confuse points and lines. Note that if $[a',b',c']$ is a nonzero scalar multiple of $[a,b,c]$, then the two triples name the same line. The line at infinity is $[0,0,1]$. Only the triple $[0,0,0]$ is disallowed.

To determine if a point (x,y,z) is on a line $[a,b,c]$, we simply need to check if the dot product $ax + by + cz$ equals zero.

From our discussion, we see that there is a duality between points and lines in the projective plane. Given any true statement about points and lines in $\mathbb{RP}^2$, when we exchange the words "point" and "line" (plus some minor grammar correction) we get another true statement. For example, the following dual statements are both true:

- Given two distinct points of the projective plane, there is exactly one line that contains both of those points.

- Given two distinct lines of the projective plane, there is exactly one point that is contained in both of those lines.

One of the celebrated results in projective geometry is the following.

Theorem 10.1 (Pappus). *Let P_1, P_2, P_3 be three distinct collinear points and let Q_1, Q_2, Q_3 be three other distinct collinear points. Let X_i be the intersection of the lines P_jQ_k and P_kQ_j where i,j,k are distinct and $1 \le i,j,k \le 3$. Then the three points X_1, X_2, X_3 are collinear.* □

Pappus's theorem is illustrated in Figure 10.1.
The dual statement to Pappus's theorem is this.

Theorem 10.2 (Dual to Pappus). *Let L_1, L_2, L_3 be three distinct concurrent lines and let M_1, M_2, M_3 be three other distinct concurrent lines. Let X_i be the line through the points of intersection $L_j \cap M_k$ and $L_k \cap M_j$ where i,j,k are distinct and satisfy $1 \le i,j,k \le 3$. Then the three lines X_1, X_2, X_3 are concurrent.* □

The dual to Pappus's theorem is illustrated in Figure 10.2.

10.2   Designing the classes `PPoint` and `PLine`

Our goal is to create C++ classes to represent points and lines in the projective plane. We call these `PPoint` and `PLine`. We need to decide how to represent these objects (i.e., what data are needed to specify the objects) as well as the methods, operators, and procedures that act on these objects.

Here are our decisions.

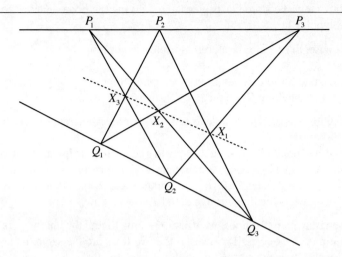

Figure 10.1:   An illustration of Pappus's theorem.

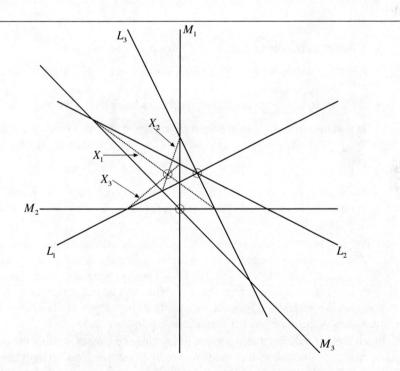

Figure 10.2:   An illustration of the dual of Pappus's theorem.

- Points are to be stored as triples (x, y, z) giving the homogeneous coordinates of the point.

  Likewise, lines are to be stored as triples, $[x, y, z]$.

- We want to be able to test points for equality (and inequality) and sort by $<$ so they can be held in C++ containers such as sets.

  Likewise, we need to be able to test lines for the same relations.

- Given two points, we want to be able to find the unique line that contains them. However, if the two points are the same, we return an invalid point; we signal this with coordinates $(0, 0, 0)$. For points x and y, the notation x+y is a good way to express the line through x and y.

  Likewise, given two lines, we want to be able to find the unique point of intersection of these lines. However, if the two lines are the same, then we return an invalid line; we signal this with the triple $[0, 0, 0]$. For lines L and M, the notation L*M is a good way to express the point of intersection.

- Given a point and line, we want to be able to determine if the point lies on the line.

- We want to be able to determine if three points are collinear.

  Likewise, we want to be able to determine if three lines are concurrent.

- We want to be able to generate a random point or line in the projective plane.

  In addition, given a line, we want to be able to choose a random point on the line. Likewise, given a point, we want to choose a random line through the point.

- We want to be able to send points and lines to output streams, writing points in the form (x,y,z) and lines in the form [x,y,z].

Notice that nearly every requirement for points has a matching requirement for lines. Thus, the code we need to write in the two instances would be nearly identical. To cut our work load in half, we exploit C++'s inheritance mechanism. We create a parent class named PObject that has two children: PPoint and PLine. As much as possible, we embed the functionality we need in the parent class, and then the children access this functionality for their own purposes.

In addition to reducing the workload in creating the classes, putting the common functionality of the classes into the parent reduces our workload in maintaining the classes. If there is an error in an algorithm, or if we create a more efficient version of the algorithm, we only need to replace the code in the parent class; we do not need to edit two separate versions.

10.3 Inheritance

Before we create the classes PPoint and PLine, we illustrate how one class is derived from another. That is, a class (let's call it Parent) is created first with certain properties. Then we create a new class (call it Child) that has all the properties of Parent plus additional properties.

For the toy example we are about to present, the Parent class houses two double real values, x and y. We provide a simple constructor to set x and y and two methods: sum() that calculates the sum x+y and print() that writes the object to the screen in the format (x,y).

The Child retains all the data and functionality of Parent but adds the following additional features. The new class has an additional data element: an integer k. It provides a new method value() that returns (x+y)*k and a new version of print() that writes the Child object to the screen in the format (x,y)*k.

Here is the code that accomplishes all these tasks.

Program 10.1: A program to illustrate inheritance.

```
1   #include <iostream>
2   using namespace std;
3
4   class Parent {
5   private:
6       double x, y;
7   public:
8       Parent(double a, double b) { x = a; y = b; }
9       double sum() const { return x+y; }
10      void print() const { cout << "(" << x << "," << y << ")"; }
11  };
12
13  class Child : public Parent {
14  private:
15      int k;
16  public:
17      Child(double a, double b, int n) : Parent(a,b) {
18          k = n;
19      }
20      double value() const { return sum()*k; }
21      void print() const { Parent::print(); cout << "*" << k; }
22  };
23
24  int main() {
25      Parent P(3., -2.);
26      P.print();
27      cout << " --> " << P.sum() << endl;
28
29      Child C(-1., 3., 5);
30      C.print();
31      cout << " --> " << C.sum() << " --> " << C.value() << endl;
32
```

```
33     return 0;
34   }
```

The `Parent` class is defined on lines 4–11. There is nothing new in this code; we have kept it short and simple to make it easy for you to read. Please look through it carefully before moving on.

The `Child` class is defined on lines 13–22 and there are several important features we need to address.

- To begin, the class `Child` is declared to be a *public subclass* of the class `Parent` on line 13:

  ```
  class Child : public Parent {
  ```

  The words `class Child` announce that we are beginning a class definition. The colon signals that this class is to be derived from another class. (Ignore the word `public` for a moment.) And the word `Parent` gives the name of the class from which this class is to be derived.

  The keyword `public` means that all the public parts of `Parent` are inherited as public parts of the derived class `Child`. The `Child` class has full access to all the public parts of `Parent` but does not have any access to the private parts of `Parent`.

  (Aside: Had we written `class Child : private Parent` then the public parts of `Parent` would become private parts of `Child`. As in the case of public inheritance, the private parts of `Public` are not accessible to `Child`.)

- On lines 14–15 we declare a private data element for `Child`: an integer `k`. The class `Child` therefore holds three data elements: `x`, `y`, and `k`.

  A method in `Child` can access `k` but not `x` or `y`. The latter are private to `Parent` and children have no right to examine their parents' private parts.

- Next comes the constructor for the `Child` class (lines 17–19). The constructor takes three arguments: real values `a` and `b` (just as `Parent` does) and additional integer value `n` (to be saved in `k`).

  What we want to do is save `a`, `b`, and `n` in the class variables `x`, `y`, and `k`, respectively. However, the following code is illegal.

  ```
  Child(double a, double b, int n) {
     x = a; y = b; k = n;
  }
  ```

  The problem is that `Child` cannot access `x` or `y`.

  Logically, what we want to do is this: first, we want to invoke the constructor for `Parent` with the arguments `a` and `b`. Second, we do the additional work special for the `Child` class, namely, assign `n` to `k`.

Look closely at line 17. Before the open brace for the method we see a colon and a call to `Parent(a,b)`. By this syntax, a constructor for a derived class (`Child`) can pass its arguments to the constructor for the base class (`Parent`).

The general form for a constructor of a derived class is this:

```
derived_class(argument list) : base_class(argument_sublist) {
    more stuff to do for the derived class;
}
```

When the derived class's constructor is invoked, it first calls its parent's constructor (passing none, some, or all of the arguments up to its parent's constructor). Once the base class's constructor completes its work, the code inside the curly braces executes to do anything extra that is required by the derived class.

- Next we implement the `value()` method (line 20). This procedure returns the quantity `(x+y)*k`. Although we have no access to x and y, the `sum()` method of `Parent` is public and so we can use that to calculate x+y. Because `sum()` is a public method of `Parent`, it is automatically a public method of `Child`. The expression `sum()*k` invokes the `sum()` method (to calculate x+y) and we multiply that by k (which is accessible to `Child`).

- Finally, we implement a `print()` method for the `Child` class (line 21). Recall that `Parent` already has a `print()` method, so this new `print()` method *overrides* the former. If `P` is an object of type `Parent` and `C` is an object of type `Child` then `P.print()` invokes the `print()` method defined on line 10 and `C.print()` invokes the one on line 21.

The new `print()` method writes the object to the computer screen in the format `(x,y)*k`. This `print()` cannot access x or y. Of course, we could add public `getX()` and `getY()` methods to `Parent`. However, there is another solution.

The `print()` method in `Parent` does most of the work already. Instead of rewriting the part of the code that prints `(x,y)`, we just need to use `Parent`'s `print()` method. The following code, however, does not work.

```
void print() const { print(); cout << "*" << k; }
```

The problem is that this code is recursive—it invokes itself. We want the second appearance of `print` to refer to `Parent`'s version. We accomplish that by prepending `Parent::` to the name of the method. The correct code for `Child`'s `print()` is this:

```
void print() const { Parent::print(); cout << "*" << k; }
```

When `Child`'s `print()` executes, it first calls `Parent`'s version of `print()` (which sends `(x,y)` to the screen). The remaining step (sending `*k`) occurs when the second statement executes.

A simple `main` follows (lines 24–34); here is the output of the program.

```
(3,-2) --> 1
(-1,3)*5 --> 2 --> 10
```

10.4 Protected class members

When we derive a new class from a base class, the public members of the base class are inherited as public members of the derived class, but the private members of the base class are inaccessible to the derived class. C++ provides a third alternative to this all-or-nothing access inheritance. In addition to public and private sections, a class may have a protected section. Both data and methods may be declared protected.

A protected member of a class becomes a private member of a derived class. A child class can access the public and protected parts of its parent, but not the private parts. A grandchild of the base class cannot access the protected parts of the base class. Let's look at an example.

Program 10.2: A program to illustrate the use of protected members of a class.

```
1  #include <iostream>
2  using namespace std;
3
4  class Base {
5  private:
6      int a;
7  protected:
8      int b;
9      int sum() const { return a+b; }
10 public:
11     Base(int x, int y) { a=x; b=y; }
12     void print() const { cout << "(" << a << "," << b << ")"; }
13 };
14
15 class Child : public Base {
16 public:
17     Child(int x, int y) : Base(x,y) { }
18     void increase_b() { b++; }
19     void print() const { Base::print(); cout << "=" << sum(); }
20 };
21
22 class GrandChild : public Child {
23 private:
24     int c;
25 public:
26     GrandChild(int x, int y, int z) : Child(x,y) { c = z; }
27     void print() const { Base::print(); cout << "/" << c; }
28 };
29
```

```
30   int main() {
31     Base       B(1,2);
32     Child      C(3,4);
33     GrandChild D(5,6,7);
34
35     B.print(); cout << endl;
36     // cout << B.sum() << endl; // Illegal, sum is protected
37
38     C.print(); cout << "  -->  ";
39     C.increase_b();
40     C.print(); cout << endl;
41
42     D.print(); cout << "  -->  ";
43     D.increase_b();
44     D.print(); cout << endl;
45
46     return 0;
47   }
```

In this program we define three classes: Base, Child, and GrandChild; each is used to derive the next.

The Base class has two data members: a private integer a and a protected integer b. The class also includes a protected method named sum, a public constructor, and a public method named print.

Class Child has no additional data members. It has a public method increase_b that increases the data member b by one. Note that it would not be possible for Child to have a similar method for increasing a. The constructor for Child passes its arguments on to its parent, Base, but then takes no further action (hence the curly braces on line 17 do not enclose any statements).

The print method for Child uses Base's print method and sum method.

Class GrandChild adds an extra private data element, c. The constructor for GrandChild passes its first two arguments to its parent's constructor and then uses the third argument to set the value of c.

The print method for GrandChild uses its grandparent's print method. Although Base::print() invokes the sum method, the GrandChild methods cannot directly call sum because it is protected in Base, hence implicitly private in Child, and hence inaccessible in GrandChild.

A main to illustrate all these ideas begins on line 30. The output of the program follows.

```
(1,2)
(3,4)=7  -->  (3,5)=8
(5,6)/7  -->  (5,7)/7
```

10.5   Class and file organization for **PPoint** and **PLine**

Because of the duality between points and lines in the projective plane, most of the data and code we use to represent these concepts in C++ are the same. If at all possible, we should avoid writing the same code twice for two reasons. First, the initial work in creating the programs is doubled. More important, maintaining the code is also made more difficult; if a change is required to the code, we need to remember to make that change twice. When we are fussing with our programs and making a number of minor modifications, it is easy to forget to update both versions. To illustrate this, we deliberately include a subtle flaw in the first version of the program; we then repair the problem once (not twice).

To this end, we define three classes: a parent class named PObject that contains as much of the code as possible and two derived classes, PPoint and PLine, that include code particular to each. These classes are defined in two files each: a header .h file and a code .cc file. Figure 10.3 illustrates the organization.

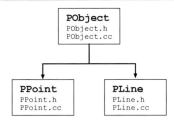

Figure 10.3:   Hierarchy of the PObject classes.

The header file PObject.h is used to declare the PObject class. Both PPoint.h and PLine.h require the directive #include "PObject.h". Programs that use projective geometry need all three. Rather than expecting the user to type multiple #include directives, we create a convenient header file that includes everything we need. We call this header file Projective.h; here it is.

Program 10.3: Header file for all projective geometry classes, Projective.h.

```
1   #ifndef PROJECTIVE_H
2   #define PROJECTIVE_H
3   #include "PPoint.h"
4   #include "PLine.h"
5   #endif
```

Notice that we do not require #include "PObject.h" because that file is already #included by PPoint.h and PLine.h. All these files have the usual structure to prevent multiple inclusion.

All told, we have seven files that implement points and lines in the projective plane.

PObject.h The header file for the PObject class.

PObject.cc The C++ code for the PObject class.

PPoint.h The header file for the PPoint class.

PPoint.cc The C++ code for the PPoint class.

PLine.h The header file for the PLine class.

PLine.cc The C++ code for the PLine class.

Projective.h The only header file subsequent programs need to include to use projective points and lines.

The decision to write the program in seven different files is based on the principle of breaking a problem down to manageable sizes and working on each piece separately. We could have packed all this work into two larger files (one .h and one .cc).

10.6 The parent class **PObject**

Data and functionality common to PPoint and PLine are implemented in the class PObject. Using the ideas presented in Section 10.2, we map out the class PObject.

Data A point or line in $\mathbb{RP}^2$ is represented by homogeneous coordinates: (x,y,z) or $[x,y,z]$. To hold these coordinates, we declare three private double variables, x, y, and z. Let us write $\langle x,y,z \rangle$ for the homogeneous coordinates of a generic projective object (either a point of a line). (See Program 10.4, line 9.)

Because $\langle 2,1,-5 \rangle$ and $\langle -4,-2,10 \rangle$ name the same object, it is useful to choose a canonical triple. One idea is to make sure that $\langle x,y,z \rangle$ is a unit vector (but then we have a sign ambiguity). The representation we use is to make the last nonzero coordinate of the triple equal to 1. The motivation for this choice is that a point in the Euclidean plane at coordinates (x,y) corresponds to the point $(x,y,1)$ in the projective plane.

If later we are unhappy with this decision, we can choose another manner to store the homogeneous coordinates. Because the coordinates are private members of PObject, we would only need to update the code for PObject.

We provide public methods getX(), getY(), and getZ() to inspect (but not modify) the values held in x, y, and z, respectively. (See Program 10.4 lines 33–35.)

We reserve $\langle 0,0,0 \rangle$ to stand for an invalid projective object. We provide a public method is_invalid() to check if an object is invalid. (See Program 10.4 lines 37–39.)

Constructors It is natural to define a constructor for a PObject with three parameters that set the homogeneous coordinates of the object. The user might invoke the constructor like this:

```
PObject P(2., 3., 5.);
```

Rather than holding this point as $\langle 2,3,5 \rangle$, we use the canonical representation $\langle 0.4, 0.6, 1 \rangle$. (See Program 10.4 lines 25–30.)

To facilitate the conversion of user-supplied coordinates to canonical coordinates, we create a private helper procedure scale. (See Program 10.4 line 10 and Program 10.5 lines 4–19.)

All C++ classes ought to define a zero-argument constructor that creates a default object of that class. In this case, a sensible choice is the object $\langle 0,0,1 \rangle$. This corresponds to the origin (as a point) or the line at infinity (as a line). (See Program 10.4 lines 21–24.)

Relations We want to be able to compare projective points (and lines) for equality, inequality, and $<$ (for sorting). To this end, we might be tempted to define operator== in the public portion of PObject like this:

```
public:
  bool operator==(const PObject& that) const {
    return ( (x==that.x) && (y==that.y) && (z==that.z) );
  }
```

Although this would give the desired result when comparing points to points or lines to lines, it would also provide the unfortunate ability to compare points to lines. Were we to use the above method for testing equality, then we might run into the following situation.

```
PPoint P(2., 3., 5.);
PLine  L(2., 3., 5.);
if (P == L) {
  cout << "They are equal." << endl;
}
```

When this code runs, the message They are equal would be written on the computer's screen.

There are at least two problems with this approach. First, the point $(2,3,5)$ and the line $[2,3,5]$ are not equal even though they have the same homogeneous coordinates. Second, lines and points should not even be comparable by ==. Code that compares a point to a line is almost certainly a bug; the best thing in this situation is for the compiler to flag such an expression as an error.

We therefore take a different approach to equality testing. We want the fundamental code that checks for equality to reside inside the PObject class because that code is common to both PPoint and PLine. We do this by declaring a protected method called equals inside PObject (see Program 10.4 line 14):

```
protected:
  bool equals(const PObject& that) const;
```

In the file PObject.cc we give the code for this method (see Program 10.5 lines 34–36):

```
bool PObject::equals (const PObject& that) const {
  return ( (x==that.x) && (y==that.y) && (z==that.z) );
}
```

Then, in the class definitions for PPoint and PLine we give the necessary operator definitions. For example, in PLine we have this:

```
public:
  bool operator==(const PLine& that) const {
    return equals(that);
  }
  bool operator!=(const PLine& that) const {
    return !equals(that);
  }
```

In this way, if (indeed, when) we decide to change the manner in which we test for equality, we only need to update PObject's protected equals method.

Similarly, rather than defining a < operator for PObject, we define a protected less method. The children can access less to define their individual, public operator< methods. (See Program 10.4 line 15 and Program 10.5 lines 38–45.)

Meet/Join Operation Given two PPoints P and Q, we want P+Q to return the line through those points. Dually, given PLines L and M, we want L*M to return the point of intersection of these lines.

In both cases, the calculations are the same: given the triples $\langle x_1, y_1, z_1 \rangle$ and $\langle x_2, y_2, z_2 \rangle$ we need to find a new triple $\langle x_3, y_3, z_3 \rangle$ that is orthogonal to the first two. To do this, we calculate the cross product:

$$\langle x_3, y_3, z_3 \rangle = \langle x_1, y_1, z_1 \rangle \times \langle x_2, y_2, z_2 \rangle.$$

Therefore, in PObject we declare a protected method called op that is invoked by operator+ in PPoint and operator* in PLine. (See Program 10.4 line 18 and Program 10.5 lines 138–148.)

Incidence Given a point and a line, we want to be able to determine if the point lies on the line. If the coordinates for these are (x, y, z) and $[a, b, c]$, respectively, then we simply need to test if $ax + by + cz = 0$.

It does not make sense to ask if one line is incident with another, so we do not make an "is incident with" method publicly available in PObject. Rather, we make a protected incident method (that calls, in turn, a private dot method for calculating dot product). (See Program 10.4 lines 11,16 and Program 10.5 lines 21–23,47–49.)

Then, PPoint and PLine can declare their own public methods that access incident. Details on this later.

Collinearity/Concurrence Are three given points collinear? Are three given lines concurrent? If the three objects have coordinates $\langle x_1, y_1, z_1 \rangle$, $\langle x_2, y_2, z_2 \rangle$, and $\langle x_3, y_3, z_3 \rangle$, then the answer is yes if and only if the vectors are linearly dependent. We check this by calculating

$$\det \begin{bmatrix} x_1 & y_1 & z_1 \\ x_2 & y_2 & z_2 \\ x_3 & y_3 & z_3 \end{bmatrix}$$

and seeing if we get zero. In PObject.h we declare a procedure dependent that takes three PObject arguments and returns true or false. We then use dependent to implement procedures collinear and concurrent for the classes PPoint and PLine (respectively). (See Program 10.4 line 44 and Program 10.5 lines 60–77.)

Random points/lines There is no way to generate a point uniformly at random in the Euclidean plane, but there is a sensible way in which we can do this for the projective plane. Recall that points in $\mathbb{RP}^2$ correspond to lines through the origin in $\mathbb{R}^3$. Thus, to select a point at random in $\mathbb{RP}^2$ we generate a point uniformly at random on the unit ball centered at the origin. An efficient way to perform this latter task is to select a vector $\mathbf{v}$ uniformly in $[-1, 1]^3$. If $\|\mathbf{v}\| > 1$, then we reject $\mathbf{v}$ and try again.

The method for generating a random line is precisely the same.

We therefore include a public method randomize that resets the coordinates of the PObject by the algorithm we just described. (See Program 10.4 line 31 and Program 10.5 lines 25–32.)

To choose a random line through a point is similar. Suppose the point is (x, y, z). The line $[a, b, c]$ should be chosen so that $[a, b, c]$ is orthogonal to (x, y, z). To do this, we find an orthonormal basis for $(x, y, z)^\perp$ that we denote $\{(a_1, b_1, c_1), (a_2, b_2, c_2)\}$. We then choose t uniformly at random in $[0, 2\pi]$. The random line $[a, b, c]$ is given by

$$a = a_1 \cos t + a_2 \sin t \qquad b = b_1 \cos t + b_2 \sin t \qquad c = c_1 \cos t + c_2 \sin t.$$

Rather than generate t and then compute two trigonometric functions, we can obtain the pair $(\cos t, \sin t)$ by choosing a point uniformly at random in the unit disk (in $\mathbb{R}^2$) and then scaling.

The technique for selecting a random point on a given line is exactly the same.

Thus, we define a protected `rand_perp` method in `PObject` (Program 10.4 line 17 and Program 10.5 lines 79–136).

The `rand_perp` method is used by `rand_point` in `PLine` and `rand_line` in `PPoint`.

Output Finally, `PObject.h` declares a procedure for writing to an output stream. The format is `<x,y,z>`. The `ostream operator<<` defined in `PObject` is overridden by like-named procedures in `PPoint` and `PLine`. (See Program 10.4 line 42 and Program 10.5 lines 51–58.)

With these explanations in place, we now give the header and code files for the class `PObject`.

Program 10.4: Header file for the `PObject` class (version 1).

```
1   #ifndef POBJECT_H
2   #define POBJECT_H
3
4   #include <iostream>
5   using namespace std;
6
7   class PObject {
8   private:
9     double x,y,z;
10    void scale();
11    double dot(const PObject& that) const;
12
13  protected:
14    bool equals(const PObject& that) const;
15    bool less(const PObject& that) const;
16    bool incident(const PObject& that) const;
17    PObject rand_perp() const;
18    PObject op(const PObject& that) const;
19
20  public:
21    PObject() {
22      x = y = 0.;
23      z = 1.;
24    }
25    PObject(double a, double b, double c) {
26      x = a;
27      y = b;
28      z = c;
29      scale();
30    }
31    void randomize();
32
33    double getX() const { return x; }
```

```
34     double getY() const { return y; }
35     double getZ() const { return z; }
36
37     bool is_invalid() const {
38       return (x==0.) && (y==0.) && (z==0.);
39     }
40   };
41
42   ostream& operator<<(ostream& os, const PObject& A);
43
44   bool dependent(const PObject& A, const PObject& B, const PObject& C);
45
46   #endif
```

Program 10.5: Program file for the `PObject` class (version 1).

```
1    #include "PObject.h"
2    #include "uniform.h"
3
4    void PObject::scale() {
5      if (z != 0.) {
6        x /= z;
7        y /= z;
8        z = 1.;
9        return;
10     }
11     if (y != 0.) {
12       x /= y;
13       y = 1.;
14       return;
15     }
16     if (x != 0) {
17       x = 1.;
18     }
19   }
20
21   double PObject::dot(const PObject& that) const {
22     return x*that.x + y*that.y + z*that.z;
23   }
24
25   void PObject::randomize() {
26     do {
27       x = unif(-1.,1.);
28       y = unif(-1.,1.);
29       z = unif(-1.,1.);
30     } while (x*x + y*y + z*z > 1.);
31     scale();
32   }
33
34   bool PObject::equals(const PObject& that) const {
35     return ( (x==that.x) && (y==that.y) && (z==that.z) );
36   }
37
38   bool PObject::less(const PObject& that) const {
39     if (x < that.x) return true;
```

```
40    if (x > that.x) return false;
41    if (y < that.y) return true;
42    if (y > that.y) return false;
43    if (z < that.z) return true;
44    return false;
45  }
46
47  bool PObject::incident(const PObject& that) const {
48    return dot(that)==0.;
49  }
50
51  ostream& operator<<(ostream& os, const PObject& A) {
52    os << "<"
53       << A.getX() << ","
54       << A.getY() << ","
55       << A.getZ()
56       << ">";
57    return os;
58  }
59
60  bool dependent(const PObject& A, const PObject& B, const PObject& C){
61    double a1 = A.getX();
62    double a2 = A.getY();
63    double a3 = A.getZ();
64
65    double b1 = B.getX();
66    double b2 = B.getY();
67    double b3 = B.getZ();
68
69    double c1 = C.getX();
70    double c2 = C.getY();
71    double c3 = C.getZ();
72
73    double det = a1*b2*c3 + a2*b3*c1 + a3*b1*c2
74               - a3*b2*c1 - a1*b3*c2 - a2*b1*c3;
75
76    return det == 0.;
77  }
78
79  PObject PObject::rand_perp() const {
80    if (is_invalid()) return PObject(0,0,0);
81
82    double x1,y1,z1;    // One vector orthogonal to (x,y,z)
83    double x2,y2,z2;    // Another orthogonal to (x,y,z) and (x1,y1,z1)
84
85    if (z == 0.) { // If z==0, take (0,0,1) for (x1,y1,y2)
86      x1 = 0;
87      y1 = 0;
88      z1 = 1;
89    }
90    else {
91      if (y == 0.) {  // z != 0 and y == 0, use (0,1,0)
92        x1 = 0;
93        y1 = 1;
94        z1 = 1;
95      }
```

```
96        else {   // y and z both nonzero, use (0,-z,y)
97           x1 = 0;
98           y1 = -z;
99           z1 = y;
100        }
101     }
102
103     // normalize (x1,y1,z1)
104     double r1 = sqrt(x1*x1 + y1*y1 + z1*z1);
105     x1 /= r1;
106     y1 /= r1;
107     z1 /= r1;
108
109     // (get x2,y2,z2) by cross product with (x,y,z) and (x1,y1,z1)
110     x2 = -(y1*z) + y*z1;
111     y2 = x1*z - x*z1;
112     z2 = -(x1*y) + x*y1;
113
114     // normalize (x2,y2,z2)
115     double r2 = sqrt(x2*x2 + y2*y2 + z2*z2);
116     x2 /= r2;
117     y2 /= r2;
118     z2 /= r2;
119
120     // get a point uniformly on the unit circle
121     double a,b,r;
122     do {
123        a = unif(-1.,1.);
124        b = unif(-1.,1.);
125        r = a*a + b*b;
126     } while (r > 1.);
127     r = sqrt(r);
128     a /= r;
129     b /= r;
130
131     double xx = x1 * a + x2 * b;
132     double yy = y1 * a + y2 * b;
133     double zz = z1 * a + z2 * b;
134
135     return PObject(xx,yy,zz);
136  }
137
138
139  PObject PObject::op(const PObject& that) const {
140
141     if (equals(that)) return PObject(0,0,0);
142
143     double c1 = y*that.z - z*that.y;
144     double c2 = z*that.x - x*that.z;
145     double c3 = x*that.y - y*that.x;
146
147     return PObject(c1,c2,c3);
148  }
```

10.7 The classes `PPoint` and `PLine`

With the code for `PObject` in place, we are ready to finish our work by writing the files for the classes `PPoint` and `PLine`. We do this work in four files: `PPoint.h`, `PPoint.cc`, `PLine.h`, and `PLine.cc` (Programs 10.6 through 10.9).

As we start to write these files, we meet a chicken-and-egg problem. Which do we define first: the class `PPoint` or the class `PLine`? For us, this is more than a philosophical conundrum. Some `PLine` methods need to refer to `PPoints`. For example, `operator*` acts on lines to produce points and `PLine`'s `rand_point` method returns a `PPoint`. Dually, some of the `PPoint` methods require the `PLine` class.

Here is how we solve this dilemma. In the `PPoint.h` file, before we give the definition of `PPoint` we have the following statement (see line 6 of Program 10.6),

```
class PLine;
```

This lets the C++ compiler know that a class named `PLine` is defined somewhere else. Therefore, when the compiler sees `PLine`, it knows that this identifier refers to some class. However, it does not know anything else about `PLine` other than the fact it is a class. This means that we may not give an inline definition for any method that refers to objects of type `PLine`. Later, in the file `PPoint.cc`, we include all the headers (conveniently with the single directive #include "Projective.h"). Therefore, the code in `PPoint.cc` can make full use of the `PLine` class.

Dually, we include the statement `class PPoint;` in the file `PLine.h`.

We focus our attention on the class `PPoint`. The analysis of `PLine` is similar.

For the class `PPoint` we have five constructors. At first, this may seem to be too many, but we show how to do this easily.

Of course, we want a three-argument constructor `PPoint(a,b,c)` that creates the point (a,b,c). We also want a zero-argument constructor `PPoint()` that creates the point $(0,0,1)$ corresponding to the origin.

It makes sense to define a two-argument constructor `PPoint(a,b)` to create the point $(a,b,1)$. In a sense, this maps the Euclidean point (a,b) to its natural correspondent in the $\mathbb{RP}^2$.

What should be the action of a single-argument constructor? The real number a can be identified with the point $(a,0)$ on the x-axis, and this in turn corresponds to $(a,0,1)$ in $\mathbb{RP}^2$. In summary, we have the following constructors and their effects.

Constructor form	Point created
`PPoint P();`	$(0,0,1)$
`PPoint P(a);`	$(a,0,1)$
`PPoint P(a,b);`	$(a,b,1)$
`PPoint P(a,b,c);`	(a,b,c)

The great news is that we can implement these four constructors with a single definition using *default parameters*.

For any C++ procedure (class method or free-standing procedure), default values can be specified. In the `.h` file, where procedures are declared, we use a syntax such as this:

```
return_type  procedure_name(type arg1 = val1, type arg2 = val2, ...);
```

Then, when the procedure is used, any missing parameters are replaced by their default values. Let's look at a concrete example. We declare a procedure named `next` that produces the next integer after a given integer. (This is a contrived example, but we want to keep things simple.) In the header file we put the following,

```
int next(int num, int step=1);
```

And in the `.cc` file, we have the code,

```
int next(int num, int step) {
  return num+step;
}
```

Notice that the argument `step` is given a default value, 1. However, the argument `num` is not given a default. It is permissible to specify only a subset of the arguments that receive default values, but if an argument has a default value, all arguments to its right must also have default values.

Notice that the optional arguments are not reiterated in the `.cc` file.

Consider the following code.

```
int a,b,c;
a = next(5);
b = next(5,1);
c = next(5,2);
```

This will set a and b equal to 6 and c equal to 7.

Alternatively, we could have given an inline definition of `next` in the header file like this:

```
inline int next(int num, int step=1) { return num+step; }
```

Returning to `PPoint`, the four constructors (with zero to three `double` arguments) can all be declared at once like this:

```
PPoint(double a=0., double b=0., double c=1.) ......
```

See line 10 of Program 10.6. The action of this constructor is simply to pass the three arguments up to its parent (line 11) and then there is nothing else to do. To show there is nothing else, we give a pair of open/close braces that enclose empty space (line 12).

The class `PPoint` needs one more constructor. Recall that `PObject` provides methods such as `rand_perp` and op that return `PObject`s. However, when these are used by `PPoint` or `PLine`, the `PObject` values need to be converted to type `PPoint` or `PLine` as appropriate. To do this `PPoint` provides a constructor that accepts a single argument of type `PObject`. Here is the simple code (see also lines 14–15 of Program 10.6).

```
PPoint(const PObject& that) :
  PObject(that.getX(), that.getY(), that.getZ()) { }
```

This code sends the x, y, and z values held in that up to the parent constructor and then does nothing else. Now, if we want to assign a PObject value to a PPoint object, we can do it in the following ways.

```
PObject X(2.,3.,5.);

PPoint  P(X);   // invoke the constructor at declaration of P

PPoint  Q;
Q = PPoint(X);  // invoke the constructor as a converter procedure

PPoint  R;
R = X;   // implicit conversion (compiler figures out what to do)
```

Here is the PPoint.h header file.

Program 10.6: Header file for the PPoint class.

```
1   #ifndef PPOINT_H
2   #define PPOINT_H
3
4   #include "PObject.h"
5
6   class PLine;
7
8   class PPoint : public PObject {
9   public:
10    PPoint(double a = 0., double b = 0., double c = 1.) :
11      PObject(a,b,c)
12      { }
13
14    PPoint(const PObject& that) :
15      PObject(that.getX(), that.getY(), that.getZ())  { }
16
17    bool operator==(const PPoint& that) const {
18      return equals(that);
19    }
20
21    bool operator!=(const PPoint& that) const {
22      return !equals(that);
23    }
24
25    bool operator<(const PPoint& that) const {
26      return less(that);
27    }
28
29    PLine operator+(const PPoint& that) const;
30
31    bool is_on(const PLine& that) const;
32
33    PLine rand_line() const;
34
35  };
```

```
36
37  ostream& operator<<(ostream& os, const PPoint& P);
38
39  inline bool
40  collinear(const PPoint& A, const PPoint& B, const PPoint& C) {
41    return dependent(A,B,C);
42  }
43
44  #endif
```

On lines 17–26 we give inline definitions of `operator==`, `operator!=`, and `operator<`. However, `operator+`, `is_on`, and `rand_line` may not be given inline because these refer to the (as yet) unknown class `PLine`. (See lines 29–33.)

The `collinear` procedure simply invokes the `dependent` procedure defined in `PObject.h`, so we give it inline here. The `inline` keyword in mandatory here because `collinear` is not a member of any class.

The parts of `PPoint` not given in `PPoint.h` are defined in `PPoint.cc` which we present next.

Program 10.7: Program file for the `PPoint` class.

```
1   #include "Projective.h"
2
3   ostream& operator<<(ostream& os, const PPoint& P) {
4     os << "(" << P.getX() << "," << P.getY() << "," << P.getZ() << ")";
5     return os;
6   }
7
8   PLine PPoint::operator+(const PPoint& that) const {
9     return PLine(op(that));
10  }
11
12  PLine PPoint::rand_line() const {
13    return PLine(rand_perp());
14  }
15
16  bool PPoint::is_on(const PLine& that) const {
17    return incident(that);
18  }
```

The code for the class `PLine` is extremely similar to that of `PPoint`. Here are the files `PLine.h` and `PLine.cc` for your perusal.

Program 10.8: Header file for the `PLine` class.

```
1   #ifndef PLINE_H
2   #define PLINE_H
3
4   #include "PObject.h"
5
6   class PPoint;
7
8   class PLine : public PObject {
```

```
9   public:
10    PLine(double a = 0., double b = 0., double c = 1.) :
11      PObject(a,b,c)
12      { }
13
14    PLine(const PObject& that) :
15      PObject(that.getX(), that.getY(), that.getZ())  { }
16
17    bool operator==(const PLine& that) const {
18      return equals(that);
19    }
20
21    bool operator!=(const PLine& that) const {
22      return !equals(that);
23    }
24
25    bool operator<(const PLine& that) const {
26      return less(that);
27    }
28
29    PPoint operator*(const PLine& that) const;
30
31    bool has(const PPoint& X) const;
32
33    PPoint rand_point() const;
34
35  };
36
37  ostream& operator<<(ostream& os, const PLine& P);
38
39  inline bool
40  concurrent(const PLine& A, const PLine& B, const PLine& C) {
41    return dependent(A,B,C);
42  }
43
44  #endif
```

Program 10.9: Program file for the `PLine` class.

```
1   #include "Projective.h"
2
3   ostream& operator<<(ostream& os, const PLine& P) {
4     os << "[" << P.getX() << "," << P.getY() << "," << P.getZ() << "]";
5     return os;
6   }
7
8   PPoint PLine::rand_point() const {
9     return PPoint(rand_perp());
10  }
11
12  PPoint PLine::operator*(const PLine& that) const {
13    return PLine(op(that));
14  }
15
16  bool PLine::has(const PPoint& that) const {
```

```
17    return incident(that);
18  }
```

10.8   Discovering and repairing a bug

With the projective point and line classes built, it is time to test our code. Here is a simple main to perform some checks.

Program 10.10: A `main` to test the $\mathbb{RP}^2$ classes.

```
1  #include <iostream>
2  #include "Projective.h"
3  #include "uniform.h"
4
5  int main() {
6    seed();
7    PPoint P;
8
9    P.randomize();
10   cout << "The random point P is " << P << endl;
11
12   PLine L,M;
13
14   L = P.rand_line();
15   M = P.rand_line();
16
17   cout << "Two lines through P are L = " << L << endl
18        << "and M = " << M << endl;
19
20   cout << "Is P on L? " << P.is_on(L) << endl;
21   cout << "Does M have P? " << M.has(P) << endl;
22
23   PPoint Q;
24   Q = L*M;
25
26   cout << "The point of intersection of L and M is Q = " << Q << endl;
27
28   cout << "Is Q on L? " << Q.is_on(L) << endl;
29   cout << "Does M have Q? " << M.has(Q) << endl;
30
31   if (P==Q) {
32     cout << "P and Q are equal" << endl;
33   }
34   else {
35     cout << "P and Q are NOT equal" << endl;
36   }
37
38   return 0;
39 }
```

When this program is run, we have the following output:

```
The random point P is (-1.32445,0.591751,1)
Two lines through P are L = [6.51303,12.8875,1]
and M = [0.871229,0.260071,1]
Is P on L? 0
Does M have P? 0
The point of intersection of L and M is Q = (-1.32445,0.591751,1)
Is Q on L? 0
Does M have Q? 0
P and Q are NOT equal
```

Much of what we see here doesn't make sense. First, the lines *L* and *M* are random lines through *P*. Yet the output indicates that *P* is on neither of these lines. Then, we generate the point *Q* at the intersection of *L* and *M*. The good news is that the coordinates of *P* and *Q* are the same: $(-5.64488, 2.562, 1)$. However, the computer still tells us that *Q* is on neither *L* nor *M*. Worse, it thinks that $P \neq Q$. *What is going on here!?*

All these problems stem from the same underlying cause: roundoff. Remember that a `double` variable is a rational approximation to a real number. Two real quantities that are computed differently may result in `double` values that are different. For example, consider this code:

```cpp
#include <iostream>
using namespace std;
int main() {
  double x = 193./191.;
  double y = 1./191.;
  y *= 193;
  if (x == y) {
    cout << "They are equal" << endl;
  }
  else {
    cout << "They are different" << endl;
    cout << "Difference = " << x-y << endl;
  }
  return 0;
}
```

Here is the output from this program.

```
They are different
Difference = -2.22045e-16
```

The computer reports that $\frac{193}{191} \neq 193 \times \frac{1}{191}$.

The equality test we created for `PObjects` checks if the three coordinates are exactly the same. We need to relax this.

The bad news is we need to rewrite some of our code to correct this problem. The great news is that we only need to repair `PObject`. The children `PPoint` and `PLine` inherit the improvements.

To begin, let us identify the places in the code for PObject where exact equality is sought.

- The equals method requires exact equality of the three coordinates.

- The incident method requires zero to be the exact result of the dot product method.

- The dependent procedure requires zero to be the exact value of the determinant.

In lieu of exact equality, we can require that the values be within a small tolerance. What tolerance should we use? We make that quantity user selectable initialized with some default value (say 10^{-12}).

To implement this idea we add a private static double variable named tolerance and define a constant named default_tolerance set to 10^{-12}.

Inside PObject we define two inline public static methods: set_tolerance and get_tolerance. Here is the revised header file.

Program 10.11: Header file for the PObject class (version 2).

```
 1  #ifndef POBJECT_H
 2  #define POBJECT_H
 3  #include <cmath>
 4  #include <iostream>
 5  using namespace std;
 6
 7  const double default_tolerance = 1e-12;
 8
 9  class PObject {
10  private:
11     double x,y,z;
12     void scale();
13     double dot(const PObject& that) const;
14     static double tolerance;
15
16  protected:
17     bool equals(const PObject& that) const;
18     bool less(const PObject& that) const;
19     bool incident(const PObject& that) const;
20     PObject rand_perp() const;
21     PObject op(const PObject& that) const;
22
23  public:
24     PObject() {
25        x = y = 0.;
26        z = 1.;
27     }
28     PObject(double a, double b, double c) {
29        x = a;
30        y = b;
31        z = c;
32        scale();
```

```
33     }
34     void randomize();
35
36     static void set_tolerance(double t) {
37       tolerance = abs(t);
38     }
39
40     static double get_tolerance() {
41       return tolerance;
42     }
43
44     double getX() const { return x; }
45     double getY() const { return y; }
46     double getZ() const { return z; }
47
48     bool is_invalid() const {
49       return (x==0.) && (y==0.) && (z==0.);
50     }
51
52  };
53
54  ostream& operator<<(ostream& os, const PObject& A);
55
56  bool dependent(const PObject& A, const PObject& B, const PObject& C);
57
58  #endif
```

Inside PObject.cc we need to declare PObject::tolerance and we give it an initial value. (See Program 10.12, line 4.)

We also need to modify the equals, incident, and dependent procedures to test for near equality instead of exact equality. You can find these modifications on lines 37–39, 52, and 80 of the new PObject.cc file which we present here.

Program 10.12: Program file for the PObject class (version 2).

```
1   #include "PObject.h"
2   #include "uniform.h"
3
4   double PObject::tolerance = default_tolerance;
5
6   void PObject::scale() {
7     if (z != 0.) {
8       x /= z;
9       y /= z;
10      z = 1.;
11      return;
12    }
13    if (y != 0.) {
14      x /= y;
15      y = 1.;
16      return;
17    }
18    if (x != 0) {
19      x = 1.;
20    }
```

```
21   }
22
23   double PObject::dot(const PObject& that) const {
24     return x*that.x + y*that.y + z*that.z;
25   }
26
27   void PObject::randomize() {
28     do {
29       x = unif(-1.,1.);
30       y = unif(-1.,1.);
31       z = unif(-1.,1.);
32     } while (x*x + y*y + z*z > 1.);
33     scale();
34   }
35
36   bool PObject::equals(const PObject& that) const {
37     double d = abs(x-that.x) + abs(y-that.y) + abs(z-that.z);
38
39     return d <= tolerance;
40   }
41
42   bool PObject::less(const PObject& that) const {
43     if (x < that.x) return true;
44     if (x > that.x) return false;
45     if (y < that.y) return true;
46     if (y > that.y) return false;
47     if (z < that.z) return true;
48     return false;
49   }
50
51   bool PObject::incident(const PObject& that) const {
52     return abs(dot(that)) <= tolerance;
53   }
54
55   ostream& operator<<(ostream& os, const PObject& A) {
56     os << "<"
57        << A.getX() << ","
58        << A.getY() << ","
59        << A.getZ()
60        << ">";
61     return os;
62   }
63
64   bool dependent(const PObject& A, const PObject& B, const PObject& C){
65     double a1 = A.getX();
66     double a2 = A.getY();
67     double a3 = A.getZ();
68
69     double b1 = B.getX();
70     double b2 = B.getY();
71     double b3 = B.getZ();
72
73     double c1 = C.getX();
74     double c2 = C.getY();
75     double c3 = C.getZ();
76
```

```
77     double det = a1*b2*c3 + a2*b3*c1 + a3*b1*c2
78                - a3*b2*c1 - a1*b3*c2 - a2*b1*c3;
79
80   return abs(det) <= PObject::get_tolerance();
81   }
82
83   PObject PObject::rand_perp() const {
84     if (is_invalid()) return PObject(0,0,0);
85
86     double x1,y1,z1;    // One vector orthogonal to (x,y,z)
87     double x2,y2,z2;    // Another orthogonal to (x,y,z) and (x1,y1,z1)
88
89     if (z == 0.) { // If z==0, take (0,0,1) for (x1,y1,y2)
90       x1 = 0;
91       y1 = 0;
92       z1 = 1;
93     }
94     else {
95       if (y == 0.) {  // z != 0 and y == 0, use (0,1,0)
96         x1 = 0;
97         y1 = 1;
98         z1 = 1;
99       }
100      else {  // y and z both nonzero, use (0,-z,y)
101        x1 = 0;
102        y1 = -z;
103        z1 = y;
104      }
105    }
106
107    // normalize (x1,y1,z1)
108    double r1 = sqrt(x1*x1 + y1*y1 + z1*z1);
109    x1 /= r1;
110    y1 /= r1;
111    z1 /= r1;
112
113    // (get x2,y2,z2) by cross-product with (x,y,z) and (x1,y1,z1)
114    x2 = -(y1*z) + y*z1;
115    y2 = x1*z - x*z1;
116    z2 = -(x1*y) + x*y1;
117
118    // normalize (x2,y2,z2)
119    double r2 = sqrt(x2*x2 + y2*y2 + z2*z2);
120    x2 /= r2;
121    y2 /= r2;
122    z2 /= r2;
123
124    // get a point uniformly on the unit circle
125    double a,b,r;
126    do {
127      a = unif(-1.,1.);
128      b = unif(-1.,1.);
129      r = a*a + b*b;
130    } while (r > 1.);
131    r = sqrt(r);
132    a /= r;
```

```
133    b /= r;
134
135    double xx = x1 * a + x2 * b;
136    double yy = y1 * a + y2 * b;
137    double zz = z1 * a + z2 * b;
138
139    return PObject(xx,yy,zz);
140 }
141
142
143 PObject PObject::op(const PObject& that) const {
144
145    if (equals(that)) return PObject(0,0,0);
146
147    double c1 = y*that.z - z*that.y;
148    double c2 = z*that.x - x*that.z;
149    double c3 = x*that.y - y*that.x;
150
151    return PObject(c1,c2,c3);
152 }
```

When the test program (Program 10.10) is run with the new PObject class, we achieve the desired results.

```
The random point P is (-0.479902,-0.616199,1)
Two lines through P are L = [0.384191,1.32364,1]
and M = [1.66531,0.32589,1]
Is P on L? 1
Does M have P? 1
The point of intersection of L and M is Q = (-0.479902,-0.616199,1)
Is Q on L? 1
Does M have Q? 1
P and Q are equal
```

The user may invoke PObject::set_tolerance(0.0); to revert to the previous behavior (exact checking).

Finally, the method we use for testing near equality can be improved. For example, we check if two projective objects are equal by computing their L^1 distance and comparing against tolerance. Alternatively, to check $\langle x_1, y_1, z_1 \rangle$ and $\langle x_2, y_2, z_2 \rangle$ for equality, we might consider a test such as this:

$$\frac{|x_1 - x_2| + |y_1 - y_2| + |z_1 - z_2|}{|x_1| + |x_2| + |y_1| + |y_2| + |z_1| + |z_2|} \le \varepsilon.$$

Whatever equality test you feel is most appropriate, it is only necessary to edit one method (equals in PObject) to implement your choice.

10.9   Pappus revisited

We close this section with a program to illustrate Pappus's Theorems and the use of the near-equality testing.

Program 10.13: A program to illustrate Pappus's theorem and its dual.

```cpp
#include "Projective.h"
#include "uniform.h"

/**
 * An illustration of Pappus's theorem and its dual
 */

void pappus() {
  seed();

  // two random lines
  PLine L1,L2;
  L1.randomize();
  L2.randomize();
  cout << "The two lines are " << endl << L1 << " and " << L2 << endl;

  // get three points on the first
  PPoint P1 = L1.rand_point();
  PPoint P2 = L1.rand_point();
  PPoint P3 = L1.rand_point();

  cout << "Three points on the first line are " << endl
       << P1 << endl << P2 << endl << P3 << endl;

  // get three points on the second
  PPoint Q1 = L2.rand_point();
  PPoint Q2 = L2.rand_point();
  PPoint Q3 = L2.rand_point();

  cout << "Three points on the second line are " << endl
       << Q1 << endl << Q2 << endl << Q3 << endl;

  // find the three pairwise intersections
  PPoint X1 = (P2+Q3)*(P3+Q2);
  PPoint X2 = (P1+Q3)*(P3+Q1);
  PPoint X3 = (P1+Q2)*(P2+Q1);

  cout << "The three points constructed are " << endl;
  cout << X1 << endl << X2 << endl << X3 << endl;

  if (collinear(X1,X2,X3)) {
    cout << "They are collinear, as guaranteed by Pappus's theorem"
         << endl;
  }
  else {
    cout << "TROUBLE! The three points are not collinear!!"
```

```
47        << endl;
48     }
49  }
50
51  void dual_pappus() {
52     // Two random points
53     PPoint A,B;
54     A.randomize();
55     B.randomize();
56     cout << "The two points are " << endl << A << " and " << B << endl;
57
58     // Three lines through the first
59     PLine L1 = A.rand_line();
60     PLine L2 = A.rand_line();
61     PLine L3 = A.rand_line();
62
63     cout << "The three lines through the first point are " << endl
64          << L1 << endl << L2 << endl << L3 << endl;
65
66     // Three lines through the second
67     PLine M1 = B.rand_line();
68     PLine M2 = B.rand_line();
69     PLine M3 = B.rand_line();
70
71     cout << "The three lines through the second point are " << endl
72          << M1 << endl << M2 << endl << M3 << endl;
73
74     // Get the three dual Pappus lines
75     PLine X1 = L2*M3 + L3*M2;
76     PLine X2 = L1*M3 + L3*M1;
77     PLine X3 = L1*M2 + L2*M1;
78
79     cout << "The three lines constructed are " << endl;
80     cout << X1 << endl << X2 << endl << X3 << endl;
81
82     if (concurrent(X1,X2,X3)) {
83       cout << "They are concurrent, as guaranteed by Pappus's theorem"
84            << endl;
85     }
86     else {
87       cout << "TROUBLE! The three lines are not concurrent!!"
88            << endl;
89     }
90  }
91
92
93  int main() {
94     double t;
95     cout << "Enter desired tolerance --> ";
96     cin >> t;
97     PObject::set_tolerance(t);
98     cout << "You set the tolerance to " << PObject::get_tolerance()
99          << endl << endl;
100
101    pappus();
102    cout << endl;
```

```
103    dual_pappus();
104
105    return 0;
106  }
```

Here are three runs of the program with the tolerance set to different values.

```
Enter desired tolerance --> 0
You set the tolerance to 0

The two lines are
[2.23943,2.19462,1] and [-0.685646,2.15228,1]
Three points on the first line are
(14.3273,-15.0755,1)
(-0.434872,-0.0119093,1)
(-0.56911,0.12507,1)
Three points on the second line are
(-2.29319,-1.19516,1)
(0.43878,-0.324843,1)
(7.8271,2.02884,1)
The three points constructed are
(-0.323743,0.01554,1)
(6.49488,5.53439,1)
(-0.0728833,0.218581,1)
TROUBLE! The three points are not collinear!!

The two points are
(0.576837,0.361625,1) and (-0.407185,0.0903103,1)
The three lines through the first point are
[-1.45089,-0.450946,1]
[0.398283,-3.4006,1]
[18.9351,-32.9691,1]
The three lines through the second point are
[2.2305,-1.01622,1]
[2.46456,0.0391258,1]
[2.51677,0.274485,1]
The three lines constructed are
[2.42584,0.116742,1]
[1.63958,0.114103,1]
[3.10284,0.119014,1]
TROUBLE! The three lines are not concurrent!!
```

```
Enter desired tolerance --> 1e-16
You set the tolerance to 1e-16

The two lines are
[0.55364,0.547428,1] and [1.05044,-0.347064,1]
Three points on the first line are
(-0.0325509,-1.79381,1)
(-0.440495,-1.38123,1)
(1.76843,-3.61523,1)
Three points on the second line are
(-2.34784,-4.22478,1)
(-0.911666,0.122021,1)
(-0.367147,1.77009,1)
The three points constructed are
```

```
(-0.421418,-0.561603,1)
(0.160804,-3.85329,1)
(-0.310677,-1.18769,1)
They are collinear, as guaranteed by Pappus's theorem

The two points are
(-15.284,42.4406,1) and (-2.5346,1.29455,1)
The three lines through the first point are
[-1.32801,-0.501814,1]
[-0.903241,-0.348843,1]
[0.00499314,-0.0217642,1]
The three lines through the second point are
[2.56928,4.25793,1]
[0.38109,-0.0263314,1]
[31.5203,60.9412,1]
The three lines constructed are
[-0.725517,-0.0128931,1]
[-9.98448,-16.6938,1]
[-0.936431,-0.392874,1]
TROUBLE! The three lines are not concurrent!!
```

```
Enter desired tolerance --> 1e-12
You set the tolerance to 1e-12

The two lines are
[0.52502,-0.458764,1] and [-0.330266,1.96863,1]
Three points on the first line are
(-2.99912,-1.25249,1)
(-1.44265,0.528772,1)
(-2.39671,-0.563076,1)
Three points on the second line are
(1.18674,-0.308875,1)
(1.38413,-0.27576,1)
(-6.11116,-1.53321,1)
The three points constructed are
(-4.23015,-0.702403,1)
(30.568,1.77536,1)
(1.20682,-0.315271,1)
They are collinear, as guaranteed by Pappus's theorem

The two points are
(15.562,5.35536,1) and (-0.432665,0.837712,1)
The three lines through the first point are
[-0.171104,0.310479,1]
[-2.21649,6.25412,1]
[-0.915338,2.47313,1]
The three lines through the second point are
[-3.3276,-2.91238,1]
[0.807493,-0.776671,1]
[-9.38979,-6.0434,1]
The three lines constructed are
[-2.58492,5.62249,1]
[-2.64901,-1.39739,1]
[-2.6087,3.01846,1]
They are concurrent, as guaranteed by Pappus's theorem
```

10.10 Exercises

10.1 Create a pair of classes named `Rectangle` and `Square`, which should be a derived subclass of `Rectangle`.

`Rectangle` should have two data members (representing the height and width) and the following methods.

- A two-argument constructor.
- Methods to get the height and width.
- Methods to change the height and width.
- Methods to report the area and perimeter.

`Square` should have a single-argument constructor (which should rely on `Rectangle`'s constructor). It should have the same methods as `Rectangle`, except that the methods to change height and width should modify *both* the height and width.

10.2 Create a class called `Parallelogram` that represents a parallelogram seated in the plane as shown in the figure.

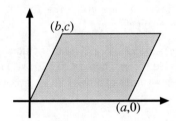

The values *a* and *c* must be nonnegative; *b* may be any real value.

The `Parallelogram` should define a zero- and three-argument constructor and methods for computing the area and perimeter of the figure.

Next, create two subclasses named `Rectangle` and `Rhombus` from the parent class `Parallelogram`. Define appropriate two-argument constructors for these subclasses.

There is no need to make a new `area()` method for the subclasses; the parent's `area()` method is efficient and serves the subclasses well. However, the `perimeter()` method in the `Parallelogram` needs to invoke the sqrt procedure to find the side length from $(0,0)$ to (b,c). However, the subclasses can find the perimeter in a more efficient manner. So, although it isn't necessary to redefine `perimeter()` for `Rhombus` and `Rectangle`, do so anyway so these procedures are more efficient in the special cases.

10.3 In Exercise 8.3 we explored C++'s inability to hold complex<double> values in a set container because the complex<double> type does not define a < operator. We resolved that issue by making a set of pair<double,double> objects that held the values (x, y) in lieu of $x + yi$.

Create an alternative solution by creating a class named mycomplex that is derived from the complex<double> type. Add to mycomplex an operator<. You also need zero-, one-, and two-argument constructors (with double arguments); these should invoke complex<double>'s constructors.

Finally, write a short main() procedure to check that mycomplex objects can be housed in a set<mycomplex> container.

10.4 Define a pair of classes named Point and Segment to represent points and line segments in the plane. The Point class should include an operator+ method; the result of P+Q is the line segment joining the points. The Segment class should include a midpoint() method that returns the Point that is midway between the end points of the line segment.

For both classes, define operator<< for writing to the screen.

10.5 In C++ it is possible to have two classes, Alpha and Beta, such that *arguments* and *return values* for methods in one class may be the type of the other class. That is, an Alpha method might return a Beta value and vice versa. (This is the case for the Point and Segment classes in Exercise 10.4.)

Can we create classes Alpha and Beta so that each contains *data members* that are of the other type?

10.6 Extend the complex<double> class to include the value ∞ (complex infinity) by creating a derived class named Complexx. This new value should interact with finite complex values in a sensible way. For example,

$$\infty + z = \infty \qquad\qquad \text{for finite } z$$
$$\infty \times \infty = \infty$$
$$z \div 0 = \infty \qquad\qquad \text{for } z \neq 0$$
$$z \div \infty = 0 \qquad\qquad \text{for finite } z$$

Some calculations with Complexx values should yield a special *undefined* value; for example, $0 \div 0$, $\infty \pm \infty$, $0 \times \infty$, ∞ / ∞, and so on.

Your class should include the operators +, - (unary and binary), *, /, ==, !=, and << (for writing to the screen).

10.7 Use the PPoint and PLine classes to write a program to illustrate Desargues' Theorem: suppose that triangles ABC and DEF are in perspective from a point O. (That is, the triples OAD, OBE, and OCF are each collinear.) Then the three points of intersection of the lines AC and DF, lines AB and DE, and lines BC and EF are collinear (see the points X, Y, and Z in Figure 10.4).

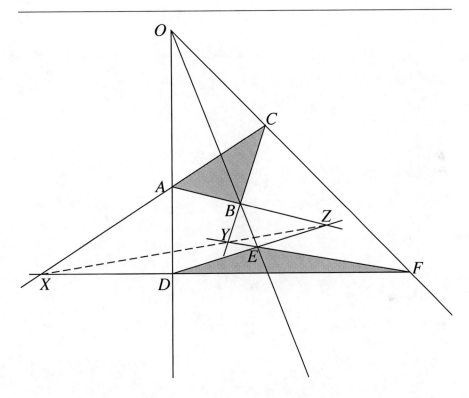

Figure 10.4:   An illustration of Desargues' Theorem.

Chapter 11

Permutations

Let n be a positive integer. A *permutation* is a one-to-one and onto function (i.e., a bijection) from the set $\{1, 2, \ldots, n\}$ to itself. The set of all permutations of this set is denoted S_n.

One way to represent a permutation π is as a list of values $[\pi(1), \pi(2), \ldots, \pi(n)]$. For example, $\pi = [1, 4, 7, 2, 5, 3, 6]$ means $\pi \in S_7$ and $\pi(1) = 1$, $\pi(2) = 4$, $\pi(3) = 7$, and so on, and $\pi(7) = 6$.

The disjoint cycle notation is an alternative way to write permutations. The permutation $\pi = [1, 4, 7, 2, 5, 3, 6]$ is written $(1)(2, 4)(3, 7, 6, 5)$. The (1) means $\pi(1) = 1$. The $(2, 4)$ means $\pi(2) = 4$ and $\pi(4) = 2$. The $(3, 7, 6, 5)$ means $\pi(3) = 7$, $\pi(7) = 6$, $\pi(6) = 5$, and $\pi(5) = 3$. In other words, $(1)(2, 4)(3, 7, 6, 5)$ means this:

$$1 \mapsto 1 \qquad 2 \mapsto 4 \mapsto 2 \qquad 3 \mapsto 7 \mapsto 6 \mapsto 5 \mapsto 3.$$

In this chapter we develop a C++ class to represent permutations. We have two goals. One is to introduce additional C++ concepts (copy constructors, destructors, and assignment operators). The other is to write a program to explore Ulam's problem.

11.1 Ulam's problem

Given a permutation $\pi \in S_n$, we may regard π as a sequence $[\pi(1), \pi(2), \ldots, \pi(n)]$. An *increasing subsequence* of π is a sequence $[\pi(i_1), \pi(i_2), \ldots, \pi(i_t)]$ where $1 \leq i_1 < i_2 < \cdots < i_t \leq n$ and $\pi(i_1) < \pi(i_2) < \cdots < \pi(i_t)$. A *decreasing* subsequence is defined analogously. A subsequence of a permutation is called *monotone* if it is either increasing or decreasing.

For example, the sequence $[1, 4, 7, 2, 5, 3, 6]$ contains the increasing subsequence $[1, 2, 3, 6]$ and the decreasing subsequence $[7, 5, 3]$. Every permutation must have a "reasonably" long monotone subsequence. This is a consequence of the Erdős–Szekeres Theorem.

Theorem 11.1 (Erdős–Szekeres). *Let k be a positive integer and let $n = k^2 + 1$. Then every permutation $\pi \in S_n$ contains a monotone subsequence of length $k + 1$.*

Informally, the result states that every permutation in S_n contains a monotone subsequence of length about $\sqrt{n}$. The proof of this result is interesting both because it is a good example of the use of the pigeonhole principle and because it leads to an algorithm for finding longest monotone subsequences of permutations.

Proof. Let $\pi \in S_n$ where $n = k^2 + 1$. For each i with $1 \le i \le n$ let (u_i, d_i) denote the lengths of the longest increasing and decreasing subsequences of π that start at position i. For example, the sequence $[1,4,7,2,5,3,6]$ gives the following values for u_i and d_i.

Index i	1	2	3	4	5	6	7
$\pi(i)$	1	4	7	2	5	3	6
u_i	4	3	1	3	2	2	1
d_i	1	2	3	1	2	1	1

The easiest way to verify this chart is to check the u_i and d_i entries starting from the right.

The key observation is that for $i \ne j$ we have $(u_i, d_i) \ne (u_j, d_j)$. To see why, let $i < j$. Then either $\pi(i) < \pi(j)$, in which case $u_i > u_j$; or else $\pi(i) > \pi(j)$, in which case $d_i > d_j$.

Suppose, for the sake of contradiction, there is a permutation π in S_n without a monotone subsequence of length $k + 1$ (where $k^2 + 1 = n$). Therefore, $1 \le u_i, d_i \le k$ for all i. Hence there are at most k^2 distinct values of (u_i, d_i) and so for some pair of indices $i \ne j$ we have $(u_i, d_i) = (u_j, d_j)$. $\Rightarrow\Leftarrow$ Therefore, π must have a monotone subsequence of length $k + 1$. $\square$

Ulam's problem concerns the longest increasing subsequence of a random permutation. Suppose a permutation π is chosen uniformly at random from S_n. That is, all permutations in S_n are equally likely, each with probability $1/n!$. What is the expected length of π's longest increasing subsequence?

If we let L_n be the random variable giving the length of a longest increasing subsequence in π, then Ulam's problem asks us to find $E(L_n)$. The Erdős–Szekeres Theorem suggests that this value should be on the order of $\sqrt{n}$. Indeed, in a celebrated paper, Hammersley[1] showed that

$$\lim_{n \to \infty} \frac{E(L_n)}{\sqrt{n}}$$

exists (and is not zero). Subsequent work in papers by Logan and Shepp[2] and by Vershik and Kerov[3] establish the value of this limit. (We give the answer at the end of this chapter.)

[1] J.M. Hammersley, A few seedlings of research, *Proceedings of the Sixth Berkeley Symposium on Mathematical Statistics and Probability*, **1** (1972), 345–394.

[2] B.F. Logan and L.A. Shepp, A variational problem for random Young tableaux, *Advances in Math.* **26** (1977), 206–222.

[3] A.M. Vershik and S.V. Kerov, Asymptotics of the Plancherel measure of the symmetric group and the limiting form of Young tables, *Soviet Math. Dokl.*, **18** (1977), 527–531.

11.2 Designing the `Permutation` class

Here is the header file, `Permutation.h` for the `Permutation` class. The overall design of the class as well as the new C++ concepts are explained after.

Program 11.1: Header file for `Permutation` class, `Permutation.h`.

```
1   #ifndef PERMUTATION_H
2   #define PERMUTATION_H
3   #include <iostream>
4   using namespace std;
5
6   class Permutation {
7   private:
8     long  n;
9     long* data;
10
11  public:
12    Permutation();
13    Permutation(long nels);
14    Permutation(long nels, long* array);
15    Permutation(const Permutation& that);
16    ~Permutation();
17
18    void swap(long i, long j);
19    void randomize();
20    void reset();
21
22    bool check() const;
23
24    long getN() const { return n; }
25    long of(long k) const;
26    long operator()(long k) const { return of(k); }
27
28    Permutation operator=(const Permutation& that);
29
30    Permutation operator*(const Permutation& that) const;
31    Permutation operator*=(const Permutation& that);
32
33    Permutation inverse() const;
34
35    bool operator==(const Permutation& that) const;
36    bool operator!=(const Permutation& that) const;
37    bool operator< (const Permutation& that) const;
38    bool isIdentity() const;
39  };
40
41  ostream& operator<<(ostream& os, const Permutation& P);
42
43  #endif
```

11.2.1 Data

To represent an element of S_n we hold the integer n in a `long` variable. We also need a way to store the values $\pi(1)$ through $\pi(n)$. For that task we have two natural choices: an array of `long` values or a container such as `vector<long>` (see Section 8.4). Although using a `vector` container would simplify some issues, we opt for an array for the following reasons. First, once we create a permutation object, we do not expect the length (n) of the permutation to change. Therefore, we do not need a `vector`'s ability to resize itself. Second, using an array compels us to learn additional features of C++.

The class `Permutation` contains two data elements in its private section such as this:

```
class Permutation {
private:
  long  n;
  long* data;
public:
  ...
};
```

The integer n holds the size of the permutation; that is, the permutation is on the integers 1 through n.

The array `data` holds the values $\pi(i)$ for $1 \le i \le n$. To make our lives easier, we let `data[1]` hold the value $\pi(1)$, `data[2]` hold $\pi(2)$, and so on. This means that we need to be sure that `data` has capacity n+1. The entry `data[0]` is wasted.[4]

To be sure that a `Permutation` object is always valid, we do not give users unfettered access to `data`. All of the methods that act on `Permutation` objects must ensure that n and `data` represent a valid permutation.

11.2.2 Constructors and destructors

The `Permutation` class contains four constructors.

- The first is a basic, zero-argument constructor `Permutation()`. This constructor creates the identity permutation (1) in S_1. (See line 12 of Program 11.1.)

- The second is a single-argument constructor `Permutation(nels)`. When `nels` is a positive integer n, this creates the identity permutation $(1)(2)\cdots(n)$ in S_n. (See line 13.)

- The third is a two-argument constructor `Permutation(nels,array)`. Here, `nels` contains a positive integer n, and `array` contains an array of $n+1$ `long`

---

[4]Note that we could use `data[0]` to store n obviating the need for the variable n. This would be efficient for the computer's memory, but inefficient for *human* memory as this would require us to remember this trick. It would also make the code harder to understand. Avoid sneaky tricks.

integers. The value of `array[k]` is $\pi(k)$ for $1 \leq k \leq n$. Thus `array` should be of size `nels+1`. (See line 14.)

The action of this constructor is to assign n from `nels` and to copy entries 1 through `nels` of `array` into `data`. This, however, is dangerous. We do not know if the data held in `array` are a valid representation of a permutation. Therefore, we include a `check` method (line 22). The `check` method returns `true` if the data held in the `Permutation` are valid. If the `check` method returns `false`, the constructor calls `reset` to reset the permutation to the identity of S_n.

See lines 20 and 22 of `Permutation.h` (Program 11.1) and lines 21–32, 44–46, and 65–85 of `Permutation.cc` (Program 11.2.)

- The fourth is a single-argument constructor `Permutation(that)` where the argument `that` is also a `Permutation` object. This is called a *copy constructor*. The new permutation created is a copy of `that`. (See line 15.)

  Until this point, we have not needed a copy constructor. Its necessity is explained shortly (Subsection 11.2.3).

In all four cases, the private data elements, n and `data`, need to be initialized by the constructor. For example, here is the portion of the file `Permutation.cc` that spells out the action of the `Permutation(nels)` constructor (copied from lines 10–19 of Program 11.2).

```
Permutation::Permutation(long nels) {
  if (nels <= 0) {
    nels = 1;
  }
  data = new long[nels+1];
  n = nels;
  for (long k=1; k<=n; k++) {
    data[k] = k;
  }
}
```

Remember that `data` is of type `long*` (an array of `long` values). Until memory has been allocated for `data`, we cannot store values in `data[1]`, `data[2]`, and so on. The allocation of storage takes place with `data = new long[nels+1];`. Once the storage has been allocated, the `data` array can be populated (see the `for` loop).

This code appears to violate our admonition that every `new` must be balanced with a corresponding `delete[]`. Where is the `delete[]` statement?

We cannot add the statement `delete[] data;` to the constructor. If we do, no sooner do we set up `data` then we would destroy it, defeating the purpose of the constructor.

What we need is for the `delete[]` instruction to execute when we are done with the object. For example, consider this bit of code.

```
void example(long n) {
  Permutation P(n);   // make a new permutation
  ...                 // do a bunch of stuff
```

```
  cout << P << endl; // print P
}
```

When the procedure `example` is invoked, a `Permutation` object `P` is created. When the procedure reaches its end, the variable `P` goes out of scope and is no longer available. If `example` were called repeatedly (say, by `main`), then we would have repeated `new` statements without any balancing `delete`s. This is a memory leak. On each invocation, a new block of memory is allocated but never released for reuse.

The solution to this problem is to create a *destructor*. A destructor is a method that the computer automatically invokes when a variable goes out of scope, typically at the end of a procedure in which it was declared.

Destructors are named as follows. The first character is a tilde and the remainder of the name is the name of the class: `~Permutation` in the current case. When declaring a destructor, no return type is specified and the argument list is empty. Here is line 16 of `Permutation.h` (Program 11.1).

```
  ~Permutation();
```

The code for `~Permutation()` is on lines 40–42 of `Permutation.cc` (see Program 11.2); we repeat that code here.

```
Permutation::~Permutation() {
  delete[] data;
}
```

Alternatively, we could have defined `~Permutation` inline in the header file in the public section of the `Permutation` class like this:

```
  ~Permutation() { delete[] data; }
```

With this destructor in place, the memory leak in the `example` procedure has been repaired. Every time the computer reaches the end of `example`, the variable `P` goes out of scope. At that point, the destructor is executed. If you wish, you may add a statement such as `cerr << "Destructor ran" << endl;` to `~Permutation`'s code; this enables you to observe the destructor when it runs.

Classes that do not dynamically allocate memory do not require destructors. The container classes (discussed in Chapter 8) have destructors; any memory they consume is freed when their variables go out of scope.

11.2.3 Copy and assign

If `a` and `b` are `long` integers, the statement `a=b;` puts a copy of the value held in `b` into `a`. Later, if we modify `b` there is no effect on `a`. Furthermore, the expression `a=b` returns a value: the common value held in `b` and now `a`. The statement `a=b=c;` is tantamount to `a=(b=c);`. The expression `b=c` is evaluated first. The value of the expression `b=c` is the value held in `c` (and now `b`). That value is then assigned to `a`. In this way, the single statement `a=b=c;` is equivalent to the pair of statements `b=c;` and `a=b;`. In summary, the statement `a=b;` has an *action* (giving `a` a copy of the value held in `b`) and a *value* (the now common value held in `a` and `b`).

If a and b are objects of a class, then the assignment operator a=b; has a predefined, default meaning in C++. The *action* is to copy each data element in b to the corresponding data element in a. The *value* of a=b; is a copy of the common value held in a and b.

For example, recall the Point class from Chapter 6. Objects of type Point contain two data elements: double variables x and y. (See Point.h, Program 6.1.) Therefore, the statement a=b; has the same effect as the two statements a.x=b.x; and a.y=b.y;. (Note: This pair of statements is illegal outside a Point method because the data elements x and y are private. However, the single statement a=b; is permissible anywhere.) The value of the statement a=b; is a copy of the value in b (and now a). Therefore, if a, b, and c are objects of type Point, the statement a=b=c; works just as do the pair of statements b=c; a=b;.

When a variable is passed to a procedure, a copy of that variable is created. For example, when we invoke a procedure such as d=gcd(a,b), copies of a and b are created and sent to the gcd procedure. (However, if a procedure's argument is set up for call by reference, then no copy is made and the function parameter is tied directly to the variable named in the corresponding position.)

Similarly, if an argument to a procedure is of an object of a certain class, then a copy of that object is created and sent to the procedure. For example, the procedure midpoint(P,Q) works on copies of P and Q. These duplicates are tacitly created by copying the x and y fields of P and Q.

These copies are made by a *copy constructor*. The default copy constructor simply replicates the data elements in the objects. It is possible to see these actions explicitly in the following program.

```
#include "Point.h"
#include <iostream>
using namespace std;

int main() {
  Point X(2,3);
  Point Y(X);
  Point Z;
  Z = Y;
  cout << X << endl;
  cout << Y << endl;
  cout << Z << endl;
  return 0;
}
```

First, the object X is created with X.x equal to 2 and X.y equal to 3. This uses the standard two-argument constructor.

Second, a copy of X is created in Y using the statement Point Y(X);. Notice that in Point.h there is no constructor with a single Point argument. C++ provides a default copy constructor. If we were to write code for the default copy constructor, it would look like this:

```
Point(const Point& that) {
  x = that.x;
  y = that.y;
```

```
}
```

Third, we create a `Point` object `z` using the zero-argument constructor. The statement `Point z;` initializes `z` to the point $(0,0)$. The next statement is `z = y;`. This invokes the default assignment operator. The action of the assignment operator is to copy the data elements of `Y` to `z`, and then return the (new) value of `z` as the result. If we were to write a program for the default assignment operator, it would look like this:

```
Point operator=(const Point& that) {
  x = that.x;
  y = that.y;
  return *this;
}
```

The last statement, `return *this;`, returns a copy of the left-hand side of the assignment statement.

For all of the classes previously considered in this book, the default copy constructor and assignment operator work fine. Indeed, we did not even mention their existence because their actions were "obvious." However, for the `Permutation` class, the default behaviors are not acceptable.

Here is the problem. Suppose that `P` and `Q` are objects of type `Permutation`. The default action of `P=Q;` is equivalent to `P.n=Q.n; P.data=Q.data;`. The first part, `P.n=Q.n;` is fine, but the second part, `P.data=Q.data;` is problematic. Recall that the `data` field of the class `Permutation` is of type `long*`. Therefore, the statement `P.data=Q.data;` does not copy `Q`'s data to `P`. Instead, the effect is for `P.data` to refer to exactly the same block of memory as `Q.data`. That would cause any subsequent changes to one of `P` or `Q` to affect the other. This is not the behavior we want. The statement `P=Q;` ought to place an independent copy of the permutation `Q` into `P`. Because the default behavior does not suit us, we need to override the default by explicitly defining an assignment operator.

In addition, the default copy constructor does not suit our purposes. Like the default assignment operator, the default copy constructor simply copies the data elements.

The default methods need to be replaced. In their stead, we declare the following methods in the public section of the `Permutation` class (see lines 15 and 28 of `Permutation.h`, Program 11.1),

```
Permutation(const Permutation& that);
Permutation operator=(const Permutation& that);
```

The code for these is given in `Permutation.cc` (Program 11.2). Let's look at the code for each.

The copy constructor code looks like this:

```
Permutation::Permutation(const Permutation& that) {
  n = that.n;
  data = new long[n+1];
  for (long k=1; k<=n; k++) data[k] = that.data[k];
}
```

Note that there is no return type because this is a constructor. The argument is a reference to a constant `Permutation` named `that`. The first action is simply to copy `that`'s n value to the new n value; this is exactly what the default copy constructor would do. Now, instead of `data=that.data;`, we allocate a new block of memory for `data` and then perform an element-by-element copy of the entries in the array `that.data` to our `data`.

The assignment operator acts in a similar manner. Here is the code.

```
Permutation Permutation::operator=(const Permutation& that) {
  delete[] data;
  n = that.n;
  data = new long[n+1];
  for (long k=1; k<=n; k++) data[k] = that.data[k];
  return *this;
}
```

The return type is `Permutation` because the expression P=Q returns a value. The argument (corresponding to the right-hand side of an assignment expression) is a reference to a `Permutation` object named `that`.

The first statement is `delete[] data;`. In the assignment statement P=Q;, the information held in P is replaced with a copy of that in Q. So we first discard the old `data` and then make a new one. (Alternatively, we could have checked if this permutation and `that` have the same value for n. If so, there would be no need to delete and then reallocate `data`.)

The rest of the code is straightforward. We copy `that.n`, allocate space for `data`, and then copy the entries in `that.data`.

Finally, we return a copy of the new value held in the left-hand side of the expression with the statement `return *this;`. Alternatively, we could have ended with `return that;` because `that` holds the same information as `*this`.

11.2.4 Basic inspection and modification methods

We provide methods for inspecting and modifying `Permutation` objects.

- `getN()` returns the value held in n. See line 24 of `Permutation.h` (Program 11.1).

- `swap(i,j)` exchanges the values $\pi(i)$ and $\pi(j)$. In conventional notation, π is replaced by $\pi \circ (i,j)$ where $1 \leq i, j \leq n$. See line 18 of `Permutation.h` and lines 48–54 of `Permutation.cc` (Program 11.2).

- `reset()` leaves n unchanged but resets the permutation to the identity permutation. See line 20 of `Permutation.h` and lines 44–46 of `Permutation.cc`.

- `randomize()` leaves n unchanged but replaces π with a permutation chosen uniformly at random from S_n. There is an efficient algorithm for doing this. For $k = 1, 2, \ldots, n$, we swap the value held in $\pi(k)$ with the value held in $\pi(j)$ where j is chosen uniform at random from $\{k, k+1, k+2, \ldots, n\}$. See line 19 of `Permutation.h` and lines 56–63 of `Permutation.cc`.

We do not provide a way to modify directly the value held in `data[i]`; to do so may result in an invalid permutation. However, we do provide a mechanism for learning the value held in `data[i]` and this is explained next.

11.2.5 Permutation operations

There are three fundamental operations we want to be able to perform with permutations. First, given a permutation π, we need to be able to obtain the value $\pi(k)$. Second, given two permutations π and σ, we want to calculate the composition $\pi \circ \tau$. And third, given a permutation π, we want to calculate π^{-1}. We consider each of these in turn.

- If P represents the permutation π, we define `P.of(k)` to return the value $\pi(k)$. A way to do this is simply to return `data[k]`. However, we need to guard against a user that submits a value k that is less than 1 or greater than n. For such improper values of `k`, the easiest thing to do is to return `k` itself.

  The declaration of `of` is on line 25 of `Permutation.h` (Program 11.1) and the code is on lines 87–90 of `Permutation.cc` (Program 11.2).

  Permutations are functions, so it would be pleasant to be able to write `P(k)` to extract the value $\pi(k)$. Fortunately, this is possible. The key to doing this is to define `operator()`. On line 26 of `Permutation.h` we see the entire inline code for this operator:

  ```
  long operator()(long k) const { return of(k); }
  ```

  The first pair of parentheses (after the keyword `operator`) shows that we are defining `P(k)` where P is a `Permutation` and `k` is a `long` integer. The second pair of parentheses encloses the argument `k`.

  Consequently, the expressions `P.of(k)` and `P(k)` are equivalent to each other.

- We define `operator*` so that `P*Q` stands for the composition of two permutations. If P and Q have the same value of `n`, the meaning of the composition is clear. In case they have different values of `n`, the resulting `Permutation` is based on the larger of the two `n`s. See the code for details (line 30 of `Permutation.h` and lines 100–111 of `Permutation.cc`).

  The first statement in the `operator*` method (line 101) finds the larger of `n` and `that.n` using an interesting C++ operator called the *trigraph*. The statement is this:

  ```
  long nmax = (n > that.n) ? n : that.n;
  ```

  The syntax for the trigraph operator `?:` is this:

  ```
  test ? val_1 : val_2
  ```

  where `test` is a Boolean expression and `val_1` and `val_2` are two expressions. The result of the trigraph depends on the value of `test`. If `test` is `true`, then the value is `val_1`; otherwise, the value is `val_2`. The statement

```
x = test ? val_1 : val_2;
```

is equivalent to the following code.

```
if (test) {
   x = val_1;
}
else {
   x = val_2;
}
```

The trigraph expression is handy if the subexpressions (`test`, `val_1`, and `val_2`) are not too complicated.

In addition to defining `operator*`, we also provide `operator*=` so we may avail ourselves of statements such as `P*=Q;`. See line 31 of `Permutation.h` and lines 113–116 of `Permutation.cc`.

- To calculate the inverse of a permutation, we provide an `inverse` method. See line 33 of `Permutation.h` and lines 118–122 of `Permutation.cc`. The statement `Q = P.inverse();` assigns the inverse of P to Q, but leaves P unchanged.

11.2.6 Comparison operators

We provide four different methods for comparing permutations. The `==` operator checks that the values of `n` are the same in both permutations, and then if the data held in both match. We do not consider the identity permutation in two different S_ns to be the same. The `!=` operator is equivalent to `!(P==Q)`.

We also provide an `operator<` to compare `Permutations`; this is convenient if we wish to create a `set` (or other sorted container) of `Permutations`.

Finally, we provide an `isIdentity` method that returns `true` if the object is the identity permutation.

See lines 35–38 of `Permutation.h` and lines 124–152 of `Permutation.cc` (Programs 11.1 and 11.2, respectively).

11.2.7 Output

We define `operator<<` for writing permutations to the computer's screen. We choose to write permutations in the disjoint cycle format. The operator is declared on line 41 of `Permutation.h` and the code is on lines 155–175 of `Permutation.cc`. The code uses a temporary array named `done` to keep track of which values have already been represented in cycles. See the code for details.

11.2.8 The code file `Permutation.c`

We close this section with the code file for the `Permutation` class.

Program 11.2: Program file for `Permutation` class.

```
1    #include "Permutation.h"
2    #include "uniform.h"
3
4    Permutation::Permutation() {
5      data = new long[2];
6      n = 1;
7      data[1] = 1;
8    }
9
10   Permutation::Permutation(long nels) {
11     if (nels <= 0) {
12       nels = 1;
13     }
14     data = new long[nels+1];
15     n = nels;
16     for (long k=1; k<=n; k++) {
17       data[k] = k;
18     }
19   }
20
21   Permutation::Permutation(long nels, long* array) {
22     if (nels <= 0) {
23       nels = 1;
24       data = new long[2];
25       data[1] = 1;
26       return;
27     }
28     n = nels;
29     data = new long[n+1];
30     for (long k=1; k<=n; k++) data[k] = array[k];
31     if (!check()) reset();
32   }
33
34   Permutation::Permutation(const Permutation& that) {
35     n = that.n;
36     data = new long[n+1];
37     for (long k=1; k<=n; k++) data[k] = that.data[k];
38   }
39
40   Permutation::~Permutation() {
41     delete[] data;
42   }
43
44   void Permutation::reset() {
45     for (long k=1; k<=n; k++) data[k] = k;
46   }
47
48   void Permutation::swap(long i, long j) {
49     if ( (i<1) || (i>n) || (j<1) || (j>n) || (i==j) ) return;
50     long a = data[i];
51     long b = data[j];
52     data[i] = b;
53     data[j] = a;
54   }
```

```
55
56   void Permutation::randomize() {
57     for (long k=1; k<n; k++) {
58       long j = unif(n-k+1)-1+k;
59       long tmp = data[j];
60       data[j] = data[k];
61       data[k] = tmp;
62     }
63   }
64
65   bool Permutation::check() const {
66     long* temp;
67
68     temp = new long[n+1];
69     for (long k=1; k<=n; k++) {
70       if ( (data[k] < 1) || (data[k] > n)) {
71         delete[] temp;
72         return false;
73       }
74       temp[k] = data[k];
75     }
76     sort(temp+1, temp+n+1);
77     for (long k=1; k<=n; k++) {
78       if (temp[k] != k) {
79         delete[] temp;
80         return false;
81       }
82     }
83     delete[] temp;
84     return true;
85   }
86
87   long Permutation::of(long k) const {
88     if ( (k<1) || (k>n) ) return k;
89     return data[k];
90   }
91
92   Permutation Permutation::operator=(const Permutation& that) {
93     delete[] data;
94     n = that.n;
95     data = new long[n+1];
96     for (long k=1; k<=n; k++) data[k] = that.data[k];
97     return *this;
98   }
99
100  Permutation Permutation::operator*(const Permutation& that) const {
101    long nmax = (n > that.n) ? n : that.n;
102    long* tmp = new long[nmax+1];
103
104    for (long k=1; k<=n; k++) {
105      tmp[k] = of(that(k));
106    }
107
108    Permutation ans(nmax,tmp);
109    delete[] tmp;
110    return ans;
```

```
111  }
112
113  Permutation Permutation::operator*=(const Permutation& that) {
114    (*this) = (*this) * that;
115    return *this;
116  }
117
118  Permutation Permutation::inverse() const {
119    Permutation ans(n);
120    for (long k=1; k<=n; k++) ans.data[data[k]] = k;
121    return ans;
122  }
123
124  bool Permutation::operator==(const Permutation& that) const {
125    if (n != that.n) return false;
126
127    for (long k=1; k<=n; k++) {
128      if (data[k] != that.data[k]) return false;
129    }
130    return true;
131  }
132
133  bool Permutation::operator!=(const Permutation& that) const {
134    return !( (*this)==that );
135  }
136
137  bool Permutation::operator<(const Permutation& that) const {
138    if (n < that.n) return true;
139    if (n > that.n) return false;
140
141    for (long k=1; k<=n; k++) {
142      if (data[k] < that.data[k]) return true;
143      if (data[k] > that.data[k]) return false;
144    }
145    return false;
146  }
147
148  bool Permutation::isIdentity() const {
149    for (int k=1; k<=n; k++) {
150      if (data[k] != k) return false;
151    }
152    return true;
153  }
154
155  ostream& operator<<(ostream& os, const Permutation& P) {
156    long n = P.getN();
157    bool* done = new bool[n+1];
158    for (long k=1; k<=n; k++) done[k] = false;
159
160    for (long k=1; k<=n; k++) {
161      if (!done[k]) {
162        os << "(" << k;
163        done[k] = true;
164        long j = P(k);
165        while (j!=k) {
166          os << "," << j;
```

```
167            done[j] = true;
168            j = P(j);
169         }
170       os << ")";
171     }
172   }
173   delete[] done;
174   return os;
175 }
```

11.3   Finding monotone subsequences

We now return to Ulam's problem: what is the expected length of a longest increasing subsequence of a random permutation? The `Permutation` class includes a `randomize` method, so we are able to generate permutations uniformly at random in S_n. Given a permutation, how do we find the length of a longest increasing (or decreasing) subsequence?

The idea for the algorithm comes from the proof of the Erdős–Szekeres Theorem (Theorem 11.1). Given a permutation $\pi \in S_n$, we define the values u_i (respectively, d_i) to be the length of a longest increasing (respectively, decreasing) subsequence of π starting at position i. To find the values u_i and d_i we work from right to left; that is, we start with $i = n$ and work our way down.

Clearly $u_n = d_n = 1$ because the longest increasing (decreasing) sequence that starts from the last position has length exactly one.

Now consider u_{n-1} and d_{n-1}. If $\pi(n-1) < \pi(n)$, then the longest increasing subsequence starting at position $n-1$ has length 2 and the longest decreasing subsequence has length 1. On the other hand, if $\pi(n-1) > \pi(n)$, then the longest increasing subsequence starting at position $n-1$ has length 1 and the longest decreasing subsequence has length 2. In other words,

$$(u_{n-1}, d_{n-1}) = \begin{cases} (2,1) & \text{if } \pi(n-1) < \pi(n), \quad \text{and} \\ (1,2) & \text{if } \pi(n-1) > \pi(n). \end{cases}$$

Suppose we have established the values u_i and d_i for $i = k+1$ through $i = n$. To find u_k, we compare $\pi(k)$ sequentially with $\pi(k+1)$, $\pi(k+2)$, and so on, up to $\pi(n)$. Among all indices $j > k$ for which $\pi(k) < \pi(j)$, find the one for which u_j is largest. We then set $u_k = u_j + 1$. If there are no indices j with $\pi(k) < \pi(j)$, then we set $u_k = 1$.

The procedure for finding d_k is analogous. Among all indices $j > k$ for which $\pi(k) > \pi(j)$, we select the j for which d_j is largest and set $d_k = d_j + 1$. If no such index j exists, we set $d_k = 1$.

Let's consider an example. For the permutation $\pi = [1,4,7,2,5,3,6]$, suppose we have calculated u_i and d_i for all $i \geq 3$. We now want to find u_2 and d_2. This is what we know so far.

Index i	1	2	3	4	5	6	7
$\pi(i)$	1	4	7	2	5	3	6
u_i	?	?	1	3	2	2	1
d_i	?	?	3	1	2	1	1

To find u_2, note that for indices $j = 3$, 5, and 7 we have $\pi(2) < \pi(j)$. Note that $u_3 = 1$, $u_5 = 2$, and $u_7 = 1$, so u_5 is largest. We therefore set $u_2 = u_5 + 1 = 3$. Indeed, the longest increasing subsequence starting at position 2 is $[4,5,6]$.

Finding d_2 is analogous. For indices $j = 4$ and 6 we have $\pi(2) > \pi(j)$. Note that $d_4 = 1$ and $d_6 = 1$, so we set $d_2 = 2$.

We create a procedure named `monotone` that takes a `Permutation` as its argument and returns a pair of `long` integers giving the lengths of the longest increasing and decreasing subsequences in the `Permutation`. Here is the header file, which we call `monotone.h`.

Program 11.3: Header file `monotone.h`.

```
1   #ifndef MONOTONE_H
2   #define MONOTONE_H
3
4   #include "Permutation.h"
5   #include <utility>
6
7   pair<long,long> monotone(const Permutation& P);
8
9   #endif
```

Note that we have `#include <utility>` because the `monotone` procedure returns a `pair`. The `utility` header file is needed to define the `pair` class.

The code for this procedure is housed in a file named `monotone.cc` that we present next.

Program 11.4: A program to find the length of the longest monotone subsequences of a permutation.

```
1    #include "monotone.h"
2
3    pair<long,long> monotone(const Permutation& P) {
4        long* up;
5        long* dn;
6        long n = P.getN();
7
8        up = new long[n+1];
9        dn = new long[n+1];
10
11       for (long k=1; k<=n; k++) {
12           up[k] = dn[k] = 1;
```

```
13      }
14
15      for (long k=n-1; k>=1; k--) {
16        for (long j=k+1; j<=n; j++) {
17          if (P(k) > P(j)) {
18            if (dn[k] <= dn[j]) {
19              dn[k] = dn[j]+1;
20            }
21          }
22          else {
23            if (up[k] <= up[j]) {
24              up[k] = up[j]+1;
25            }
26          }
27        }
28      }
29
30      long up_max = 1;
31      long dn_max = 1;
32      for (long k=1; k<=n; k++) {
33        if (up_max < up[k]) up_max = up[k];
34        if (dn_max < dn[k]) dn_max = dn[k];
35      }
36
37      delete[] up;
38      delete[] dn;
39
40      return make_pair(up_max,dn_max);
41    }
```

The arrays up and dn are used to hold the sequences u_i and d_i.

Finally, we need a main to generate random permutations repeatedly and calculate the average lengths of longest increasing and decreasing subsequences. Here is such a program.

Program 11.5: A program to illustrate Ulam's problem.

```
1   #include "Permutation.h"
2   #include "uniform.h"
3   #include "monotone.h"
4   #include <iostream>
5   using namespace std;
6
7   int main() {
8     long n;
9     long reps;
10    seed();
11
12    cout << "Enter n (size of permutation) --> ";
13    cin >> n;
14    cout << "Enter number of repetitions   --> ";
15    cin >> reps;
16
17    Permutation P(n);
18
```

```
19    long sum_up = 0;
20    long sum_dn = 0;
21
22    for (long k=0; k<reps; k++) {
23       P.randomize();
24       pair<long,long> ans;
25       ans = monotone(P);
26       sum_up += ans.first;
27       sum_dn += ans.second;
28    }
29
30    cout << "Average length of longest increasing subsequence is "
31         << double(sum_up)/double(reps) << endl;
32
33    cout << "Average length of longest decreasing subsequence is "
34         << double(sum_dn)/double(reps) << endl;
35
36    return 0;
37 }
```

Here is the result of running this program for permutations in S_n for n equal to 10,000 for one hundred iterations.

```
Enter n (size of permutation) --> 10000
Enter number of repetitions    --> 100
Average length of longest increasing subsequence is 192.31
Average length of longest decreasing subsequence is 192.94
```

For this case, we observe $E(L_n)/\sqrt{n} \approx 1.9$. In fact,

$$\lim_{n\to\infty} \frac{E(L_n)}{\sqrt{n}} = 2.$$

11.4   Exercises

11.1 Devise a procedure or method to find the order of a permutation. (For a permutation π, the *order* of π is the least positive integer k so that π^k is the identity.)

11.2 Create a class name `Counted` that can keep track of the number of objects of type `Counted` which exist at any point in the program. That is, the `Counted` class should include a `static` method named `count` that returns the number of `Counted` objects currently in memory.

For example, consider the following `main()`.

```
#include "Counted.h"
#include <iostream>
using namespace std;

int main() {
```

```
Counted X, Y;
Counted* array;
array = new Counted[20];

cout << "There are " << Counted::count()
     << " Counted objects in memory" << endl;

delete[] array;

cout << "And now there are " << Counted::count()
     << " Counted objects in memory" << endl;

return 0;
}
```

This should produce the following output.

```
There are 22 Counted objects in memory
And now there are 2 Counted objects in memory
```

(It's hard to imagine a mathematical reason we might want to keep track of how many objects of a given sort are being held in memory, but this is a useful debugging trick to see if there is a memory leak.)

11.3 Redo Exercise 8.5 (page 152) without using the container classes of the Standard Template Library. That is, the parts of the partition should be held in a conventional C++ array and maintained in sorted order. This change requires you to create a copy constructor, an assignment operator, and a destructor.

Replace the get_parts method by an operator[]; if P is a Partition, then the expression P[k] should give the kth part of the partition.

11.4 Let n be a positive integer. Suppose that n points are placed in the unit cube $[0,1]^3$. Write a procedure to find the size s of a largest subset of these n points such that their coordinates satisfy

$$x_1 \le x_2 \le x_3 \le \cdots \le x_s$$
$$y_1 \le y_2 \le y_3 \le \cdots \le y_s$$
$$z_1 \le z_2 \le z_3 \le \cdots \le z_s$$

Suppose the points are placed in the unit cube independently and uniformly at random. Use your procedure to conjecture the expected size of the largest such set.

11.5 Create a class SmartArray that behaves as a C++ array on long values, but that allows any integer subscript. That is, a SmartArray is declared like this:

```
SmartArray X(100);
```

This creates an object X that has the same behavior as if it were declared long X[100]; but allows indexing beyond the range 0 to 99. An index of

k outside this range is replaced by k mod 100. In this way, X[-1] refers to the last element of the array (in this case X[99]).

Hint: The hard part for this problem is enabling the expression X[k] to appear on the left side of an assignment. Suppose you declare the indexing operator like this:

```
long operator[](long k);
```

Then you cannot have an expression such as X[2]=4;. The trick is to declare the operator like this:

```
long& operator[](long k);
```

11.6 Create a class to represent linear fractional transformations. These are functions of the form

$$f(z) = \frac{az+b}{cz+d}$$

where $z, a, b, c, d \in \mathbb{C}$.

Be sure to include an operator() for evaluating a linear fractional transformation at a given complex number and an operator* for the composition of two transformations.

11.7 Create a Path class that represents a polygonal path in the plane (i.e., an ordered sequence of Points in $\mathbb{R}^2$). Include the following methods:

- A default constructor that creates an empty path.
- A one-argument constructor that creates a path containing a single point.
- An operator+ that concatenates two paths, or concatenates a path and a point. (Be sure to take care of both Path+Point and Point+Path.)
- An operator[] to get the kth point on the path.

Chapter 12

Polynomials

In this chapter we create a C++ class to represent polynomials. The coefficients of these polynomials can be from any field K such as the real numbers, the complex numbers, or $\mathbb{Z}_p$. One way to do this is to create several different polynomial classes depending on the coefficient field. A better solution is to learn how to use C++ templates.

12.1 Procedure templates

Imagine that we often need to find the largest of three quantities in our programming. We create a procedure named `max_of_three` like this:

```
long max_of_three(long a, long b, long c) {
  if (a<b) {
    if (b<c) return c;
    return b;
  }
  if (b<c) return c;
  return b;
}
```

Then, the expression `max_of_three(3,6,-2)` evaluates to 6.

If we also want to find the maximum of three real values, we create another version of `max_of_three`:

```
double max_of_three(double a, double b, double c) {
  if (a<b) {
    if (b<c) return c;
    return b;
  }
  if (b<c) return c;
  return b;
}
```

Pythagorean triples (see Chapter 7) can also be compared by <, so if we want to compare `PTriple` objects, we create yet another version of `max_of_three`:

```
PTriple max_of_three(PTriple a, PTriple  b, PTriple c) {
  if (a<b) {
    if (b<c) return c;
```

```
      return b;
    }
    if (b<c) return c;
    return b;
}
```

At some point, we may want to apply `max_of_three` to another class. We could write more and more versions of this procedure, each with a different first line to accommodate the additional types.

The result is many different incarnations of the `max_of_three` procedure. This is legal in C++; one can have multiple procedures with the same name provided the types or number of arguments are different for each version. The procedure `max_of_three(double,double,double)` happily coexists with the version of `max_of_three` that acts on arguments of type `PTriple`. It is annoying that we need to create the exact same procedure over and over again for different types of arguments. Furthermore, if we want to change the algorithm for `max_of_three`, we need to edit all the different versions.

Fortunately, there is a better way. It is possible to write a single version of `max_of_three` that automatically adapts to the required situation using templates. Here is the template for the `max_of_three` procedure.

Program 12.1: The header file containing the template for the `max_of_three` procedure.

```
1   #ifndef MAX_OF_THREE_H
2   #define MAX_OF_THREE_H
3
4   template <class T>
5   T max_of_three(T a, T b, T c) {
6     if (a<b) {
7       if (b<c) return c;
8       return b;
9     }
10    if (b<c) return c;
11    return b;
12  }
13
14  #endif
```

There are two points to notice.

- The procedure is given, in full, in the header file, `max_of_three.h`. This is standard for templates. One does not separate the declaration in the header file from the code in the `.cc` file.

  In addition, the keyword `inline` is not needed.

- The procedure begins with `template <class T>`. This sets up the letter `T` to stand for a type.

  The remainder of the code is exactly what we had before with `T` standing for `long`, `double`, `PTriple`, or any other type. The letter `T` acts as a "type variable."

One may think of a template as a procedure schema. When the C++ compiler encounters `max_of_three` in a program, it uses the template to create the appropriate version.

For example, here is a `main` that utilizes the adaptable nature of `max_of_three`.

```
#include "max_of_three.h"
#include <iostream>
using namespace std;

int main() {
  double x = 3.4;
  double y = -4.1;
  double z = 11.2;

  long a = 14;
  long b = 17;
  long c = 0;

  cout << "The max of " << x << ", " << y << ", " << z << " is "
       << max_of_three(x,y,z) << endl;

  cout << "The max of " << a << ", " << b << ", " << c << " is "
       << max_of_three(a,b,c) << endl;

  return 0;
}
```

The first invocation of `max_of_three` has three `double` arguments; the compiler uses the `max_of_three` template with `T` equal to `double` to create the appropriate code. The second invocation has three `long` arguments, and as before, the compiler uses the template to create the appropriate version of the procedure.

The output of this `main` is as we expect.

```
The max of 3.4, -4.1, 11.2 is 11.2
The max of 14, 17, 0 is 17
```

The `max_of_three` template works for any type `T` for which < is defined. If < is not defined for the type, the compiler generates an error message. For example, consider the following `main`.

```
#include <iostream>
#include <complex>
#include "max_of_three.h"
using namespace std;

int main() {
  complex<double> x(3,0);
  complex<double> y(-2,3);
  complex<double> z(1,1);

  cout << "The max of " << x << ", " << y << ", " << z << " is "
       << max_of_three(x,y,z) << endl;

  return 0;
}
```

When we attempt to compile this program, we get the following error messages (your computer might produce different error messages).

```
max_of_three.h: In function 'T max_of_three(T, T, T) [with T =
   std::complex<double>]':
main.cc:12:    instantiated from here
max_of_three.h:6: error: no match for 'std::complex<double>& <
   std::complex<double>&' operator
max_of_three.h:7: error: no match for 'std::complex<double>& <
   std::complex<double>&' operator
max_of_three.h:10: error: no match for 'std::complex<double>& <
   std::complex<double>&' operator
```

The compiler is complaining that it cannot find an `operator<` that takes two arguments of type `complex<double>` at lines 6, 7, and 10 in the file `max_of_three.h`; this is precisely where the `<` expressions occur.

It is possible to create templates with multiple type parameters. Such templates look like this:

```
template <class A, class B>
void do_something(A arg1, B arg2) { ... }
```

When the compiler encounters code such as

```
long alpha = 4;
double beta = 4.5;
do_something(alpha, beta);
```

it creates and uses a version of `do_something` in which `A` stands for `long` and `B` stands for `double`.

12.2   Class templates

12.2.1   Using class templates

In Section 2.7 we showed how to declare C++ variables to hold complex numbers. After the directive `#include <complex>`, we have statements such as these:

```
complex<double> w(-3.4, 5.1);
complex<long>   z(4, -7);
```

The first declares a complex variable `w` in which the real and imaginary parts are `double` real numbers and the second declares `z` to be a Gaussian integer (`long` integer real and imaginary parts).

The class `complex` is, in fact, a class template. By specifying different types between the `<` and `>` delimiters, we create different classes. For example, we could use the `Mod` type (see Chapter 9):

```
#include <complex>
#include "Mod.h"
using namespace std;

int main() {
  Mod::set_default_modulus(11);
  complex<Mod> z(4,7);

  cout << "z = " << z << endl;
  cout << "z squared = " << z*z << endl;

  return 0;
}
```

This program calculates $(4+7i)^2$ where 4 and 7 are elements of $\mathbb{Z}_{11}$ where we have $(4+7i)^2 = -33+56i = i$. This is confirmed when the program is run.

```
z = (Mod(4,11),Mod(7,11))
z squared = (Mod(0,11),Mod(1,11))
```

Likewise, `vector` is a class template. To create a `vector` that contains integers, we use `vector<long>`. To create a `vector` containing complex numbers with real coefficients, the following mildly baroque construction is required,

```
vector<complex<double> > zlist;
```

Note the space between the two closing > delimiters; the space is mandatory here. If it were omitted, the >> would look like the operator we use in expressions such as `cin >> x` causing angst for the compiler. For better readability, you may prefer this:

```
vector< complex<double> > zlist;
```

The `pair` class template takes two type arguments. To create an ordered pair where the first entry is a real number and the second is a Boolean, we write this:

```
pair<double,bool> P;
```

Using class templates is straightforward. The template is transformed into a specific class by adding the needed type argument(s) between the < and > delimiters. The main pitfall to avoid is supplying a type that is incompatible with the template. For example, it would not make sense to declare a variable to be of type `complex<PTriple>`.

By the way: The use of < and > delimiters in `#include <iostream>` is unrelated to their use in templates.

12.2.2 Creating class templates

Now that we have examined how to use class templates we are led to the issue: How do we create class templates? The technique is similar to that of creating procedure templates. To demonstrate the process, we create our own, extremely limited version of the `complex` template that we call `mycomplex`. The template resides in a file named `mycomplex.h`; there is no `mycomplex.cc` file. Here is the header file containing the template.

Program 12.2: The template for the `mycomplex` classes.

```
1   #ifndef MY_COMPLEX_H
2   #define MY_COMPLEX_H
3
4   #include <iostream>
5   using namespace std;
6
7   template <class T>
8   class mycomplex {
9
10  private:
11     T real_part;
12     T imag_part;
13
14  public:
15     mycomplex<T>() {
16        real_part = T(0);
17        imag_part = T(0);
18     }
19
20     mycomplex<T>(T a) {
21        real_part = a;
22        imag_part = T(0);
23     }
24
25     mycomplex<T>(T a, T b) {
26        real_part = a;
27        imag_part = b;
28     }
29
30     T re() const { return real_part; }
31     T im() const { return imag_part; }
32
33  };
34
35  template<class T>
36  ostream& operator<<(ostream& os, const mycomplex<T>& z) {
37     os << "(" << z.re() << ") + (" << z.im() << ")i";
38     return os;
39  }
40
41  #endif
```

The overall structure of the class template is this:

```
template <class T>
class mycomplex {
  // private and public data and methods
};
```

Let's examine this code.

- The initial `template <class T>` serves the same purpose as in the case of template procedures. This establishes `T` as a parameter specifying a type. When we declare a variable with a statement like this

```
mycomplex<double> w;
```

the T stands for the type `double`. See lines 7–8.

- The class template has two private data fields: `real_part` and `imag_part`. These are declared as type T. Thus if w is declared type `mycomplex<double>` then `w.real_part` and `w.imag_part` are both type `double`. See lines 10–12.

- Three constructors are given. The same name is used for these constructors: `mycomplex<T>`. (The `<T>` is optional here; the constructors could be named simply `mycomplex` with the `<T>` implicitly added.)

  The zero-argument constructor creates the complex number $0 + 0i$. Note that the value 0 is explicitly converted to type T. This requires that type T be able to convert a zero into type T. If `some_type` is a class that does not have a single-argument numerical constructor, then declaring a variable to be a `mycomplex<some_type>` causes a compiler error. See lines 15–18.

  The single- and double-argument constructors follow (lines 20–28). They are self-explanatory.

- We provide the methods `re()` and `im()` to retrieve a `mycomplex` variable's real and imaginary parts. Note that the return type of these methods is T. See lines 30–31.

- The last part of the file `mycomplex.h` is the `operator<<` procedure for writing `mycomplex` variables to the computer's screen. This procedure is not a member of the `mycomplex` class template, so it needs a separate introductory `template <class T>` in front of the procedure's definition. See line 35.

  The next line looks like the usual declaration of the `<<` operator. The second argument's type is a reference to a variable of type `mycomplex<T>`.

  After this comes the inline code for the procedure. The keyword `inline` is optional because all templates must be defined inline. (If we want to include the keyword `inline`, it would follow `template <class T>` and precede `ostream&`.)

Of course, to make this a useful template, we would need to add methods/procedures for arithmetic, comparison, exponentiation, and so forth.

We could create a different version of `mycomplex` in which the real and imaginary parts are allowed to be different types. The code would look like this.

Program 12.3: A revised version of `mycomplex` that allows real and imaginary parts of different types.

```
1  template <class T1, class T2>
2  class mycomplex {
3
```

```
 4  private:
 5    T1 real_part;
 6    T2 imag_part;
 7
 8  public:
 9    mycomplex() {
10      real_part = T1(0);
11      imag_part = T2(0);
12    }
13
14    mycomplex(T1 a) {
15      real_part = a;
16      imag_part = T2(0);
17    }
18
19    mycomplex(T1 a, T2 b) {
20      real_part = a;
21      imag_part = b;
22    }
23
24    T1 re() const { return real_part; }
25    T2 im() const { return imag_part; }
26
27  };
28
29  template<class T1, class T2>
30  ostream& operator<<(ostream& os, const mycomplex<T1,T2>& z) {
31    os << "(" << z.re() << ") + (" << z.im() << ")i";
32    return os;
33  }
```

If we use this alternative `mycomplex` template, variable declarations require the specification of two types, such as this:

```
mycomplex<long, double> mixed_up_z;
```

12.3   The `Polynomial` class template

Our goal is to create a C++ class template to represent polynomials over any field K and we call this class template `Polynomial`. Consider the following declarations,

```
Polynomial<double>              P;
Polynomial< complex<double> > Q;
Polynomial<Mod>                 R;
```

These are to create polynomials in $\mathbb{R}[x]$, $\mathbb{C}[x]$, and $\mathbb{Z}_p[x]$, respectively. (Of course, we need `#include <complex>` and `#include "Mod.h"`.)

12.3.1 Data

We need to store the coefficients of the polynomial. For this, we use a `vector` variable (see Section 8.4) named `coef`. The coefficient of x^k is held in `coef[k]`. We therefore require the directive `#include <vector>` in the file `Polynomial.h`.

We hold the degree of the polynomial in a `long` variable named `dg`. The degree is the largest index k such that `coef[k]` is nonzero. In the case where the polynomial is equal to zero, we set `dg` equal to -1.

See lines 11–12 in Program 12.4.

12.3.2 Constructors

A basic, zero-argument constructor produces the zero polynomial. This is accomplished by a call to a `clear()` method that resizes `coef` to hold only one value, sets that value to zero, and sets `dg` equal to -1. See lines 23–25 and 78–81 of Program 12.4.

A single-argument constructor is used to produce a constant polynomial; that is, a polynomial in which $c_k = 0$ for all $k \geq 1$. This constructor may be used explicitly or implicitly to convert scalar values to polynomials. For example, consider this code:

```
Polynomial<double> P(5.);
Polynomial<double> Q;
Polynomial< complex<double> > R;
Q = 6;
R = -M_PI;
```

The first line creates the polynomial $p(x) = 5$ using the constructor explicitly. The polynomials $q(x) = 6$ and $r(x) = (-\pi + 0i)$ both use the constructor implicitly. Notice that for both `Q` and `R` there is also a conversion of the right-hand side of the assignment into another numeric type. The 6 is an integer type and is cast into `double` for storage in `Q`. Similarly, `-M_PI` is a real number and this needs to be converted to a complex type for storage in `R`.

To do this, we allow the argument to be of any type and then cast that argument to type `K`. Let's look at the full code for this constructor (lines 27–33):

```
template <class J>
Polynomial<K>(J a) {
  coef.clear();
  coef.resize(1);
  coef[0] = K(a);
  deg_check();
}
```

This constructor is a template within a template! (The outer template has the structure `template <class K> Polynomial { ... };`.) Thus, in code such as

```
Polynomial< complex<double> > P(val);
Polynomial< complex<double> > Q;
Q = val;
```

the variable `val` may be of any type.

There is, of course, a catch. We assign `coef[0]` with the value `K(a)`. This is an explicit request to convert the value `a` to type `K`. This is fine if we are converting a `long` to a `double` or a `double` to a `complex<double>`, but fails when `a` is not convertible to type `K`, for example, if `K` were type `double` and `a` were type `PTriple`.

Notice the call to the private method `deg_check()`. This method scans the data held in `coef` to find the last nonzero value and resets `dg` accordingly. Various operations on polynomials might alter their degree (e.g., addition of polynomials, changing coefficients, etc.) and this method makes sure `dg` holds the correct value.

Next we have a three-argument constructor (lines 35–43). To create the polynomial $ax^2 + bx + c$ one simply invokes `Polynomial<K>(a,b,c)`. As before, the three arguments need not be type `K`; it is enough that they be convertible to type `K`. For example, consider this code.

```
long           a = 7;
complex<double> b(4.,-3.);
double         c = M_PI;

Polynomial< complex<double> > P(a,b,c);
```

This creates a polynomial `P` equal to $7x^2 + (4 - 3i)x + \pi$.

Finally, it is useful to be able to create a polynomial given an array of coefficients. The constructor on lines 45–60 takes two arguments: the size of an array and an array of coefficients. The array is declared as type `J*`; that is, an array of elements of type `J`. The only requirement on `J` is that values of that type must be convertible to type `K`.

12.3.3 Get and set methods

We include an assortment of methods to inspect and modify the coefficients held in a `Polynomial`.

- The `deg()` method returns the degree of the polynomial. (See line 62.)

- The `get(k)` method returns the coefficient of x^k. In the case where k is invalid (negative or greater than the degree), we return zero. (See lines 64–67.)

  As an alternative to `get(k)` we provide `operator[]` (line 69). For a polynomial `P`, both `P[k]` and `P.get(k)` have the same effect. However, we have not set up `operator[]` to work on the left-hand side of an assignment; we cannot change the kth coefficient by a statement such as `P[k]=c;`.

- The `isZero()` method returns true if the polynomial is identically zero. See line 89.

- The `set(idx,a)` method sets the coefficient `coef[idx]` equal to the value a. See lines 71–76.

Some care must be taken here. First, if `idx` is negative, no action is taken. If `idx` is greater than the degree, we need to expand `coef` accordingly. Also, this method might set the highest coefficient to zero, so we invoke `deg_check()`.

- The `clear()` method sets all coefficients to zero. See lines 78–82.

- The `minimize()` method frees up any wasted memory held in `coef`. It is conceivable that a polynomial may at one point have large degree (causing `coef` to expand) and then later have small degree. Although the size of `coef` grows and shrinks with the degree, the capacity of `coef` would remain large. This method causes the `vector` method `reserve` to be invoked for `coef`. See lines 84–87.

- Finally, we have the `shift(n)` method. See lines 91–105.

  If n is positive, this has the effect of multiplying the polynomial by x^n. Each coefficient is shifted upwards; the coefficient of x^k before the shift becomes the coefficient of x^{k+n} after.

  Shifting with a negative index has the opposite effect. Coefficients are moved to lower powers of x. Coefficients shifted to negative positions are discarded.

The polynomial P	$4x^2 + 6x - 2$
After P.shift(2)	$4x^4 + 6x^3 - 2x^2$
After P.shift(-1)	$4x + 6$.

Notice that we give the argument `n` a default value of 1 (see line 91). Thus `P.shift()` is the same as `P.shift(1)`.

12.3.4 Function methods

Polynomials are functions and to use them as such we provide a method named `of`. Invoking `P.of(a)` evaluates the polynomial $p(x)$ at $x = a$. See lines 109–118.

An efficient way to evaluate a polynomial such as $3x^4 + 5x^3 - 6x^2 + 2x + 7$ at $x = a$ is this:

$$\left(\left(\left((3 \times a) + 5\right) \times a + (-6)\right) \times a + 2\right) \times a + 7.$$

In addition to the `of` method, we provide an `operator()` method (line 120). This way we can express function application in the natural way `P(a)` in addition to `P.of(a)`.

Because polynomials are functions, they may be composed and the result is again a polynomial. We use the same method names, `of` and `operator()`, for polynomial composition. For polynomials P and Q, the result of `P.of(Q)` (and also `P(Q)`) is the polynomial $p(q(x))$. See lines 122–135.

These methods depend on the ability to do polynomial arithmetic, and we describe those methods below (Subsection 12.3.6).

12.3.5 Equality

To check if two polynomials are equal, we make sure they have the same degree and that corresponding coefficients are equal. The operators == and != are given on lines 139–149.

12.3.6 Arithmetic

We provide methods for the usual arithmetic operators: addition, subtraction, multiplication, and division (quotient and remainder). See lines 153ff.

Each of the basic operators is defined like this:

```
Polynomial<K> operator+(const Polynomial<K>& that) const { ... }
```

If P and Q are of type `Polynomial<double>`, then the expression P+Q invokes this method with `that` referring to Q. However, the expression P+5 also engages this method; implicitly the 5 is cast to a polynomial via `Polynomial<double>(5)`. However, the expression 5+P cannot be handled by this method (because 5 is not a polynomial). Therefore, we also provide procedure templates such as this:

```
template <class J, class K>
Polynomial<K> operator+(J x, const Polynomial<K>& P) {
  return P + K(x);
}
```

See lines 291–298.

In addition to the usual operators + - * / % we provide their operate/assign cousins: += -= *= /= %=. We also give methods for unary minus and exponentiation (to a nonnegative integer power).

With the exception of division, the code for these various operators is reasonably straightforward. Division requires a bit more attention.

As in the case of integers, division of polynomials produces two results: the quotient and the remainder. Let $a(x)$ and $b(x)$ be polynomials with $b \neq 0$. Then there exist polynomials $q(x)$ and $r(x)$ for which

$$a(x) = q(x)b(x) + r(x) \qquad \text{and} \qquad \deg r(x) < \deg b(x).$$

Furthermore, q and r are unique. For example, if $a(x) = 3x^2 + x - 2$ and $b(x) = 2x + 1$, then $q(x) = \frac{3}{2}x - \frac{1}{4}$ and $r(x) = -\frac{7}{4}$.

We define A/B and A%B to be the quotient and remainder, respectively, when we divide $a(x)$ by $b(x)$.

To this end, we define the procedure `quot_rem(A,B,Q,R)` (see lines 300–321) to find the quotient and remainder. The `operator/` and `operator%` make use of `quot_rem` to do their work.

Please note that the division methods require that K be a field. If K is only a commutative ring (e.g., `long` integers), then most of the class template works fine, but the division methods do not.

12.3.7 Output to the screen

The operator `<<` for writing to the computer's screen appears on lines 269–289. This operator writes the polynomial $5x^3 - x + \frac{1}{2}$ like this:

```
(5)X^3 + (-1)X + (0.5)
```

The procedure is smart enough to omit terms whose coefficient is zero, to omit the superscript 1 on the linear term, and to omit the variable altogether on the constant term. However, it does not convert `(-1)X` into the more legible $-$ `X`, or even $-$ `(1)X`. We might think that it is possible for the code to check if a coefficient is negative and modify behavior accordingly. However, for some fields K (such as $\mathbb{C}$ and $\mathbb{Z}_p$) this does not make sense.

12.3.8 GCD

Let $a(x)$ and $b(x)$ be polynomials. A *common divisor* of $a(x)$ and $b(x)$ is a polynomial $c(x)$ with the property that there exist polynomials $s(x)$ and $t(x)$ so that $a(x) = c(x)s(x)$ and $b(x) = c(x)t(x)$. A *greatest common divisor* of $a(x)$ and $b(x)$ is a common divisor of highest possible degree.

Two polynomials may have several different greatest common divisors because any nonzero scalar multiple of a greatest common divisor is again a greatest common divisor.

To settle on a specific meaning for $\gcd[a(x), b(x)]$ we choose the greatest common divisor $d(x)$ whose leading coefficient is 1. (A polynomial whose leading coefficient is 1 is called *monic*.) Any two nonzero polynomials have a unique monic greatest common divisor.

The `gcd` procedure for two polynomials is given on lines 323–336. This procedure uses the helper methods `is_monic()` and `make_monic()`; the former checks if the polynomial is monic and the latter transforms the polynomial into a monic polynomial by dividing by the leading coefficient. See lines 256–265.

We use the Euclidean algorithm (described in Section 3.3) to calculate the gcd of two polynomials.

In addition to the usual gcd procedure, we provide an extended version. Given polynomials $a(x)$ and $b(x)$, the extended gcd procedure finds $d(x) = \gcd[a(x), b(x)]$ as well as polynomials $s(x)$ and $t(x)$ so that $d(x) = a(x)s(x) + b(x)t(x)$. See lines 337–376.

12.3.9 The code

Here is the listing of `Polynomial.h`. This includes the `Polynomial<K>` class template as well as associated procedure templates (`operator<<`, `gcd`, etc.).

Program 12.4: Header file for the `Polynomial` class template.

```
1   #ifndef POLYNOMIAL_H
2   #define POLYNOMIAL_H
```

```
3
4    #include <vector>
5    #include <iostream>
6    using namespace std;
7
8    template <class K>
9    class Polynomial {
10   private:
11     vector<K> coef;
12     long       dg;
13
14     void deg_check() {
15       dg = -1;
16       for (long k=0; k< long(coef.size()); k++) {
17         if (coef[k] != K(0)) dg = k;
18       }
19       coef.resize(dg+1);
20     }
21
22   public:
23     Polynomial<K>() {
24       clear();
25     }
26
27     template <class J>
28     Polynomial<K>(J a) {
29       coef.clear();
30       coef.resize(1);
31       coef[0] = K(a);
32       deg_check();
33     }
34
35     template <class J, class JJ, class JJJ>
36     Polynomial<K>(J a, JJ b, JJJ c) {
37       coef.clear();
38       coef.resize(3);
39       coef[2] = K(a);
40       coef[1] = K(b);
41       coef[0] = K(c);
42       deg_check();
43     }
44
45     template <class J>
46     Polynomial<K>(long array_size, const J* array) {
47       if (array_size < 0) {
48         coef.clear();
49         coef.resize(1);
50         coef[0] = K(0);
51         dg = -1;
52         return;
53       }
54       coef.clear();
55       coef.resize(array_size);
56       for (long k=0; k<array_size; k++) {
57         coef[k] = K(array[k]);
58       }
```

```
59      deg_check();
60    }
61
62    long deg() const { return dg; }
63
64    K get(long k) const {
65      if ( (k<0) || (k>dg) ) return K(0);
66      return coef[k];
67    }
68
69    K operator[](long k) const { return get(k); }
70
71    void set(long idx, K a) {
72      if (idx < 0) return;
73      if (idx+1 >long(coef.size())) coef.resize(idx+1);
74      coef[idx] = a;
75      deg_check();
76    }
77
78    void clear() {
79      coef.resize(1);
80      dg = -1;
81      coef[0] = K(0);
82    }
83
84    void minimize() {
85      deg_check();
86      coef.reserve(dg+1);
87    }
88
89    bool isZero() const { return (dg < 0); }
90
91    void shift(long n = 1) {
92      if (n==0) return;
93
94      if (n<0) {
95        for (long k=0; k<=dg+n; k++) coef[k] = coef[k-n];
96        for (long k=dg+n+1; k<=dg; k++) coef[k] = K(0);
97        deg_check();
98        return;
99      }
100
101      coef.resize(n+dg+1);
102      for (long k=dg; k>=0; k--) coef[k+n] = coef[k];
103      for (long k=0; k<n; k++) coef[k] = K(0);
104      dg += n;
105    }
106
107    // FUNCTION APPLICATION //
108
109    K of(K a) const {
110      if (dg <= 0) return coef[0];
111      K ans;
112      ans = K(0);
113      for (long k=dg; k>=0; k--) {
114        ans *= a;
```

```
115        ans += coef[k];
116      }
117      return ans;
118    }
119
120    K operator()(K a) const { return of(a); }
121
122    Polynomial<K> of(const Polynomial<K>& that) const {
123      if (dg <= 0) return Polynomial<K>(coef[0]);
124
125      Polynomial<K> ans(K(0));
126      for (long k=dg; k>=0; k--) {
127        ans *= that;
128        ans += Polynomial<K>(coef[k]);
129      }
130      return ans;
131    }
132
133    Polynomial<K> operator()(const Polynomial<K>& that) const {
134      return of(that);
135    }
136
137    // COMPARISON //
138
139    bool operator==(const Polynomial<K>& that) const {
140      if (dg != that.dg) return false;
141      for (long k=0; k<=dg; k++) {
142        if (coef[k] != that.coef[k]) return false;
143      }
144      return true;
145    }
146
147    bool operator!=(const Polynomial<K>& that) const {
148      return !(*this == that);
149    }
150
151    // ARITHMETIC //
152
153    Polynomial<K> operator+(const Polynomial<K>& that) const {
154      Polynomial<K> ans;
155      long dmax = (dg > that.dg) ? dg : that.dg;
156      ans.coef.resize(dmax+1);
157      for (long k=0; k<=dmax; k++) {
158        ans.coef[k] = get(k) + that.get(k);
159      }
160      ans.deg_check();
161      return ans;
162    }
163
164    Polynomial<K> operator+=(const Polynomial<K>& that) {
165      (*this) = (*this) + that;
166      return *this;
167    }
168
169    Polynomial<K> operator-() const {
170      Polynomial<K> ans;
```

```
171      ans.coef.resize(dg+1);
172      for (long k=0; k<=dg; k++) ans.coef[k] = -coef[k];
173      ans.deg_check();
174      return ans;
175    }
176
177    Polynomial<K> operator-(const Polynomial<K>& that) const {
178      Polynomial<K> ans;
179      long dmax = (dg > that.dg) ? dg : that.dg;
180      ans.coef.resize(dmax+1);
181      for (long k=0; k<=dmax; k++) {
182        ans.coef[k] = get(k) - that.get(k);
183      }
184      ans.deg_check();
185      return ans;
186    }
187
188    Polynomial<K> operator-=(const Polynomial<K>& that)  {
189      (*this) = (*this) - that;
190      return *this;
191    }
192
193    Polynomial<K> operator*(const Polynomial<K>& that) const {
194      Polynomial<K> ans;
195      if (isZero() || that.isZero()) return ans;
196      long dans = dg + that.dg;
197      ans.coef.resize(dans+1);
198      for (long k=0; k<=dans; k++) {
199        K c(0);
200        for (long j=0; j<=k; j++) {
201          if ((j<=dg) && (k-j<=that.dg)) c += coef[j]*that.coef[k-j];
202        }
203        ans.coef[k] = c;
204      }
205      ans.deg_check();
206      return ans;
207    }
208
209    Polynomial<K> operator*=(const Polynomial<K>& that)  {
210      *this = (*this) * that;
211      return *this;
212    }
213
214    Polynomial<K> pow(long k) const {
215      if (k==0) return Polynomial<K>(1);
216      if (k==1) return *this;
217
218      if (k%2 == 0) {
219        long half_k = k/2;
220        Polynomial<K> ans;
221        ans = (*this).pow(half_k);
222        ans *= ans;
223        return ans;
224      }
225
226      long half_k = (k-1)/2;
```

```
227      Polynomial<K> ans;
228      ans = (*this).pow(half_k);
229      ans *= ans;
230      ans *= *this;
231      return ans;
232    }
233
234    Polynomial<K> operator/(const Polynomial<K>& that) const {
235      Polynomial<K> Q,R;
236      quot_rem(*this, that, Q, R);
237      return Q;
238    }
239
240    Polynomial<K> operator/=(const Polynomial<K>& that) {
241      *this = (*this)/that;
242      return *this;
243    }
244
245    Polynomial<K> operator%(const Polynomial<K>& that) const {
246      Polynomial<K> Q,R;
247      quot_rem(*this, that, Q, R);
248      return R;
249    }
250
251    Polynomial<K> operator%=(const Polynomial<K>& that) {
252      (*this) = (*this) % that;
253      return *this;
254    }
255
256    void make_monic() {
257      if (dg < 0) return;
258      K lead = coef[dg];
259      for (long j=0; j<=dg; j++) coef[j] /= lead;
260    }
261
262    bool is_monic() const {
263      if (dg < 0) return false;
264      return coef[dg] == K(1);
265    }
266
267 };    // end of Polynomial<K> class template
268
269 template <class K>
270 ostream& operator<<(ostream& os, const Polynomial<K>& P) {
271   if (P.deg() <= 0) {
272     os << "(" << P[0] << ")";
273     return os;
274   }
275   for (long k=P.deg(); k>=0; k--) {
276     if (P[k] != K(0)) {
277       if (k < P.deg()) os << " + ";
278       os << "(" << P[k] << ")";
279       if (k>1) {
280         os << "X^" << k;
281         continue;
282       }
```

```
283         if (k==1) {
284            os << "X";
285         }
286      }
287   }
288   return os;
289 }
290
291 template <class J, class K>
292 Polynomial<K> operator+(J x, const Polynomial<K>& P)
293 { return P + K(x); }
294
295 template <class J, class K>
296 Polynomial<K> operator-(J x, const Polynomial<K>& P)
297 { return (-P) + K(x); }
298
299 template <class J, class K>
300 Polynomial<K> operator*(J x, const Polynomial<K>& P)
301 { return P * K(x); }
302
303 template <class K>
304 void quot_rem(const Polynomial<K>& A,
305               const Polynomial<K>& B,
306               Polynomial<K>& Q,
307               Polynomial<K>& R) {
308   Q.clear();
309   R.clear();
310
311   Polynomial<K> AA (A);   // copy of A
312
313   while (AA.deg() >= B.deg()) {
314     long k = AA.deg()-B.deg();
315     Polynomial<K> BB = B;
316     BB.shift(k);
317     K a_lead = AA[AA.deg()];
318     K b_lead = BB[BB.deg()];
319     for (long j=0; j <= BB.deg(); j++) BB.set(j, BB[j]*a_lead/b_lead);
320     AA -= BB;
321     Q.set(k,a_lead/b_lead);
322   }
323   R = A - Q*B;
324 }
325
326 template <class K>
327 Polynomial<K> gcd(const Polynomial<K>& A, const Polynomial<K>& B) {
328   if (B.isZero()) {
329     if (A.is_monic()) return A;
330     Polynomial<K> AA(A);
331     AA.make_monic();
332     return AA;
333   }
334
335   Polynomial<K> C;
336   C = A%B;
337   return gcd(B,C);
338 }
```

```
339
340    template <class K>
341    Polynomial<K> gcd(const Polynomial<K>& A,
342                     const Polynomial<K>& B,
343                           Polynomial<K>& S,
344                           Polynomial<K>& T)        {
345      Polynomial<K> D; // holds the answer
346
347      // If A and B are both 0, set S=T=0 and return 0.
348      if (A.isZero() && B.isZero()) {
349        S.clear();
350        T.clear();
351        return D;
352      }
353
354      // If A is not 0 but B is, D = A/a_lead, S = a_lead, T = 0
355      if (B.isZero()) {
356        D = A;
357        K a_lead = A[A.deg()];
358        D.make_monic();
359        S = Polynomial<K>(K(1)/a_lead);
360        T.clear();
361        return D;
362      }
363
364      // Neither A nor B is zero, so we recurse
365
366      Polynomial<K> Q;
367      Polynomial<K> R;
368      quot_rem(A,B,Q,R);
369
370      Polynomial<K> SS;
371      Polynomial<K> TT;
372
373      D = gcd(B,R,SS,TT);
374
375      S = TT;
376      T = SS - Q*TT;
377
378      return D;
379    }
380
381    #endif
```

12.4 The GCD problem revisited

In Chapter 3 we considered the question (phrased imprecisely here): What is the probability that two positive integers are relatively prime? We found that the answer is $1/\zeta(2)$. Here we consider a similar problem: What is the probability that two randomly chosen polynomials are relatively prime?

To be more precise, let B_d denote the set of all polynomials in $\mathbb{Z}_2[x]$ of degree less than d; there are 2^d such polynomials $a_{d-1}x^{d-1} + a_{d-2}x^{d-2} + \cdots + a_1x + a_0$ where the a_js are 0 or 1. Let p_d denote the probability the two polynomials, chosen uniformly and independently from B_d are relatively prime. What can we say about p_d as $d \to \infty$?

To formulate a conjecture, we write a program to evaluate p_d by direct enumeration. This is the approach we used in Section 3.5. With luck, modest values of d will lead us to the answer. The overall structure of the program is this:

1. Ask the user to input d.

2. Build an array containing all the polynomials in B_d.

3. For all $i < j$, count the number of times the ith and jth polynomials are relatively prime.

4. From this count, we learn the numerator of p_d. Divide by 2^{2d} to find the answer.

Of these, the most difficult part is the construction of the list in step 2. To generate this table efficiently, we observe that there is a natural one-to-one correspondence between d-digit binary numbers and polynomials in B_d, illustrated here with $d = 6$.

$$
\begin{aligned}
000000 &\leftrightarrow 0 \\
000001 &\leftrightarrow 1 \\
000010 &\leftrightarrow x \\
000011 &\leftrightarrow x+1 \\
000100 &\leftrightarrow x^2 \\
000101 &\leftrightarrow x^2+1 \\
&\vdots \\
111111 &\leftrightarrow x^5+x^4+x^3+x^2+x+1
\end{aligned}
$$

Integer values are stored in computers in binary, so our first step is to write a procedure to convert integers into polynomials:

$$b_{d-1}b_{d-2}\ldots b_1b_0 \mapsto b_{d-1}x^{d-1} + b_{d-2}x^{d-2} + \cdots + b_1x + b_0 \in B_d$$

We call this procedure `long2poly`. Here is its header file.

Program 12.5: Header file `long2poly.h`.

```
1  #ifndef LONG_TO_POLY_H
2  #define LONG_TO_POLY_H
3
4  #include "Polynomial.h"
```

```
 5   #include "Mod.h"
 6
 7   const long max_bits = 31;
 8
 9   Polynomial<Mod> long2poly(long m);
10
11   #endif
```

This header defines a constant `max_bits` that sets an upper bound on d; this value is based on the size of a `long` integer on the computer on which this program is to be run.

The procedure takes a `long` integer argument and returns a `Polynomial<Mod>`. To write this program, we want to access the individual bits of the integer argument, `m`. The way we do this is to check if `m` is even or odd, and then set b_0 accordingly. We then divide `m` by 2, check if the result is even or odd, and then set d_1. We continue in this fashion until `m` is zero. Here is the code.

Program 12.6: Code file for the `long2poly` procedure.

```
 1   #include "long2poly.h"
 2
 3   Polynomial<Mod> long2poly(long m) {
 4     Polynomial<Mod> ans;
 5
 6     long j = 0;
 7     while (m != 0) {
 8       ans.set(j, Mod(m,2));
 9       m /= 2;
10       j++;
11     }
12
13     return ans;
14   }
```

Next, we need a main to implement the exhaustive algorithm.

Program 12.7: Main program for the GCD revisited problem.

```
 1   #include "Polynomial.h"
 2   #include "Mod.h"
 3   #include "long2poly.h"
 4
 5   using namespace std;
 6
 7   int main() {
 8     long d;
 9     cout << "Enter degree bound --> ";
10     cin >> d;
11
12     if ( (d<1) || (d>max_bits) ) {
13       cerr << "Please choose d between 1 and " << max_bits << endl;
14       return 0;
15     }
16
```

```
17    long bound = 1<<d;
18
19    Polynomial<Mod> *list;
20
21    list = new Polynomial<Mod>[bound];
22
23    cerr << "Generating polynomials ... ";
24    for (long k=0; k<bound; k++) {
25      list[k] = long2poly(k);
26    }
27    cerr << "done! " << endl << bound
28        << " polynomials of degree less than "
29        << d <<" generated" << endl;
30
31    long count = 0;
32    const Polynomial<Mod> one(Mod(1,2));
33
34    for (long i=0; i<bound-1; i++) {
35      for (long j=i+1; j<bound; j++) {
36        if( gcd(list[i],list[j]) == one ) count++;
37      }
38    }
39
40    count = 2*count + 1;
41
42    cout << count << " out of " << bound*bound
43        << " pairs are relatively prime" << endl;
44
45    cout << count / (double(bound) * double(bound)) << endl;
46
47    return 0;
48  }
```

Finally, when the program is run, we see the following.

```
Enter degree bound --> 10
Generating polynomials ... done!
1024 polynomials of degree less than 10 generated
524289 out of 1048576 pairs are relatively prime
0.500001
```

The formulation of a conjecture, and its proof,[1] are left as an exercise for the reader.

---

[1]For a proof via generating functions, see S. Corteel, C. Savage, H. Wilf, D. Zeilberger, A pentagonal number sieve, *Journal of Combinatorial Theory, Series A* **82** (1998) 186–192. Recently, Art Benjamin and Curtis Bennett have found a bijective proof (submitted for publication).

12.5   Working in binary

The `long2poly` procedure used a trick to convert a `long` integer m into polynomials $p(x)$ in $\mathbb{Z}_2[x]$. We set the constant coefficient of $p(x)$ based on the parity of m, and then we divided m by 2 (keeping only the integer part). We then repeated this technique to set higher and higher coefficients until m vanished. In short, we used division arithmetic to read off the base-2 digits of m.

In other words, we used a *mathematical trick* to find the binary representation of m. However, the binary is already present inside the computer; it is more efficient to work directly with that. C++ provides a family of operators for working directly on the bits in the binary form of integers.

12.5.1   Signed versus unsigned integers

Integers are stored inside the computer in binary. The number 20 is represented internally as 0000000000010100.

In this, and subsequent examples, we assume the integers are held as `short` types; on my computer these are two bytes (16 bits) long. Other integer types may have 32 or 64 bits.

The storage of negative integers is mildly counterintuitive. The leftmost bit is known as the *sign bit*. If this bit is 1, the number represented is negative. However, -20 is *not* represented as 1000000000010100. Look closely at the correct internal representation of -20 and 19:

Value	Binary representation
-20	1111111111101100
19	0000000000010011

For a positive integer n, the binary representation of n is just the usual base-2 representation. However, the binary representation of $-n$ is formed by complementing the bits of $n - 1$. Of course, zero is represented by an all-zero binary number.

This manner of storing negative values is known as the *twos complement* representation. This representation is used for the sake of computational efficiency.

The integer types (`char`, `short`, `int`, and `long`) all have variants that restrict their range to nonnegative values. These variant types prepend the word `unsigned` to the type name. For example:

```
unsigned short x;
```

To illustrate the difference, suppose we have two variables x and y declared thus:

```
unsigned short x;
short y;
```

Suppose both of these hold the bits 1111111111111111. In this case, the value of x is 65,535 ($2^{16} - 1$) whereas the value of y is -1.

12.5.2 Bit operations

C++ provides six operators for working with the binary representation of integers.

Bitwise and For integer variables x and y, the expression x&y is the bitwise and of x and y. That is, the kth bit of x&y is 1 if and only if the kth bits of both x and y are 1. Here is an example.

x	0100001101100000
y	0001000111101101
x&y	0000000101100000

The bitwise and operation & should not be confused with the Boolean and operation &&. You should use && only with bool values.

Bitwise or Similar to bitwise and, the operation x|y gives the bitwise or of x and y. That is, the kth bit of x|y is 0 if and only if the kth bits of both x and y are 0. Here is an example.

x	0100001101100000
y	0001000111101101
x|y	0101001111101101

The bitwise or operation | should not be confused with the Boolean or operation ||. You should use || only with bool values.

Exclusive or The expression x^y gives the bitwise exclusive or of x and y. That is, the kth bit of x|y is 0 if and only if exactly one of the kth bits of both x and y is 1. Here is an example.

x	0100001101100000
y	0001000111101101
x^y	0101001010001101

Bitwise not The expression ˜x interchanges 1s and 0s in x. That is, the kth bit of ˜x is 1 if and only if the kth bit of x is 0. Here is an example.

x	0100001101100000
˜x	1011110010011111

The bitwise not operation ˜ should not be confused with the Boolean not operation !. You should use ! only with bool values.

Left shift The expression x<<n (where n is a nonnegative integer) shifts the bits of x to the left n steps. The right-hand side of the result is filled with 0s. Any bits in the highest n positions are lost. Here is an example.

x	0100001101100000
x<<5	0110110000000000

The symbol << is the same one we use for writing to the console, as in the statement `cout << x << endl;`. C++ is able to distinguish between these cases by inspecting the types of objects on either side of the << symbol.

The expression x<<n is equivalent to multiplying x by 2^n (unless bits are lost at the left).

Right shift The expression x>>n (where n is a nonnegative integer) shifts the bits of x to the right n places. Bits in the lower n places are lost. The vacated positions on the left are filled in with 0s or with 1s depending on the situation:

- If x is an `unsigned` integer type, 0s are inserted at the left.

- If x is a signed integer type and x is nonnegative, 0s are inserted at the left.

- If x is a negative integer, then 1s are inserted at the left. Here are some examples.

short x	0010010010001010
x>>5	0000000100100100
unsigned short y	1000110010110111
y>>5	0000010001100101
short z	1000110010110111
z>>5	1111110001100101

The right shift operator >> uses the same symbol we use for keyboard input, for example, `cin >> x;`. As with left shift, C++ distinguishes these cases by inspecting the types of the objects on either side of the >> symbol.

All six of these operators can be combined with the assignment operator, =. The expression x &= y is equivalent to x = (x&y), and so on.

Bit operations can be combined to perform operations that would be difficult with standard mathematical operators. For example, suppose we want to set the kth bit of x to 1; the following code does the trick: `x |= (1<<k);`. If we want to set that bit to zero, we do this: `x &= ~(1<<k);`.

12.5.3 The `bitset` class template

Using integer types to represent a list of binary values is efficient, but presents two drawbacks. First, this technique is limited to the size of an integer on your computer; if you want a list of, say, 200 bits, there is no integer type with that capacity. Second, using bit manipulation can result in obfuscated code. Human beings find statements such as `x&=~(1<<k);` difficult to understand. (The statement sets the kth bit of x to zero.) If your problem requires high speed for short lists of bits, then bit manipulation of integer types may be your best option. However, there are two other choices of which you should be aware.

The first option, also discussed in Section 8.4, is to use `vector<bool>` variables; these are adjustable arrays of true/false values. To set the kth bit of such an array equal to zero (`false`), we use the considerably clearer statement `x[k]=0;` or `x[k]=false;`. Variables of type `vector<bool>` use memory efficiently, can be easily resized, and provide convenient access to their elements. However, the bitwise operations (such as `&`, `~`, `>>`, etc.) cannot be used with variables of type `vector<bool>`. If one wished to interchange all the 0s and 1s held in `x`, the statement `x=~x;` does not work. Instead, one would need to write a `for` loop to change the bits one by one:

```
for (int k=0; k<x.size(); k++) {
  x[k] = !x[k];
}
```

The second option is to use a `bitset`. A `bitset` object is a fixed size repository of bits. To use variables of type `bitset`, start with `#include <bitset>` and declare variables like this:

```
bitset<100> x;
```

This sets up `x` as a list of 100 bits. Notice that `bitset` is a template but its argument is a number, not a type; we explain how to do this later in this section. The important point is that this number is a constant, not a variable. The following code is illegal.

```
int n;
cout << "Enter number of bits --> ";
cin >> n;
bitset<n> x;   // illegal constructor, n is not a constant
```

The size of the `bitset` must be declared when you write your program, not while the program is running.

Here is a list of the various methods and operators available for `bitset`s.

- **Constructors**. The standard constructor has the form

  ```
  bitset<number> var;
  ```

  where `number` is a specific positive integer. This may be a `const int` defined earlier in the code, or an explicitly typed value, such as 100. The variable `var` holds `number` bits, and at this point these are all 0s.

  One may construct from an `unsigned long` integer value. For example,

  ```
  bitset<20> x(39);
  ```

  sets `x` to 00000000000000100111 (the binary representation of 39).

  One may also construct from a C++ `string` object (these are discussed later in Section 14.2). The constructor

  ```
  bitset<20> x(string("10110001"));
  ```

  sets `x` to 00000000000010110001. The type `string` is required; don't use `bitset<20> x("10110001"));`.

  Finally, a copy constructor is available:

```
bitset<20> y(x);
```

makes y a copy of x. Note that x must also be a bitset<20> and may not be a bitset of any other size.

- **Inspection methods.** These are methods one can use to learn information about the bits held in a bitset. Suppose x is a bitset<100>:

  - x.size() returns the number of bits that x holds (in this example, 100).
  - x.any() returns true if at least one bit in x is a 1.
  - x.none() returns true if all of the bits are 0.
  - x.count() returns the number of 1s in x.
  - x.test(k) returns true if the *k*th bit of x is a 1. Of course, k must be at least 0 and less than x.size().

- **Bit manipulation methods.** The following methods may be used to alter the value held in the bits of a bitset. Suppose x is a bitset<100>:

  - x.set() sets all of x's bits to 1.
  - x.set(k) sets the *k*th bit to 1.
  - x.set(k,b) sets the *k*th bit base on the value held in the integer variable b. If b is zero, the *k*th bit of x is set to 0; otherwise it is set to 1.
  - x.reset() sets all of x's bits to 0.
  - x.reset(k) sets bit *k* to 0.
  - x.flip() swaps 0 and 1 values in every position of x. For example, suppose x holds 1110001110; after the statement x.flip();, it now holds 0001110001.
  - x.flip(k) flips the *k*th bit of x.

- **Comparison operators.** If x and y are both bitsets of the same size, then we may compare them with the usual expressions x==y and x!=y.

- **Bit operators.** The standard bitwise operators (&, |, ^, ~, <<, >>) and their assignment variants (&=, |=, ^=, ~=, <<=, >>=) may be used on a pair of bitsets of the same size.

- **Array style access.** In addition to the methods described above, individual elements of a bitset may be accessed using square brackets. The expression x[k] is the *k*th element of x. The expression x[k] may appear on the right or the left of an assignment statement such as x[4]=~x[10];.

In addition, the expression x[5].flip() is equivalent to x.flip(5); both of these toggle the fifth bit of x.

- **Input/output**. Objects of type `bitset` can be written to the computer's screen and read from the keyboard.

The statement `cin >> x;` reads a sequence of 0s and 1s from the keyboard. At most `x.size()` bits are read. If fewer bits are read (before reaching a character other than 0 or 1), the left bits are filled with zeros.

The statement `cout << x;` prints x to the screen. The highest bit (in position `x.size()-1`) is printed first and the lowest bit, `x[0]`, is printed last.

Here is a short program that illustrates these ideas.

```
#include <bitset>
#include <iostream>
using namespace std;

int main() {
  bitset<10> x;
  cout << "Enter bits -> ";
  cin >> x;
  cout << "x = " << x << endl;
  for (int k=0; k<10; k++) {
    cout << "x[" << k << "] = " << x[k] << endl;
  }
  return 0;
}
```

Here is a sample run of this program.

```
Enter bits -> 1101
x = 0000001101
x[0] = 1
x[1] = 0
x[2] = 1
x[3] = 1
x[4] = 0
x[5] = 0
x[6] = 0
x[7] = 0
x[8] = 0
x[9] = 0
```

12.5.4 Class templates with non-type arguments

We have seen a variety of templates, and in nearly all cases the arguments to the template, given between the < and > delimiters, are C++ types. For example, the `complex` class template is completed with a numeric type (e.g., `complex<double>`) and our `max_of_three` procedure template may use any type arguments that can be compared with < (e.g., three `PTriple` values).

The exception is the `bitset` class template. Here, the template is completed by specifying an unsigned integer value. How is this done?

A typical class template is defined in a header file like this:

```
template <class T>
MyTemplateClass {
  . . . . .
};
```

where the data and methods in `MyTemplateClass` may refer to variables of type `T`. Variables are declared with statements such as this.

```
MyTemplateClass<double> x;
```

However, the template parameters (the arguments between < and > delimiters) need not be classes. For example, we could create a class template such as this:

```
template <long N>
AnotherClass {
  . . . . .
};
```

The parameter `N` may appear in the data and methods of `AnotherClass` wherever a constant `long` integer might rightly go. For example, `AnotherClass` might include a data member that is an array declared like this:

```
private:
  double coordinates[N];
```

Note that `AnotherClass<10>` and `AnotherClass<11>` are different classes (although based on the same template).

Template classes may have multiple parameters that may be classes or specific values as in this example:

```
template <class T, double X, int N, class S>
ComplicatedClass {
  . . . . .
};
```

A declaration based on this template would look like this:

```
ComplicatedClass<int, -3.5, 17, double> x;
```

12.6 Exercises

12.1 Exercise 7.3 (page 126) asks you to create a procedure to find the median of a list of real (`double`) numbers. Make a new version that can handle numbers of any type by using a template. The procedure should not modify the array.

12.2 Exercise 11.5 (page 233) asks you to create the class `SmartArray` that allows arbitrary indexing (e.g., a -1 index returns the last element of the array). In that problem, a `SmartArray` contains `long` integers. Create a new version of `SmartArray` that can hold values of any given type. For example, to declare a `SmartArray` to hold 20 `double` values, you would type this:

```
SmartArray<double> X(20);
```

12.3 Create a `derivative` procedure that finds the derivative of a `Polynomial`.

12.4 Create a root-finding procedure for polynomials based on Newton's method. Given a polynomial p and initial guess x_0, the procedure should solve $p(x) = 0$ using the iteration

$$x_{k+1} = x_k - \frac{p(x_k)}{p'(x_k)}.$$

How many iterations should be performed? You may either let the user set the number of iterations or a desired tolerance ε so that $|p(x)| < \varepsilon$.

Be sure to address the following issues.

- The polynomial may be either real or complex.

- The initial x_0 may be either real or complex.

- The roots might not be simple (i.e., the roots might have multiplicity greater than 1).

12.5 The `pair` class template (defined in the `utility` header) is a handy mechanism for creating ordered pairs; see Section 8.5. Using `pair` as an inspiration, create a `triple` class template that represents an ordered three-tuple (x, y, z) where the three elements may be of any type (including different types from each other). Make the data fields `public` and name them `first`, `second`, and `third`.

Remember to define an `operator<` that orders the triples lexicographically.

In addition, provide a `make_triple` procedure procedure that is analogous to `make_pair`.

12.6 Create a `RationalFunction` class template to represent rational functions. [A *rational function* is a quotient of two polynomials, $p(x)/q(x)$.] The coefficients may be real, complex, or from $\mathbb{Z}_p$ for some prime p. Include the basic operations $+$, $-$, $\times$, $\div$.

12.7 Write a program to print out all subsets of the set $\{1, 2, \ldots, n\}$. Do this with an integer variable that steps from 0 to $2^n - 1$ and convert that value into a set.

Part III

Topics

Chapter 13

Using Other Packages

Good news!

The first good news is that we have covered nearly all of the C++ concepts you need to know for mathematical work. All that remains is a more extensive discussion on getting information in and out of your programs (covered next in Chapter 14).

The second good news is we are ready to stand on the shoulders of giants. If you are reading this book, chances are you are not an expert in computer programming. So, the next best thing is to have an expert assistant to create C++ classes for you. And you do! Thanks to the ubiquity of C++, there are classes available for many types of work including number theory, algebraic geometry, optimization, quaternions, combinatorics, cup products for finite groups, and more. Many of these packages are available for free (for noncommercial use) over the Web.

In this chapter we introduce a few of these packages.

13.1 Arbitrary precision arithmetic: The GMP package

The C++ `long` type can accommodate integer values in a finite range (see Section 2.1). For work in number theory or cryptography, one needs to handle integers with hundreds or thousands of digits. Or perhaps we want to work with rational values, but `double` variables are unacceptable because they do not hold exact values. The solution to this problem is to create C++ classes for handling arbitrarily large integers and exact rational numbers.

The creation of such classes takes a lot of work, and making the algorithms efficient takes a great deal of skill. Fortunately, programmers with a great deal of skill have done the hard work of creating the GNU Multiple Precision Library (called GMP for short). (We describe version 4.1.4.)

The GMP library is available on the Web at `http://www.swox.com/gmp/`, but it may already be available on your computer (it is often included on Linux systems). The underlying GMP library is created for programming in C, but it includes good support for C++ programming. If GMP has not already been installed on your computer, you can download it from the Web and follow the installation instructions (see the file `install` or the manual in PDF format). If you run into trouble, find a friendly computer scientist for some assistance.

The GMP package provides four important C++ classes:

- `mpz_class` for arbitrary precision integers,

- `mpq_class` for exact rational numbers,

- `mpf_class` for floating point numbers, and

- `gmp_randclass` for generating large random numbers.

To use these classes, your program needs to include a header file:

```
#include <gmpxx.h>
```

Variables are declared in the usual manner. For example, to declare x to be an arbitrary precision integer, use this:

```
mpz_class x;
```

This initializes x with the value 0. To give it a large value, the following does *not* work.

```
x = 30984729387509874398572345435342326626245985;    // this fails
```

The problem is that C++ is not able to deal with that large value as one of its basic types. Instead, you can type this:

```
x = "30984729387509874398572345435342326626245985";  // this works
```

The `mpz_class` type can convert character arrays (containing digits).
     The `mpz_class` can be initialized using other bases. For example,

```
mpz_class x("12321423112312321312314001200001213", 5);
```

initializes x with a base-5 value.
     The usual arithmetic and comparison operators work just as expected. Multiplication of extremely large integers is accomplished using sophisticated efficient algorithms.

The rational type, `mpq_class`, is constructed either from two integer arguments

```
mpq_class a(n_1,n_2);
```

or else a character array, such as this:

```
mpq_class b("53490875234097/1134381");
```

Rationals can also be constructed from `double` values. Indeed, the GMP types can be converted to and constructed from nearly any numeric type.
     Rational `mpq_class` objects have a `canonicalize()` method. The purpose of this method is to clear any common factors between numerator and denominator. It's a good idea to invoke this method after creating an `mpq_class` value.

The C++ features of the GMP package are a supplement to the C base that forms the bulk of GMP. Some GMP procedures require some fancy footwork to be used in C++. For example, to find the greatest common divisor of two `mpz_class` integers one uses the procedure `mpz_gcd`. This procedure takes three arguments of type `mpz_t`—not `mpz_class`. The `mpz_class` objects contain an `mpz_t` type value internally and to access that internal value one uses the `get_mpz_t()` method. Here's how this all works.

First, we set up our variables:

```
mpz_class a = "47825100";
mpz_class b = "55431225";
mpz_class d;    // place to hold the answer
```

Then we call the `mpz_gcd` method like this:

```
mpz_gcd(d.get_mpz_t(), a.get_mpz_t(), b.get_mpz_t());
```

Now d holds the gcd of a and b.

An object of type `gmp_randclass` is a random number generator. This object offers a choice of pseudo random number generator algorithms and allows the user to set the seed. For example, to create a new random number generator with a default algorithm and seeded from the system clock, we write this:

```
gmp_randclass X(gmp_randinit_default);
X.seed(time(0));
```

To extract a random value from the generator, we use the `get_z_bits` method and specify the size (in bits) of the result. For example, `X.get_z_bits(100)` returns a 100-bit random integer.

Here is a program that illustrates these ideas.

Program 13.1: A program to illustrate the use of the GMP package.

```
1   #include <gmpxx.h>
2   #include <iostream>
3
4   using namespace std;
5
6   int main() {
7     mpz_class a, b;
8
9     a = "54098745908347598037452";
10    b = "44523409864";
11
12    cout << "a = " << a << endl;
13    cout << "b = " << b << endl;
14    cout << "a*b = " << a*b << endl;
15    cout << "a/b = " << a/b << endl;
16    cout << "a%b = " << a%b << endl;
17    cout << "a+b = " << a+b << endl;
18    cout << "a-b = " << a-b << endl;
19    cout << "b-a = " << b-a << endl;
```

```
20
21    mpz_class d;
22    mpz_gcd(d.get_mpz_t(), a.get_mpz_t(), b.get_mpz_t());
23    cout << "gcd(a,b) = " << d << endl;
24
25
26    cout << "Is a < b? " << ((a<b) ? "Yes" : "No");
27    cout << endl;
28
29    mpq_class r(a,b);
30    r.canonicalize();
31
32    cout << "As a rational number, a/b is " << r << endl;
33
34    mpq_class q(-1.125);
35    cout << "q = " << q << endl;
36
37    gmp_randclass X(gmp_randinit_default);
38    X.seed(time(0));
39    cout << "Here's a random number: " << X.get_z_bits(100) << endl;
40
41    return 0;
42 }
```

Compiling this program can be tricky. The computer needs to find the following items,

- The header file gmpxx.h, and

- The libraries[1] gmp and gmpxx.

If the GMP header files are installed in a standard location, no special steps are required for the compiler to locate the header file gmpxx.h. No special steps are needed if the file gmpxx.h resides in the same directory (folder) as the program that #includes it. Otherwise, the compiler needs to be informed of the header file's location. The precise manner in which this is done can vary between compilers. For example, with the g++ compiler (a popular choice), one specifies the directory in which gmpxx.h resides with the -I command line option. (See the full example below as well as Appendix A.)

It is also necessary to tell the compiler to use the libraries gmp and gmpxx (these are distinct from the header file gmp.h). For example, with the g++ compiler, this is accomplished with a pair of -l command line options (see below). In the case where these libraries are installed in a nonstandard location, it may be necessary to tell the computer where to find these libraries. This is accomplished with the -L command line option in g++.

For example, to compile the gmp-tester.cc program (Program 13.1) above, I use the following command,

```
g++ gmp-tester.cc -I/sw/include -L/sw/lib -lgmp -lgmpxx
```

---

[1] A library contains compiled code needed by the GMP package.

On my computer, the `gmpxx.h` header file is in the directory `/sw/include` (specified by the `-I` option), the libraries are located in `/sw/lib` (specified by the `-L` option), and the `gmp` and `gmpxx` libraries are named explicitly (by the pair of `-l` options).

Here is the output of the program.

```
a = 54098745908347598037452
b = 44523409864
a*b = 2408660637205753086401397418226528
a/b = 1215062953929
a%b = 4181881796
a+b = 54098745908392121447316
a-b = 54098745908303074627588
b-a = -54098745908303074627588
gcd(a,b) = 4
Is a < b? No
As a rational number, a/b is 13524686477086899509363/11130852466
q = -9/8
Here's a random number: 52132270408518643545513754888
```

13.2 Linear algebra

13.2.1 Two-dimensional arrays in C++

One-dimensional arrays in C++ are mildly difficult to use. One needs either to give the explicit size of the array in advance (e.g., `long vals[10];`) or else to declare the array via a pointer (`long *vals;`), allocate space (`vals = new long[n];`), and then release the memory when finished (`delete[] vals;`). We need to remember that the first element of an array is indexed as `vals[0]` and there is no protection against accessing data beyond the bounds of the array. There is no convenient way to extend the size of the array. The `vector` class template solves many of these problems.

It is possible to use multidimensional arrays in C++, but the difficulties are compounded. We describe such arrays here briefly in order to convince you that you do *not* want to use these tricky structures and, instead, will avail yourself of some of the options we discuss below.

To declare a two-dimensional array of a given size, we use a statement such as this:

```
long vals[4][10];
```

This establishes a variable named `vals` as a 4×10-array of `long` integer values. Please note that the following statement is incorrect: `long vals[4,10];`. Unfortunately, it is valid C++ (don't bother trying to figure out what it does) so the compiler won't catch this error.

To access an element of this array, we use the syntax `vals[i][j]` where `i` is between 0 and 3, and `j` is between 0 and 9. The expression `vals[i,j]` is incorrect.

It is possible to declare three- (or higher-) dimensional arrays. For example,

```
long vals[4][10][19];
```

declares `vals` to be a $4 \times 10 \times 19$-array of `long` values. The i,j,k-entry of this array is given by `vals[i][j][k]` where i,j,k have the expected constraints.

Passing multidimensional arrays to procedures is possible, but the syntax is tricky. Procedures that receive such arrays need to specify the size of the array they expect to receive.

Fortunately, there are better alternatives. The two that we present have the added advantage of providing linear algebraic capabilities (matrix multiplication, eigenvalue calculation, etc.).

13.2.2 The TNT and JAMA packages

The U.S. government's National Institute of Standards and Technology (NIST) has created a pair of packages for working with one-, two-, and three-dimensional arrays and to perform standard linear algebra functions thereon. The packages are the *Template Numerical Toolkit* (TNT) and the *C++ Java Matrix Library* (JAMA).[2] These are available for free download from the following Web site,

```
http://math.nist.gov/tnt/index.html
```

You need to download the TNT and JAMA packages separately. Be sure to download the documentation as well; this consists of a collection of Web pages that describe the various classes and procedures. (We describe TNT version 1.2.4 and JAMA version 1.2.2.)

Installation is easy because these packages are simply collections of `.h` header files. They can be copied to any convenient location on your computer (including, e.g., the directory containing your program that uses these files).

The most important classes in the TNT package are `Array1D` (for vectors) and `Array2D` (for matrices[3]). To use these, we need the directive `#include "tnt.h"` and the statement `using namespace TNT;` at the beginning of our program. The classes and procedures in the TNT class are in their own namespace. If we did not use this statement, then we would need to prepend `TNT::` to the names of TNT classes and procedures (e.g., `TNT::Array2D` in lieu of `Array2D`).

To declare an array of elements of type, say, `double`, we use statements like these:

```
Array2D<double> M(5,10);
Array1D<double> v(9);
Array2D<double> X;
```

---

[2] This library was developed first for the Java programming language and subsequently ported to C++.

[3] You may discover that the TNT package includes classes called `Matrix` and `Vector`. Do not use these. They are *deprecated* classes; this means they are obsolete and their inclusion in the package is only to support people who used older versions of TNT.

Here, M is a 5×10-array (matrix) of double values, v is a length-9 array (vector), and X is an as-yet unsized matrix.

To access elements of an array we use the usual C++ conventions. For the variables M and v above, the syntax is as follows,

- M[i][j] where $0 \leq i < 5$ and $0 \leq j < 10$, and

- v[k] where $0 \leq k < 9$.

To learn the size of an array, the methods dim1() and dim2() give the number of rows and columns (if more than one-dimensional) in the array.

Read this carefully: The assignment operator for TNT array classes (e.g., B=A;) works in a manner that is different from other classes we have encountered thus far. After this assignment not only are A and B arrays of the same size and not only do they hold the same values, they now refer to the exact same data. In other words, modifications to one of A or B results in change in the other. (We say that A and B provide different *views* to the same underlying data.) If we wish to make an independent copy of A in B, we need to use the copy method like this:

B = A.copy();

Here is a program that illustrates the difference between the statements B=A; and B=A.copy();.

Program 13.2: Assignment versus copying in the TNT package.

```
1   #include <iostream>
2   #include "tnt.h"
3
4   using namespace std;
5   using namespace TNT;
6
7   int main() {
8     Array2D<double> A(3,3), B, C;
9
10    for (int i=0; i<3; i++)
11      for (int j=0; j<3; j++)
12        A[i][j] = (i+1) + 0.1*(j+1);
13
14    cout << "Original matrix A:" << endl << A << endl;
15
16    B = A;          // Now B and A share data
17    C = A.copy();   // C is an independent copy of a
18
19    B[1][1] = 888;
20    A[0][2] = 999;
21
22    cout << "Now A is this " << endl << A << endl;
23    cout << "and B is this " << endl << B << endl;
24    cout << "and C is this " << endl << C << endl;
25
26    return 0;
27  }
```

Here is the output of the program.

```
Original matrix A:
3 3
1.1 1.2 1.3
2.1 2.2 2.3
3.1 3.2 3.3

Now A is this
3 3
1.1 1.2 999
2.1 888 2.3
3.1 3.2 3.3

and B is this
3 3
1.1 1.2 999
2.1 888 2.3
3.1 3.2 3.3

and C is this
3 3
1.1 1.2 1.3
2.1 2.2 2.3
3.1 3.2 3.3
```

(Notice that `Array2D` objects can be written to the computer's screen using the usual `<<` operator. The first line of output is a pair of integers giving the number of rows and columns; the array follows one row at a time.)

One implication of this unusual assignment operator behavior is that when TNT arrays are passed to procedures, the procedure does not receive an independent copy. Suppose we have a procedure that looks like this:

```
void proc(Array2D<double> X) { ... }
```

It appears that this procedure is ready to receive its argument using call by value. From this we might infer (incorrectly) that if `proc` modifies `X`, there would be no effect on the matrix sent to this procedure. However, because assignment does not make a true copy (but only an alternative view of the same data), modifications do cause changes in the array on which this procedure was invoked.

The solution to this issue is simple: **Do not use call by value for TNT arrays.** This is a good practice in general because sending large objects as parameters of procedures is inefficient. Instead, use call by reference, and include the keyword `const` when you need to certify that the procedure does not modify its argument. Thus, the procedure should look like this:

```
void proc(const Array2D<double>& X) { ... }
```

Alternatively, if your procedure is not limited to arrays of `double` values but may be used on a broader assortment of types, use a template:

```
template <class T>
void proc(const Array2D<T>& X { ... }
```

For example, here is a header file named `trace.h` that provides a template procedure to calculate the trace of an `Array2D` matrix:

Program 13.3: A template to calculate the trace of an `Array2D` matrix.

```
1   #ifndef TRACE_H
2   #define TRACE_H
3
4   #include "tnt.h"
5
6   template <class T>
7   T trace(const TNT::Array2D<T>& X) {
8     T sum = T(0);
9     for (int k=0; k<X.dim1() && k<X.dim2(); k++) {
10      sum += X[k][k];
11    }
12    return sum;
13  }
14
15  #endif
```

The TNT package provides rudimentary array arithmetic. If A and B are arrays of the same type and same size (e.g., both `Array2D<long>` and both $m \times n$) then the following operations are available.

```
A+B     A+=B     A-B     A-=B     A*B     A*=B     A/B     A/=B
```

In each case, the arithmetic is performed element by element. If A, B, and C are `Array2D<double>` objects, and if A and B are the same size (both $m \times n$), then `C=A*B;` sets C to be an $m \times n$-array in which `C[i][j]` is `A[i][j]*B[i][j]`. It is *not* matrix multiplication.

Matrix multiplication is available via the `matmult` procedure. Matrix multiplication can only be performed on `Array2D` arrays of the appropriate shapes. The statement `C = matmult(A,B);` sets C to the appropriate size (if A is $m \times n$ and B is $n \times p$, then C is $m \times p$) and sets the elements of C appropriately:

$$C[i][j] = \sum_{k=0}^{n-1} A[i][k]*B[k][j]$$

The TNT package does not provide scalar multiplication of arrays, nor does it provide matrix–vector (`Array2D` times `Array1D`) multiplication. However, it is not difficult to create template procedures to perform these tasks. For example, here is a template for scalar–vector multiplication.

```
#include "tnt.h"
using namespace TNT;
template <class T>
void scalar_vector_multiply(T s,
                            const Array1D<T>& vec,
                            Array1D<T>& ans) {
  ans = Array1D<T>(vec.dim1()); // resize ans
  for (int i=0; i<vec.dim1(); ++i) {
```

```
    ans[i] = s*vec[i];
  }
}
```

More advanced linear algebra functions for real matrices are provided by the JAMA package. This package provides the following classes:

- JAMA::Cholesky — Given a real, symmetric, positive definite matrix A, find a lower triangular matrix L so that $A = LL^T$. Header file: jama_cholesky.h.

- JAMA::Eigenvalue — Given a real, square matrix A find the eigenvalues and eigenvectors of A. If these are complex, the real and imaginary parts of the eigenvalues are accessed separately. (This class does not use the complex<> types.) Header file: jama_eig.h.

- JAMA::LU — Given a real, $m \times n$-matrix (with $m \geq n$) A, find a lower triangular matrix L and an upper triangular matrix U so that LU is a (row permutation of) A. Header file: jama_lu.h.

- JAMA::QR — Given a real, $m \times n$-matrix (with $m \geq n$) A, find an $m \times n$ orthogonal matrix Q and an $n \times n$ upper triangular matrix R so that $A = QR$. Header file: jama_qr.h.

- JAMA::SVD — Given a real, $m \times n$ matrix A (with $m \geq n$), find an $m \times n$ orthogonal matrix U, an $n \times n$ diagonal matrix Σ, and an $n \times n$ orthogonal matrix V so that $A = U\Sigma V^T$. Header file: jama_svd.h.

The design philosophy for all these classes is the same. To find, say, the eigenvalues of a matrix A, we do not use a procedure to which A is passed. Instead, we create a JAMA::Eigenvalue object like this:

```
#include "tnt.h"
#include "jama_eig.h"
using namespace TNT;
using namespace JAMA;

int main() {
  Array2D<double> A(10,10);
  // assign values to the elements of A
  ....
  Eigenvalue<double> eigs(A);
  ....
}
```

This code loads the necessary headers and sets up a matrix A. We then create an Eigenvalue object named eigs. The matrix A is passed to the constructor.

The eigenvalues and eigenvectors are now embedded inside the Eigenvalue object named eigs. To extract the information we use one of the access methods provided by the Eigenvalue class:

- `eigs.getRealEigenvalues()` returns an `Array1D` object containing the real parts of the matrix's eigenvalues.

- `eigs.getImagEigenvalues()` returns an `Array1D` object containing the imaginary parts of the eigenvalues.

- `eigs.getV()` returns an `Array2D` (matrix) whose columns contain the eigenvectors of the matrix. (Refer to the documentation to see how `Eigenvalue` handles complex eigenvectors.)

There are other linear algebraic entities one might like to compute including the rank or determinant of a matrix, or the solution to the linear system $Ax = \mathbf{b}$. The JAMA package provides tools for doing this. To find them, one needs to browse the documentation for the various classes (`Cholesky`, `Eigenvalue`, etc.), but it's not too hard to guess where these might lie. Some examples:

- *To find the determinant of a matrix*: It's easy to find the determinant of a matrix from its *LU*-factorization. Naturally enough, the `LU` class includes a `det()` method. For example, if `A` is a square, `Array2D<double>` matrix, the following code finds its determinant,

```
LU alu(A);
cout << "The determinant of A is " << alu.det() << endl;
```

- *To solve a linear system* $Ax = \mathbf{b}$: Again, this is often found through the *LU*-factorization, and the `LU` class contains two `solve()` methods: one that solves $Ax = \mathbf{b}$ and another that solves $AX = B$ (where $\mathbf{x}$ and $\mathbf{b}$ are vectors, and X and B are matrices).

  The `QR` class also provides `solve()` methods; if the system is overdetermined, this `solve` gives a least-squares solution.

  In the special case that A is symmetric and positive definite, a solution to the linear system can also be found through the Cholesky factorization. The class `Cholesky` has a method named `is_spd()` to check if a matrix is symmetric and positive definite, and a pair of `solve()` methods for solving linear systems.

  Generally, it is numerically unwise to solve $Ax = \mathbf{b}$ by inverting A and calculating $A^{-1}\mathbf{b}$. However, if we wish to find the inverse of a square matrix A, we can create an identity matrix I and solve $AX = I$.

  See the documentation for details.

- *To find the rank of a matrix*: The rank of A is the number of nonzero singular values. So, to find the rank of a matrix, use the `SVD` class's `rank()` method:

```
Array2D<double> A;
...
SVD sing_vals(A);
cout << "The rank of A is " << sing_vals.rank() << endl;
```

The calculation of the rank of a matrix is difficult if the matrix is ill conditioned. To check a matrix's condition number, use the `cond()` method, also found in the SVD class.

We illustrate the use of the TNT and JAMA packages with the following program. Recall that a *Hilbert matrix* is an $n \times n$-matrix whose i, j-entry is $1/(i + j - 1)$ (in the usual notation in which the upper left corner contains the $1, 1$-entry). Hilbert matrices are invertible, but notoriously ill conditioned. Hence, finding their inverses is numerically unstable. The following program creates a Hilbert matrix (whose size is specified by the user), finds its inverse from its *LU*-factorization, calculates its eigenvalues, and (because Hilbert matrices are symmetric and positive definite) finds its Cholesky factorization.

Program 13.4: A program to illustrate the TNT and JAMA packages with calculations on a Hilbert matrix.

```
1   #include "tnt.h"
2   #include "jama_lu.h"
3   #include "jama_eig.h"
4   #include "jama_cholesky.h"
5   #include <iostream>
6   using namespace std;
7   using namespace TNT;
8   using namespace JAMA;
9
10  // A program to demonstrate the use of TNT and JAMA packages for
11  // linear algebra work.
12
13  int main() {
14      // Get the size of the matrix from the user
15      cout << "Enter Hilbert matrix size --> ";
16      int n;
17      cin >> n;
18
19      if (n < 2) {
20          cerr << "Please enter a value greater than 1" << endl;
21          return 1;
22      }
23
24      // H holds the n-by-n Hilbert matrix and eye holds the n-by-n
25      // identity matrix.
26      Array2D<double> H(n,n);
27      Array2D<double> eye(n,n);
28
29      // Fill in the entries in H
30      for (int i=0; i<n; i++) {
31          for (int j=0; j<n; j++) {
32              H[i][j] =1./(i+j+1);
33          }
34      }
35
36      // Set up identity matrix
37      for (int i=0; i<n; i++) eye[i][i] = 1;
```

```
38
39    // Print out H
40    cout << "H = " << endl;
41    cout << H << endl;
42
43    // Use the LU decomposition's solve method to find the inverse of H.
44    LU<double> HLU(H);
45    Array2D<double> Hinv = HLU.solve(eye);
46    cout << "H inverse = " << endl;
47    cout << Hinv<< endl;
48
49    // Use the LU decomposition of H to calculate its determinant
50    cout << "The determinant of H is " << HLU.det() << endl;
51
52    // Use the Eigenvalue class to find the eigenvalues of H
53    Eigenvalue<double> Heig(H);
54    cout << "The eigenvalues of H are " << endl;
55    Array1D<double> eig_vals;
56    Heig.getRealEigenvalues(eig_vals);
57    for (int i=0; i<n; i++) cout << eig_vals[i] << " ";
58    cout << endl << endl;
59
60    // Use the Cholesky class to find a matrix C so that C*C' = H
61    Cholesky<double> Hchol(H);
62    Array2D<double> C = Hchol.getL();
63    cout << "The Cholesky matrix for H is " << endl;
64    cout << C << endl;
65
66    // Construct C's transpose
67    Array2D<double> Ctrans(n,n);
68    for (int i=0; i<n; i++) {
69      for (int j=0; j<n; j++) {
70        Ctrans[i][j] = C[j][i];
71      }
72    }
73
74    // Check that C*C' gives H
75    cout << "C * C' = " << endl;
76    cout << matmult(C,Ctrans) << endl;
77
78    return 0;
79  }
```

Here is the output of a typical run.

```
Enter Hilbert matrix size --> 5
H =
5 5
1 0.5 0.333333 0.25 0.2
0.5 0.333333 0.25 0.2 0.166667
0.333333 0.25 0.2 0.166667 0.142857
0.25 0.2 0.166667 0.142857 0.125
0.2 0.166667 0.142857 0.125 0.111111

H inverse =
5 5
```

```
25 -300 1050 -1400 630
-300 4800 -18900 26880 -12600
1050 -18900 79380 -117600 56700
-1400 26880 -117600 179200 -88200
630 -12600 56700 -88200 44100

The determinant of H is 3.7493e-12
The eigenvalues of H are
3.28793e-06 0.000305898 0.0114075 0.208534 1.56705

The Cholesky matrix for H is
5 5
1 0 0 0 0
0.5 0.288675 0 0 0
0.333333 0.288675 0.0745356 0 0
0.25 0.259808 0.111803 0.0188982 0
0.2 0.23094 0.127775 0.0377964 0.0047619

C * C' =
5 5
1 0.5 0.333333 0.25 0.2
0.5 0.333333 0.25 0.2 0.166667
0.333333 0.25 0.2 0.166667 0.142857
0.25 0.2 0.166667 0.142857 0.125
0.2 0.166667 0.142857 0.125 0.111111
```

13.2.3 The `newmat` **package**

Another package designed specifically for linear algebra is the `newmat` package. This package was developed by Robert Davies and is available for free from his Web site, http://www.robertnz.net/. (We describe version 11, beta.)

This package needs to be compiled on your computer; the instructions on how to do this are included in the download. Once compiled, you will find a library (named `libnewmat.a`, or something similar) and several `.h` header files. You may copy these to another location if you desire.

All programs that use the `newmat` package need a `#include "newmat.h"` directive. If the program requires more than the basic features of the package, other headers may need to be included. For example, to use `<<` to print matrices to `cout`, be sure to include the `newmatio.h` header.

When you compile your own program, you need to tell the compiler where to find the header files (e.g., with the `-I` option), where to find the library (e.g., with the `-L` option), and to link with the `newmat` library (e.g., with `-lnewmat`). See Appendix A.1.4 for more information about the `-L` and `-l` options.

The `newmat` package provides several different classes for holding matrices; in all cases, the matrices hold `double` real values.[4] The fundamental matrix type is

---

[4]It is possible to switch this to `float` values if you prefer.

called `Matrix` and its constructor takes two arguments giving the number of rows and columns. Here is an example,

```
Matrix A(2,3);
```

Now `A` is a 2×3-matrix. It is also possible to declare a matrix without specifying its size, for example, `Matrix A;`.

Elements of a matrix are specified using the standard mathematical convention: the element in row `i` and column `j` of a matrix `A` is `A(i,j)`. Here, `i` (respectively, `j`) is at least 1 and at most the number of rows (respectively, columns) of `A`. The expression `A(i,j)` may appear on either side of an assignment. (It is also possible to have `newmat` use the C++ multidimensional array notation `A[i][j]` in which case the subscripts begin at zero, but this is not recommended.)

The `Matrix` constructor does not specify values for its elements; it is the programmer's responsibility to handle that. So, for example, to create a 10×5-matrix of zeros, do this:

```
Matrix Z(10,5);
Z = 0.;
```

In addition to `Matrix`, the `newmat` package provides several other classes for holding matrices including the following,

```
SquareMatrix,
UpperTriangularMatrix,
LowerTriangularMatrix,
SymmetricMatrix,
DiagonalMatrix,
IdentityMatrix,
```

and various banded matrices. For vectors, the package provides `ColumnVector` and `RowVector`. These alternatives use the computer's memory efficiently. In addition, some procedures require specific types of matrices. For example, the `EigenValues` procedure may only be used on symmetric matrices and the eigenvalues are returned in a diagonal matrix:

```
SymmetricMatrix A(n);  // n-by-n symmetric matrix
...                    // load values into A
DiagonalMatrix D;      // place to hold the eigenvalues
EigenValues(A,D);      // find the eigenvalues of A
```

The `SymmetricMatrix` class is clever about assigning values to elements. Consider this code.

```
#include <iostream>
#include "newmat.h"
#include "newmatio.h"
using namespace std;

int main() {
  SymmetricMatrix A(3);
  A(1,2) = 3;
  A(3,2) = 2;
  cout << A << endl;
```

```
    return 0;
}
```

(Note: The inclusion of the `iostream` header must precede the `newmatio.h` header.)
The output of this program is this.

```
0.000000  3.000000  0.000000
3.000000  0.000000  2.000000
0.000000  2.000000  0.000000
```

Because A is symmetric, setting `A(i,j)` to some value automatically sets `A(j,i)` to
the same value as well.

The `newmat` packages provide a rich assortment of matrix operations. Here is a
brief description of some of them.

`B=A;`	Make a copy of A in B.
`A.t()`	Return the transpose of A.
`A.i()`	Return the inverse of A.
`A+B, A+=B;`	Matrix addition.
`A-B, A-=B;`	Matrix subtraction.
`A*B, A*=B;`	Matrix multiplication.
`A|B, A|=B;`	Horizontal concatenation.
`A&B, A&=B;`	Vertical concatenation.
`A==B, A!=B`	Equality/inequality testing.

Considerable care was taken in designing these operations. For example, to solve
the linear system $Ax = \mathbf{b}$, the statement `x = A.i() * b;` does not invert the matrix,
but finds the solution by a better method. However, if A and B are $n \times n$-matrices and x
is an n-vector, the expression `A*(B*x)` evaluates much more quickly than `(A*B)*x`.
If you type the ambiguous `A*B*x` it is not clear in what order the computer does the
calculation.

Scalar–matrix calculations (including scalar–vector) behave as you might expect.
If s is type `double` and A is type `Matrix`, then `A+s` (or `s+A`) gives a new matrix
whose i,j entry is `A(i,j)+s`. The statement `A+=s;` adds the value in s to all the
elements in A. The operations `-`, `*`, and `/` behave analogously.

The `nrows()` and `ncols()` methods give the number of rows and columns, respectively, of a matrix. There is a variety of methods for modifying the shape (and
type) of a matrix including `ReSize()`, `AsColumn()`, `AsRow()`, `AsDiagonal()`,
and so on. These are spelled out in the documentation.

The `newmat` package provides a rich assortment of procedures for important linear
algebraic problems including: Cholesky factorization of symmetric matrices, *QR*-
factorization, singular value decomposition, symmetric matrix eigenvalue decomposition, fast Fourier (and other trigonometric) transforms, determinant, trace, various
matrix norms, dot product of vectors, and so on.

To illustrate some of these, here is a program that finds the determinant, trace,
eigenvalues, and inverse of a Hilbert matrix (compare with Program 13.4).

Program 13.5: A program to illustrate the `newmat` package with calculations on a Hilbert matrix.

```
1    #include <iostream>
2    #include "newmat.h"
3    #include "newmatio.h"
4    #include "newmatap.h"
5
6    using namespace std;
7
8    int main() {
9      cout << "Enter size of Hilbert matrix --> ";
10     int n;
11     cin >> n;
12
13     SymmetricMatrix H(n);
14     for (int i=1; i<=n; i++) {
15       for (int j=i; j<=n; j++) {
16         H(i,j) = 1. / double(i+j-1);
17       }
18     }
19
20     cout << "The Hilbert matrix is " << endl << H << endl;
21
22     cout << "Its det is " << H.Determinant() << endl;
23
24     cout << "Its inverse is " << endl << H.i() << endl;
25
26     DiagonalMatrix D;
27     EigenValues(H,D);
28
29     cout << "The eigenvalues of the Hilbert matrix are" << endl;
30     cout << D << endl;
31
32     cout << "Trace calculated two ways: " << H.Trace() << " and "
33       << D.Trace() << endl;
34
35     return 0;
36   }
```

Here is an output of the program in which the user specifies a Hilbert matrix of size 5.

```
Enter size of Hilbert matrix --> 5
The Hilbert matrix is
1.000000 0.500000 0.333333 0.250000 0.200000
0.500000 0.333333 0.250000 0.200000 0.166667
0.333333 0.250000 0.200000 0.166667 0.142857
0.250000 0.200000 0.166667 0.142857 0.125000
0.200000 0.166667 0.142857 0.125000 0.111111

Its det is 3.7493e-12
Its inverse is
25.000000 -300.000000 1050.000000 -1400.000000 630.000000
-300.000000 4800.000000 -18900.000000 26880.000000 -12600.000000
```

```
1050.000000 -18900.000000 79380.000000 -117600.000000 56700.000000
-1400.000000 26880.000000 -117600.000000 179200.000000 -88200.000000
630.000000 -12600.000000 56700.000000 -88200.000000 44100.000000

The eigenvalues of the Hilbert matrix are
0.000003
 0.000306
  0.011407
   0.208534
    1.567051

Trace calculated two ways: 1.7873 and 1.7873
```

13.3 Other packages

There are many more packages available on the Web than we can possibly describe in detail. We conclude this chapter by listing a few additional packages that might be of interest. A well-chosen selection of keywords typed in a search engine is likely to turn up additional options.

- *Seldon* is a full-feature linear algebra package, available for free online here:

  `http://www.osl.iu.edu/research/mtl/`

- The *Matrix Template Library* is another package for working with matrices. Visit the Web site for more information and a free download:

  `http://www.osl.iu.edu/research/mtl/`

- The *Computational Geometry Algorithms Library* or *CGAL* for short is a large library of C++ classes and procedures for geometry work. Use it to find everything from convex hulls to Voronoi diagrams to minimum enclosing ellipses.

  CGAL is available for free from `www.cgal.org`.

- For graph theory work, try *Boost*, available for free from `www.boost.org`.

  Read the installation instructions carefully. The first step is to build a helper program called `bjam` that is used to direct the building of `boost` itself.

- For computational number theory, consider *LiDIA*. Based on the GMP package (see Section 13.1), it gives programmers the ability to do calculations in finite fields and on elliptic curves, use lattice reduction algorithms, perform linear algebra calculations, and more. It is available for free:

  `http://www.informatik.tu-darmstadt.de/TI/LiDIA/`

- The *Library of Efficient Data Structures and Algorithms*, also known as *LEDA*, is a commercial package for work in graph theory, geometry, cryptography, and more. More information, including pricing, is available here:

  `http://www.algorithmic-solutions.com/enleda.htm`

- Interested in computing the homology group (over $\mathbb{Z}$ or $\mathbb{Z}_p$) of a chain complex? Consider the C++ software available from *CHomP*—the Computational Homology Program. See their Web site for more information and a free download:

  `http://www.math.gatech.edu/~chom/`

13.4 Exercises

13.1 The base ten representation of 100! ends with a long string of zeros; it is a classic problem to find how many. Solve this problem with C++ using the GMP package.

13.2 Write a program that fills a two-dimensional array with Pascal's triangle. That is, the n,k-entry is $\binom{n}{k}$. [If $n < k$, set $\binom{n}{k} = 0$.] Build the table to include rows/columns 0 through 20. Generate each row from the previous row using the identity $\binom{n}{k} = \binom{n-1}{k-1} + \binom{n-1}{k}$.

What modification to your program would be necessary were you to increase the size of the table to 100×100?

Chapter 14

Strings, Input/Output, and Visualization

For the most part, mathematical work does not involve manipulation of character data. Nonetheless, it is useful to have a general understanding of how C++ handles character data, how to convert text to numbers, how to use command line arguments, and how to read and write data in files. We also show how to modify the formatting of output (e.g., how to increase the number of digits printed after the decimal point). We illustrate many of these ideas by creating a class to parse files one line at a time, and break those lines into individual words.

C++ has two ways to handle character data: arrays of `char` values and in objects of type `string`. The `char` arrays are a legacy of C++'s roots in the C programming language. It is necessary to understand their basics, but their use should be avoided where possible. The newer `string` variables are easier to use. We begin with a brief introduction to `char` arrays.

Textual and numerical output are important, but there are times when graphical output is especially insightful. We close this chapter with a discussion of how to draw pictures in C++.

14.1 Character arrays

Character (or text) data consist either of individual characters (letters, numerals, punctuation) or of lists of characters. The C++ data type `char` holds a single character from the Latin character set (26 lower- and uppercase letters, numerals, white space, punctuation). These are known as the ASCII characters and are the only characters with which this book deals. Computers can also deal with a richer set of glyphs (from accented Latin letters to Chinese characters) using a system called *unicode*; this system is beyond our scope.

An ordered list of characters is generally called a *character string*. For example, the words `Hello Gauss` are such a list comprising 11 characters. As mentioned, C++ has two principal ways to work with character strings: as null-terminated `char` arrays and as objects of the class `string`. The character array representation is a primitive scheme inherited from the language C. It is useful for writing messages to the screen and other simple chores. The moment one wishes to do any manipulation of characters (e.g., concatenate two strings), the C++ `string` class makes programming much easier. We begin by discussing the basics of character arrays and then introduce the `string` class in the next section.

In a statement such as `cout<<"The answer is "<<x<<endl;`, the characters enclosed in quotation marks form a *character array*. That is, the sequence of letters is a C++ object of type `char*`: an array whose elements are type `char`.

In this book we have used character arrays exclusively for writing messages to the computer's screen, but it is possible to hold such arrays in variables. For example:

```
const char* word = "Hello";
```

This creates an array of characters named `word`. The individual characters can be accessed using the usual array notation. For example, `word[0]` is the first character of the array, that is, H.

It is surprising to learn that the length[1] of the array `word` (as declared above) is six (even though *Hello* is a five-letter word). Character arrays in C++ are *null terminated*; this means that after the last character of the string, the numerical value 0 is appended to mark the end of the string. Figure 14.1 illustrates how the contents of the variable `word` are stored inside the computer's memory. The array is held at some

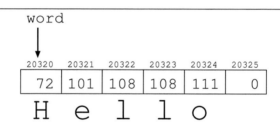

Figure 14.1:   Illustrating a null-terminated character array.

location in memory (arbitrarily set at 20320 in the figure); the variable `word` holds that memory location. The elements of the array, `word[0]` through `word[5]`, are stored as ASCII values. The letter H has ASCII value 72, hence that's what is stored in `word[0]`. The subsequent values are 101, 108, 108, and 111 corresponding to the letters e, l, l, and o. Finally, `word[5]` contains the numerical value 0 (which does not correspond to any printable character) and this marks the end of the character array.

There are procedures for processing character array data; here are a few examples.

- `strlen` gives the length of the character string (not including the terminating 0).

---

[1]The length of the character string is 5, but the length of the array that supports that string is 6. The C/C++ procedure `strlen` applied to the character array `"Hello"` returns the value 5.

- `strcpy` and `strncpy` are used for copying one character array to another.

- `strcat` and `strncat` are used to concatenate two character arrays; that is, if `s1` is `"Good"` and `s2` is `"Morning"`, their concatenation is `"GoodMorning"`.

- `strcmp` and `strncmp` give a total ordering on the set of character arrays; this is useful for sorting a list of character strings or determining if two strings are equal.

On some compilers, you may need the directive `#include <cstring>` in order to use these procedures.

Using character arrays and their associated procedures is awkward and error prone. In nearly all cases, it is much simpler to use the C++ `string` class instead. So, rather than delve into the details of these procedures, we turn to the friendlier and more powerful `string` class.

14.2 The `string` class

For any character processing tasks beyond the most basic, use C++'s `string` class. To use `string` objects, you might need the directive `#include <string>` (the header is optional with some compilers).

The `string` class contains a large number of features; in this section we discuss those with the greatest utility in mathematical work.

14.2.1 Initialization

Variables of type `string` can be declared in several ways; here are the most basic versions:

- `string s;` declares s to be an empty character string.

- `string s("Hello");` declares s to be a character string containing the letters `Hello`. In lieu of `"Hello"` we can use any null-terminated character array, such as this:

  ```
  const char* word = "Gauss";
  string s(word);
  ```

  Please note: It is not permissible to use a single `char` value as an argument to a `string` constructor. Thus, `string s('j');` is illegal. Instead, use `string s("j");`. Alternatively, the following is permissible.

  ```
  string s;
  s = 'j'; // this is OK
  ```

- `string s(s2);` initializes s with a copy of the string s2.

- `string s(s2,idx);` initializes s with a copy of the portion of string s2 that starts at position `idx`.

- `string s(s2,idx,len);` initializes s with a copy of the portion of string s2 that starts at position `idx` and runs for `len` characters. For example,

```
string s("Mathematics");
string t(s,2,3);
cout << t << endl;
```

writes the word `the` on the computer's screen. Note that the 0th character of s is M.

- `string s(reps, ch);` where `ch` is a character and `reps` is a nonnegative integer. This initializes s with `reps` copies of the character `ch`. For example,

```
string snore(10,'z');
cout << z << endl;
```

writes `zzzzzzzzzz` on the screen.

In addition to setting a `string`'s value when it is declared, we may modify its value using an assignment statement such as any of these:

```
s = "Gauss";
s = 'g';
s = other;   // other is another string object
```

14.2.2 Fundamental operations

The most basic operation one can perform on `string` objects is concatenation; the result of concatenating strings s1 and s2 is a new string comprising the characters of s1 immediately followed by the characters in s2. Concatenation of `strings` in C++ is denoted by the addition operator, +. For example, consider this code:

```
string s1("Hello ");
string s2("Gauss");
cout << s1+s2 << endl;
```

This prints `Hello Gauss` on the computer's screen. The + operation can be used to combine strings with single characters or with character arrays. Here are some examples.

```
string s("Hello");
cout << s + " Gauss" << endl;   // writes "Hello Gauss"

string s1 = "good";
string s2 = "luck";
cout << s1 + ' ' + s2 << endl; // writes "good luck"
```

One `string` can be appended to the end of another using the += operation:

```
string s("Carl Friedrich");
s += ' ';        // append a space
s += "Gauss";    // append the last name
cout << s << endl;
```

writes `Carl Friedrich Gauss` on the screen.

Characters may be inserted into the middle of a string using the `insert` method. The statement `s.insert(pos,t);` inserts the string `t` into `s` just before character `s[pos]`. The statement `s.insert(0,t);` inserts the characters at the beginning of `s`. Here is an example.

```
string s("CarlGauss");
s.insert(4," Friedrich ");
cout << s << endl;
```

writes `Carl Friedrich Gauss` on the screen.

In the statement `s.insert(pos,t);` the variable `t` may be either a `string` or a character array. It may not be type `char`.

```
s.insert(3,"x");   // allowed
s.insert(3,'x');   // forbidden
```

The `erase` method is used for deleting characters from a string. The statement `s.erase(pos);` deletes all characters from position `pos` to the end of the string. For example,

```
string s = "abcdefghijklmnopqrstuvwxyz";
s.erase(5);
cout << s << endl;
```

writes `abcde` to the screen. (Remember, for the original string, `s[5]` is `f`.)

The statement `s.erase(pos,nchars);` deletes `nchars` of `s` starting at `s[pos]`. For example,

```
string s = "abcdefghijklmnopqrstuvwxyz";
s.erase(5,3);
cout << s << endl;
```

writes `abcdeijklmnopqrstuvwxyz` to the screen.

A portion of a string can be modified using the `replace` method. The statement `s.replace(pos,nchars,new_chars);` deletes `nchars` starting at position `pos`, and then inserts `new_chars` in place of the missing portion. The `new_chars` may be either a `string` or a `char*` character array, and may be of any length. Here is an example:

```
string s = "abcdefghijklmnopqrstuvwxyz";
s.replace(5,16,"...");
cout << s << endl;
```

This writes `abcde...vwxyz` to the screen.

The substr method is used to extract a substring of a string; it does not modify the string. The expression s.substr(pos) returns the substring of s starting with character s[pos] through to the end of s. More generally, the expression s.substr(pos,nchars) returns the substring of s starting with s[pos] up to, but not including, s[pos+nchars]. Here is an example.

```
string s = "abcdefghijklmnopqrstuvwxyz";
cout << s.substr(5,3) << endl;
```

This writes fgh to the screen.

The length of a string can be ascertained using either s.size() or s.length(). One can test if the string is an empty string with s.empty() which returns true if the length of s is zero.

Square brackets can be used to access a given character in a string (either for reading or for modification). In consonance with C++ principles, s[0] is the first character of s. This code

```
string s = "good luck";
s[2] = 'l';
cout << s << endl;
```

writes gold luck on the screen. The value between the square brackets must be nonnegative and less than the length of the string. Alternatively, the at method may be used: s.at(k) gives character number k of the string s (i.e., s[k]).

Two string objects may be compared using the standard C++ comparison operations: ==, !=, <, <=, >, and >=. The ordering is mostly lexicographic. However, all uppercase letters precede lowercase letters. The following program illustrates the ordering of string values; note that sort implicitly relies on the < operator.

Program 14.1: A program to illustrate the sorting of string values.

```
1   #include <iostream>
2   using namespace std;
3
4   const int NWORDS = 7;
5
6   int main() {
7     string words[NWORDS];
8     words[0] = " zebra";
9     words[1] = "ant eater";
10    words[2] = "Aaron";
11    words[3] = "aardvark";
12    words[4] = "Baltimore";
13    words[5] = "anteater";
14    words[6] = "BREAKFAST";
15
16    sort(words, words+NWORDS);
17
18    for (int k=0; k<NWORDS; k++) {
19      cout << words[k] << endl;
20    }
```

```
21
22     return 0;
23   }
```

Here is the output of this program.

```
 zebra
Aaron
BREAKFAST
Baltimore
aardvark
ant eater
anteater
```

In a dictionary, *aardvark* precedes *Aaron*, but C++ sorts these the other way around because uppercase A comes before all lowercase letters. The space character precedes all letters, hence ⎵zebra is first in the sorted output.

14.2.3 Searching

The `string` class provides methods for searching for substrings. The most basic of these is `find`. The method is invoked with an expression such as `s.find(pat)` where `s` is a `string` and `pat` is a `string` or a `char*`. It returns the location of the first occurrence of `pat` in `s`. The following code prints the number 8 on the screen.

```
string s = "abcdefghijklmnopqrstuvwxyz";
cout << s.find("ijk") << endl;
```

What happens if `find` cannot find `pat`? The answer is complicated. To begin, we need to explain that `find` returns a value of type `std::string::size_type`. (If you include the statement `using namespace std;` then you may drop the prefix `std::`.) In most cases, we save the value returned by `find` in a variable for further processing. To do this, we use code such as this:

```
string s = "I feel like a louse";
string pat = "eel";
string::size_type idx;    // or std::string::size_type idx;
idx = s.find(pat);
```

In this case, `idx` is set equal to 3.

If, however, `pat` is set to `"house"`, then `find` returns a special value named `std::string::npos`. (The prefix `std::` may be omitted if we have the statement `using namespace std;`.) Here is a short program that illustrates how to use `find`.

```
#include <iostream>
using namespace std;

int main() {
  string s = "Mississippi";
  string pat;
  cout << "Enter substring --> ";
  cin >> pat;
```

```
    string::size_type idx;      // or std::string::size_type idx;
    idx = s.find(pat);

    if (idx != string::npos) {
      cout << "The substring \"" << pat
           << "\" was found at position " << idx << endl;
    }
    else {
      cout << "The substring \"" << pat << "\" was not found"
           << endl;
    }
    return 0;
}
```

Here are two runs of the program.

```
Enter substring --> ssi
The substring "ssi" was found at position 2
```

```
Enter substring --> sse
The substring "sse" was not found
```

Closely related to `find` is the `rfind` method. The statement `s.rfind(pat)` returns the index of the last occurrence of `pat` in `s`, or `string::npos` if `pat` cannot be found.

Two additional `string` searching methods are provided: `find_first_of` and `find_last_of`. The expression `s.find_first_of(pat)` searches the string `s` for a character that is found in `pat` and returns its index. If none of the characters in `pat` is present in `s`, then `string::npos` is returned. Here is a program to illustrate how this works.

```
#include <iostream>
using namespace std;

int main() {
  string s = "Mississippi";
  string pat;
  cout << "Enter substring --> ";
  cin >> pat;

  string::size_type idx;      // or std::string::size_type idx;
  idx = s.find_first_of(pat);

  if (idx != string::npos) {
    cout << "One of the characters \"" << pat
         << "\" was found at position " << idx << endl;
  }
  else {
    cout << "None of the characters\"" << pat << "\" was found"
         << endl;
  }
  return 0;
}
```

Here are two executions of this code.

```
Enter substring --> aeiouy
One of the characters "aeiouy" was found at position 1
```

```
Enter substring --> wxyz
None of the characters"wxyz" was found
```

The expression `s.find_last_of(pat)` method gives the index of the last character in `s` that is also in `pat`, or `string::npos` if no such character exists.

14.2.4 Converting between `string` and `char*` types

Conversion from a null-terminated character array (type `char*`) to a `string` is easy. If `word` is a `char*` (character array) and `s` is a string, the assignment `s = word;` does the job. Alternatively, we can convert `word` to a `string` when `s` is declared:

```
string s(word);
```

Finally, we can write `string(word)` to convert `word` into a `string`.

The conversion from a `string` to a `char*` is more complicated. The `string` class includes a method called `c_str` for this purpose. The expression `s.c_str()` returns a pointer to an unmodifiable, null-terminated character array with the same contents as `s`.

```
string s = "Leonhard Euler";
const char* word = s.c_str();
cout << word << endl;
```

Notice that `word` is declared `const`; omitting this keyword results in an error. If you need to do further processing on the character array returned by `c_str`, you need to copy the characters into another `char*` array and work on that copy.

Fortunately, one rarely needs to convert a `string` to a `char*`. The exception is when we wish to use a procedure that takes a `char*` argument, but no `string` alternative is available. (For an example, see Exercise 14.1.)

14.3   Command line arguments

In all the programs we have presented thus far, data are entered into the program using a prompt/response paradigm:

```
cout << "Enter n --> ";
cin >> n;
```

An alternative mechanism for sending a few values to a program is to use *command line arguments*. For example, a greatest common divisor program, named `gcd`, would be invoked from the terminal by typing gcd 289 51. The arguments are sent to main as character arrays, "289" and "51". The main procedure then needs to convert these to integers, send those values to a gcd procedure, and print the result. Here is how all of this is accomplished.

The first step is to declare main in a different manner. Thus far in this book, main has been always declared as int main(). In this version, no arguments are sent to main and an integer value is to be returned. The alternative declaration for main specifies arguments:

```
int main(int argc, char** argv) { ... }
```

The first argument is an int value that specifies the number of arguments typed on the command line. The name of the program itself is considered an argument, so this number is always at least one. The name of this argument is not required to be argc (for "argument count") but this convention is nearly universal, so you are encouraged to follow suit.

The second argument, named argv, is an array of character arrays (hence the double star). This need not be named argv, but this name is also nearly universally used for this purpose.

When main is invoked, this array is populated as follows: the character array in argv[0] is the name of the program. The arrays argv[1] through argv[argc-1] are the other arguments on the command line. An example makes this clear.

Program 14.2: A program that illustrates how to access command line arguments in a main.

```
1  #include <iostream>
2  using namespace std;
3
4  int main(int argc, char** argv) {
5    for (int k=0; k<argc; k++) {
6      cout << "argv[" << k << "] is " << argv[k] << endl;
7    }
8    return 0;
9  }
```

Suppose this program is compiled and the executable is called test-main. If the program is run with the command line

```
./test-main one two 3 negative-four
```

the following output results.

```
argv[0] is ./test-main
argv[1] is one
argv[2] is two
argv[3] is 3
argv[4] is negative-four
```

Note that in this case, argc equals 5 accounting for the name of the program and the four additional arguments passed to the program.

The command line arguments are sent to `main` as character arrays. Often, we want to convert these values to integer or double values. Unfortunately, the following does not work.

```
int main(int argc, char** argv) {
  int n1;
  n1 = argv[1]; // INCORRECT
  cout << "n1 = " << n1 << endl;
  return 0;
}
```

There are two problems—one minor and one serious—with the line flagged with the comment INCORRECT. The minor issue is that we did not check if `argc` is at least 2; if `argc` is only 1, then `argv[1]` is not a valid element of the `argv` array. The serious error is that the statement `n1 = argv[1];` does not convert the character array into an integer. Even if `argv[1]` holds a valid representation of a decimal integer, say `"89"`, the statement does not convert the character array into the expected integer value, 89. Unfortunately, on some compilers, this might not be an error.[2]

To convert a character array to the numerical value it represents, use one of the following procedures (these are built in to C++).

- `atoi(word)` converts the character array in `word` to an `int` value. Thus, if `word` holds `"-51"`, then `atoi(word)` returns the value -51.

- `atol(word)` converts the character array in `word` to a `long` integer value.

- `atof(word)` converts the character array in `word` to a `float` value.

- `atod(word)` converts the character array in `word` to a `double` value.

Here is a sample program to illustrate their use.

Program 14.3: A program to calculate the gcd of two values specified on the command line.

```
1   #include <iostream>
2   #include "gcd.h"
3
4   using namespace std;
5
6   int main(int argc, char** argv) {
7     if (argc != 3) {
8       cerr << "Usage: " << argv[0] << " n1 n2" << endl;
9       cerr << "to find the gcd of n1 and n2" << endl;
10      return 1;
11    }
12
13    long n1 = atol(argv[1]);
```

---

[2]This program only generates a warning on my compiler. On my computer, the output of this program is n1 = -1073743042 because the *address* of argv[1], when converted to a signed integer, is -1073743042.

```
14    long n2 = atol(argv[2]);
15    cout << gcd(n1,n2) << endl;
16
17    return 0;
18 }
```

Notice that lines 7–11 check that the appropriate number of arguments are given to the program; if not, the program prints an error message and a reminder of how it should be used. Here is a sample session using this program.

```
$ ./gcd 51 289
17

$ ./gcd 5
Usage: ./gcd n1 n2
to find the gcd of n1 and n2

$ ./gcd hello Gauss
0

$
```

Notes: The dollar sign is the computer's shell prompt (not something the user types and not considered a command line argument). The first invocation of the program is properly formatted and the result is typed on the screen. The second invocation has an incorrect number of command line arguments; the program detects this and prints the error message. In the final invocation of the program the arguments ought to be numbers, but instead we send nonsense (`hello` and `Gauss`). We request that these be converted to `long` values (lines 13–14). However, `atol`, unable to recognize these character arrays as representations of numbers, returns the value 0. A more sophisticated program could examine the contents of `argv[1]` and `argv[2]` to see if they held properly formatted numbers; if not, an error message would be generated.

14.4   Reading and writing data in files

14.4.1   Opening files for input/output

The input/output objects `cin`, `cout`, and `cerr` are designed for transferring data from the computer's keyboard or to the computer's screen.[3] Often, however, we want to read data from a file (i.e., a document) or to write the results of our computation into a file (for later processing, or inclusion in a report or email message).

---

[3]On most computers it is possible to redirect these input/output streams so that data, that would normally be written to the screen, are sent to a file (or another program) instead.

Fortunately, it is not difficult to declare *input* and *output streams*. These are objects like `cin` and `cout`, but rather than being associated with the keyboard or the screen, they are associated with a file on the computer's hard drive.

Computer files are of two sorts: *plain text* (or ASCII) and *binary*. Plain text files contain only the ordinary characters (of the sort that can be held in a `char` variable); that is, lower- and uppercase Latin letters, numerals, punctuation, and blank space. They do not contain letters from other character sets (e.g., Chinese) or other types of data. Examples of plain text files are `.cc` and `.h` files for programming, TₑX and LᴬTₑX files, PostScript documents, and `.html` Web pages. Binary files, on the other hand, contain characters beyond the ASCII set or other types of data (including images, sounds, etc.). Examples of binary files include Microsoft Word documents, multimedia files (from `.jpg` photographs to `.wmv` video), PDF documents, and executable programs (built from your C++ code).

We focus our attention solely on reading and writing plain text files. Although C++ is capable of dealing with binary files, it is more complicated to handle such data. For special situations, you may be able to find publicly available C++ procedures for reading and writing specific types of data (e.g., `.jpg` files).

To read and write from files, include the directive `#include <fstream>` at the beginning of your program. The `fstream` header defines two important classes: `ifstream` for *input file stream* and `ofstream` for *output file stream*.

The first step is to declare an object of type `ifstream` or `ofstream`. In both cases, we provide a single argument giving the name of the file to be read/written; the argument is a character array (type `char*`). The constructors look like this:

```
ifstream my_in("input_file");
ofstream my_out("output_file");
```

The first sets up `my_in` to read data from a file named `input_file` and the second writes data to a file named `output_file`. Before we do either of these, there are a few important cautionary notes.

- The file `input_file` might not exist or might not be readable by your program (e.g., if you do not have sufficient privileges to read that file). So, before attempting to read data from that file, we perform the following simple test.

  ```
  if (my_in.fail()) {
    cerr << "Unable to read the file input_file" << endl;
    return 1;
  }
  ```

  Input streams contain methods named `fail` and `good`. If (and only if) the stream is in a good state, then `good()` returns `true` and `fail()` returns `false`. Thus, if the file cannot be opened, `my_in.fail()` returns `true`.

  It is important to do a test such as this or else the rest of your program may run into trouble.

  A program also uses the `good` and `fail` methods to detect when an input file has been exhausted; see Section 14.4.3.

- Likewise, it might not be possible for the program to write to `output_file` (the disk might be locked, another program might be using the file, or your program may lack sufficient privileges to write a file in the particular directory). To test if the output file was opened successfully, use code such as this:

```
if (my_out.fail()) {
  cerr << "Unable to write to the file output_file" << endl;
  return 1;
}
```

- For output, please be aware that **opening an existing file for output completely erases the file**. There's no second chance. The file is *not* moved to the "trash" or recoverable in any way.

  These is an alternative way to open an output file that does not overwrite the existing file. We may open an output file so that data written to that file are appended to the end of the file. If this is what is desired, use the following constructor.

```
ofstream my_out("output_file", ios::app);
```

A file stream may be associated with a file after it is declared using the `open` method. Here is an example.

```
ifstream my_in;
ofstream my_out;
// intervening code
my_in.open("input_file");
my_out.open("output_file");
```

Alternatively, to append data to an output file, use this statement:

```
my_out.open("output_file", ios::app);
```

Generally, it is not necessary to close a file—the file associated with a stream is automatically closed when the stream goes out of scope. However, there are times when we need to close a file explicitly. In that case, we use the `close()` method.

One instance when we would use the explicit `open` and `close` methods is when the command line arguments name files that we want to process. Consider the following example.

Program 14.4: A program the processes files specified on the command line.

```
1  #include <iostream>
2  #include <fstream>
3  using namespace std;
4
5  int main(int argc, char** argv) {
6    ifstream in;
7
8    for(int k=1; k<argc; k++) {
9      in.open(argv[k]);
```

```
10      if (in.fail()) {
11        cerr << "*** Unable to process file " << argv[k] << endl;
12      }
13      else {
14        cerr << "Working on file " << argv[k] << endl;
15        // do whatever we need to do with the file named in argv[k]
16        // in >> variables; etc; etc;
17      }
18      in.close();
19      in.clear();
20    }
21    return 0;
22 }
```

The main `for` loop is bracketed by calls to `in.open` and `in.close` (see lines 9 and 18). Each command line argument is supposed to name a file. We try to open the file for input; if this is not successful (line 10) we print an error message and move on. Otherwise, we process the file in whatever way would be appropriate, presumably until we reach the end of that file.[4] We then close the file (line 18).

Before we step to the next file we invoke the `ifstream`'s `clear()` method. This resets any error conditions triggered by the `ifstream`, and there are two likely error conditions that would arise in this program: inability to open the file and reaching the end of the file. After one of these events occurs, the expression `in.good()` yields the value `false` (and `in.fail()` yields `true`) until we cancel the error condition with `in.clear()`.

14.4.2 Reading and writing

Once the stream object is declared and we have tested that the file has been successfully opened, we can use the usual `<<` (for `ofstream`) and `>>` (for `ifstream`) operators for writing/reading the file. Here is an example.

Program 14.5: A program that illustrates writing data to a file.

```
1  #include <iostream>
2  #include <fstream>
3  using namespace std;
4
5  int main() {
6    const char* output_file_name = "example.out";
7
8    ofstream my_out(output_file_name);
9    if (my_out.fail()) {
10      cerr << "Unable to open the file " << output_file_name
11           << " for writing." << endl;
12      return 1;
13    }
14
```

---

[4]End of file detection is explained in Subsection 14.4.3.

```
15    for (int k=1; k<=10; k++) {
16       my_out << k << " ";
17    }
18    my_out << endl;
19
20    return 0;
21 }
```

After this program is compiled and run, a file named `example.out` is created and its contents look like this:

```
1 2 3 4 5 6 7 8 9 10
```

14.4.3 Detecting the end of an input file

When reading data from a file, a program might not know a priori how many data are in the file. For example, the file may contain many integers and it's the program's job to sum those integers. To do this, the program repeatedly requests input (using the >> operator) until it reaches the end of the file.

The question is, how does a program tell when it has reached the end of an input file? The solution is to use the `ifstream`'s `good` and `fail` methods.

When a file has been exhausted, the expression `ifstream.fail()` yields the value `true`. The following program illustrates how to use this idea; it sums the numbers it finds in the file `example.out` generated by Program 14.5.

Program 14.6: A program that sums the integer values it finds in a file.

```
1  #include <iostream>
2  #include <fstream>
3  using namespace std;
4
5  int main() {
6    const char* input_file_name = "example.out";
7
8    ifstream my_in(input_file_name);
9    if (my_in.fail()) {
10      cerr << "Unable to open the file " << input_file_name
11          << " for input." << endl;
12      return 1;
13    }
14    int n;
15    int sum = 0;
16    while (true) {
17      my_in >> n;
18      if (my_in.fail()) break;
19      sum += n;
20    }
21    cout << "The sum of the numbers is " << sum << endl;
22
23    return 0;
24 }
```

Focus your attention on lines 16–20. The loop is controlled by the construction `while (true) {...}`. The loop runs forever until the `break` on line 18 is reached. When `my_in.fail()` is evaluated one of two things happens: either (a) the program has successfully read an integer into `n` or else (b) there are no more values left in the file to be found (because we have reached the end of the file). In case (a), `my_in.fail()` evaluates to `false`. However, in case (b), it evaluates to `true`. Therefore, the loop continues as long as the input is successful. Once the end of the input file is reached, the loop terminates and the sum of the values in the file is written to the computer screen. The output of this program looks like this:

```
The sum of the numbers is 55
```

14.4.4 Other methods for input

The `>>` operator handles most input needs. When handling character data, however, it is sometimes useful to be able to deal with single characters and with full lines of text.

The `get` method is used to read a single character from an input stream. Here's an example.

```
char ch;
cin.get(ch);
if (cin.good()) {
   cout << "We read the character " << ch << endl;
}
else {
   cout << "No more input available" << endl;
}
```

The statement `cin.get(ch)` reads a single character from the stream `cin` and stores the result in the variable `ch`.

The expression `cin.get(ch)` is not equivalent to `cin >> ch`. The former reads the next character available no matter what, but the `>>` statement skips any white space before reading a character. Consider this program.

```
#include <iostream>
using namespace std;

int main() {

   char ch;
   cout << "Type something -->";
   cin >> ch;
   cout << "We read the character '" << ch << "'" << endl;
   cin.get(ch);
   cout << "We read the character '" << ch << "'" << endl;

   return 0;
}
```

Here are some sample executions of the code.

```
Type something -->123
We read the character '1'
We read the character '2'
```

```
Type something -->    123
We read the character '1'
We read the character '2'
```

```
Type something --> 1  2  3
We read the character '1'
We read the character ' '
```

In the first execution, the `cin >> ch;` statement reads the character 1 and then the `cin.get(ch);` reads the character 2. The same thing happens in the second execution because `cin >> ch;` skips the white spaces before the 1. However, in the third run, the statement `cin.get(ch);` reads the space character immediately following the 1.

There is also a `put` method for output streams; it is used to write a single character. If `ch` is a `char` variable, the statement `cout.put(ch);` is tantamount to `cout << ch;`.

Suppose `word` is a `string` variable. The statement `cin >> word;` skips white space before reading data into the variable `word`, and then stops as soon as additional white space is encountered. For example, the user types

```
    Suppose $f$ is    continuous.
```

then the statement `cin >> word;` puts `Suppose` into the variable `word`. If the statement is executed repeatedly, it would subsequently save the string `$f$`, then `is`, and then `continuous.` into `word`.

Sometimes it is useful to read a full line of text into a `string` variable. There are two ways to do this. In both cases the procedure invoked is named `getline`.

We may write `cin.getline(buffer, nchars);` where `buffer` is a character array (type `char*`) and `nchars` is a positive integer. This statement reads at most `nchars` characters from `cin` and loads them into the character array `buffer`. The reading stops either when the end of the line is encountered or `nchars` have been read. Here is how this method might be used in a program.

```
const int MAX_LINE = 10000;
char buffer[MAX_LINE];
cout << "Type a line: ";
cin.getline(buffer,MAX_LINE);
cout << "You typed: " << buffer << endl;
```

The drawbacks to this form of `getline` are (a) one needs to know a priori an upper bound on the number of characters in a line and (b) the characters are saved in a character array.

The following alternative version of `getline` is more convenient. The statement `getline(cin,theLine);` (where `theLine` is a `string` variable) reads characters from `cin` until reaching the end of the line; the characters are saved in `theLine`.

Both forms of `getline` take an optional additional argument: a `char` value specifying a delimiter character. Then, instead of reading to the end of the line, `getline` reads until the delimiter character is encountered. For example, the statement `getline(cin,theLine,'/');` reads characters into `theLine` until a / character is encountered.

14.5 String streams

C++ provide a means to treat character strings in a manner akin to file streams. Objects of the classes `istringstream` and `ostringstream` may be used in the same manner as input and output file streams, but their data come from (or go to) a `string` embedded in the object. The use of these classes requires a `#include <sstream>` directive.

An `istringstream` is initialized with a `string` or a `char*` array. For example,

```
string line("angle 70.3 degrees");
istringstream is(line);
```

We can now use `is` just as we would any other input stream object. The subsequent code

```
string s1;
is >> s1;
double x;
is >> x;
string s2;
is >> s2;
```

places the string `angle` into `s1`, the value 70.3 into `x`, and the string `degrees` into `s2`.

It's important that the variable receiving data from an `istringstream` via the `>>` operator be of the appropriate type. No error is reported in this situation, but the contents of the variable are unpredictable.

An `ostringstream` behaves in the same manner as an output stream, but the data it is sent are saved into a `string`, not a file. We declare an `ostringstream` variable without any arguments like this:

```
ostringstream os;
```

and then we send data using the usual `<<` operator: `os << k;`. After we have finished putting data into `os` we extract the string we built using the `str()` method. The following code illustrates these ideas.

Program 14.7: A program to illustrate the use of string streams.

```
1   #include <iostream>
2   #include <sstream>
3   using namespace std;
4
5   int main() {
6       ostringstream os;
7
8       os << "The base-ten digits are";
9       for (int k=0; k<=9; k++) os << " " << k;
10      os << endl;
11      cout << os.str();
12
13      string words(" 3.9    10    hello     good bye");
14      istringstream is(words);
15      double x;
16      is >> x; // 3.9
17      int n;
18      is >> n; // 10
19      string s;
20      is >> s; // hello
21      long k;
22      is >> k; // good
23      cout << "x = " << x << ", n = " << n << ", s = " << s
24              << ", and k = " << k << endl;
25
26      return 0;
27  }
```

Here is the output from the program.

```
The base-ten digits are 0 1 2 3 4 5 6 7 8 9
x = 3.9, n = 10, s = hello, and k = 0
```

14.6 Formatting

The statement `cout << x;` prints the value held in x on the computer's screen. If x is a `double` variable, the output may look like one of these.

Value	Comment
0.142857	decimal value of $1/7$
1.42857	$10/7$
0.00142857	$1/700$
-0.142857	$-1/7$
2.5e+11	2.5×10^{11}
1	one

Notice that the number of decimal places displayed varies, but in each case at most six significant digits are given (not counting leading zeros). Also observe that positive numbers do not have a leading + sign, and that the decimal point is dropped when x holds an integer value.

In most cases, the default behavior of `cout << x;` is adequate for mathematical purposes. However, we may wish to print more than six significant figures or print out a table with figures nicely arranged in columns. To accomplish these, we need to modify the default behavior of the output stream.

A convenient way to do this is through the use of *manipulators* that we send to output streams using the << operator. Before we may use manipulators, we need the directive `#include <iomanip>` at the beginning of the program.

14.6.1 Setting precision

By default, real numbers sent to an output stream (such as `cout`) are printed with six decimal digits of precision. If we want more (or fewer) we use the `setprecision` manipulator. Its use looks like this:

```
#include <iomanip>
...
cout << exp(1.) << endl;
cout << setprecision(10);
cout << exp(1.) << endl;
```

The output of this code looks like this:

```
2.71828
2.718281828
```

14.6.2 Showing all digits

Increasing the precision of the output stream does not necessarily result in additional digits being printed. For example, the statement

```
cout << setprecision(10) << 1.0 << endl;
```

just prints 1 on the screen. The manipulator `showpoint` coerces the printing of the decimal point and the extra digits. The code

```
cout << setprecision(10) << 1.0 << endl;
cout << showpoint << 1.0 << endl;
```

results in the following output.

```
1
1.000000000
```

To restore the default behavior, send the `noshowpoint` object to the output stream.

14.6.3 Setting the width

Setting the precision of the output does not directly determine the number of characters typed to the screen. A leading minus sign, the position of the first nonzero digit, whether the number is to appear in scientific notation, and whether `showpoint` is in effect all influence the number of characters that `cout << x` prints. This variability can wreak havoc with any attempt to line up the output in neat columns.

By default, `cout << x;` prints x in exactly as much space as required; no extra space is padded.

The `setw` manipulator provides a mechanism that guarantees the number of characters `cout << x;` prints. The statement

```
cout << setw(20) << x;
```

prints the value stored in x in a field that is 20 characters wide. If `cout << x` would normally produce fewer than 20 characters, then, by default, the value is printed right justified in the 20-character region. The code

```
cout << '|' << setw(20) << exp(1.) << '|' << endl;
```

results in this output.

```
|             2.71828|
```

It is possible for the printed value to appear left or right justified within the amount of space specified by `setw`. The manipulators controlling this are named `left` and `right`.

```
cout << '|' << setw(20) << left  << exp(1.) << '|' << endl;
cout << '|' << setw(20) << right << exp(1.) << '|' << endl;
```

```
|2.71828             |
|             2.71828|
```

Unlike the `setprecision` and `showpoint` manipulators, the effect of `setw` does not persist between outputs. Once output has been sent, the effect of `setw` is immediately canceled and the default behavior is restored. The code

```
cout << '|' << setw(20) << exp(1.) << '|' << M_PI << '|' << endl;
```

gives the following result.

```
|             2.71828|3.14159|
```

When using `setw`, the extra characters typed are spaces. However, this can be changed with the `setfill` manipulator. Here is an example.

```
string hi("Hello Gauss");
cout << setfill('-');
cout << setw(20) << hi << endl;
cout << setw(20) << left << hi << endl;
```

```
---------Hello Gauss
Hello Gauss---------
```

14.6.4 Other manipulators

There are several additional manipulators available in C++; here we mention some of them that you might find useful.

- `cout << showpos`: This causes nonnegative numbers to be prepended with a + sign. To restore the default behavior, use `noshowpos`.

- `cout << scientific`: This forces real numbers (types `float` and `double`) to appear in scientific notation. To restore default behavior (real values are either printed in decimal or scientific notation depending on their value) use the statement `cout << setiosflags(ios::floatfield);`.

  The mantissa of the real value is separated from the power of 10 by the letter e. The case of the letter e can be modified using the manipulators `uppercase` and `nouppercase`.

- `cout << fixed`: This forces real numbers to be printed in decimal notation (and not scientific). For example:

```
cout << exp(20.0) << endl;
cout << fixed;
cout << exp(20.0) << endl;
```

```
4.85165e+08
485165195.409790
```

  Restore default behavior with `cout<<setiosflags(ios::floatfield);`.

- `cout << boolalpha`: By default, `bool` values are printed as 0 or 1. After applying the `boolalpha` manipulator, these are printed as `false` and `true`. Restore the default behavior with `cout << noboolalpha;`.

- `cout << dec`, `cout << oct`, and `cout << hex`: Integer values are normally printed in base ten, but may also be printed in base eight (octal) or sixteen (hexadecimal). Use these manipulators to select the desired base.

14.7 A class to parse files

We close this chapter by presenting a class for parsing files. In applied work, mathematicians often are given files containing data. It can be an annoying chore simply to read the file into a program. For example, a file might contain geometric information about a large molecule. The input file specifies the molecule with various kinds of lines:

- Type and location of atoms: These lines have the following format:

  ```
  ATOM atom_number symbol x-coord y-coord z-coord
  ```

- Chemical bonds: These lines have the format:

  ```
  BOND atom_number atom_number
  ```

- Comments: These lines begin with a # and are followed by arbitrary text.

- Blank lines.

Such an input file might look like this:

```
# Data acquired from the ACME Molecule Machine and
# saved in directory /shared/molecules/specimen-4.
ATOM 1 C    0 0 0
ATOM 2 H    1 0 0
ATOM 3 Br   0 -1 0
ATOM 4 C    0 0 1.2

#Here are the bonds

BOND 1 2
BOND 1 4
BOND 1 4
     # Note double bond between the carbons
BOND 1 3
```

The `LineParser` class we present in this section is a device that reads a file (such as the molecule description file above) one line at a time, and then breaks the line into individual words. The class provides the following methods.

- The constructor `LineParser(file_name)`: This creates a new `LineParser` object that reads data from the file named in the `char*` array `file_name`.

- A method `read()` that reads a line from the input file and breaks it into individual words. This method returns `true` if it is able to read a line. Let's call the last line processed by `read()` the *current line*.

- A method `show_line()` that returns the current line.

- A method `show_line_number()` that gives the line number of the current line.

- A method `num_words()` that reports the number of separate words found on the current line.

- A square brackets operator to get the words on the line. If `LP` is a `LineParser` object, `LP[k]` returns the *k*th word of the current line.

Here are the header and program files for the `LineParser` class followed by a `main` program that illustrates the use of this class.

Program 14.8: Header file for the `LineParser` class.

```cpp
1   #ifndef LINE_PARSER_H
2   #define LINE_PARSER_H
3   #include <fstream>
4   #include <vector>
5   #include <string>
6   using namespace std;
7
8   class LineParser {
9   private:
10    ifstream       in;
11    string         theLine;
12    long           lineNumber;
13    int            nWords;
14    vector<string> words;
15    void           parse();
16
17  public:
18    LineParser(const char* file_name);
19    bool read();
20    string show_line() const { return theLine; }
21    long   show_line_number() const { return lineNumber; }
22    int    num_words() const { return nWords; }
23    string operator[](int k) const { return words[k]; }
24  };
25
26  #endif
```

Program 14.9: Program file for the `LineParser` class.

```cpp
1   #include "LineParser.h"
2   #include <sstream>
3   #include <iostream>
4   using namespace std;
5
6   LineParser::LineParser(const char* file_name) {
7     in.open(file_name);
8     if (!in) {
9       cerr << "WARNING: Unable to open " << file_name << endl;
10    }
11    lineNumber = 0;
12    theLine = "";
13  }
14
15  bool LineParser::read() {
16    getline(in,theLine);
17    bool result = in.good(); // check if getline succeeded
18    if (result) {
19      lineNumber++;
20      parse();
21    }
22    return result;
23  }
```

```
24
25  void LineParser::parse() {
26    words.clear();
27    words.resize(100);
28    istringstream s(theLine);
29    nWords = 0;
30    string tmp;
31    while (s>>tmp) {
32      if (words.size() < unsigned(nWords)) words.resize(nWords+10);
33      words[nWords] = tmp;
34      nWords++;
35    }
36  }
```

Program 14.10: A program to demonstrate the use of the `LineParser` class.

```
1   #include "LineParser.h"
2   #include <iostream>
3   using namespace std;
4
5   int main(int argc, char** argv) {
6
7     if (argc != 2) {
8       cerr << "Usage: " << argv[0] << " filename" << endl;
9       return 1;
10    }
11
12    LineParser LP(argv[1]);
13
14    while (LP.read()) {
15      cout << "Line " << LP.show_line_number() << "\t\t\""
16           << LP.show_line() << "\"" << endl << endl;
17      for (int k=0; k<LP.num_words(); k++) {
18        cout << "Word " << k << " is \t\"" << LP[k] << "\"" << endl;
19      }
20      cout << "-------------------------------------------------"
21           << endl;
22    }
23
24    return 0;
25  }
```

When the `main` program is run on the following four-line file

```
What's it all about,          Alfie?
Catch 22

Nothing on the previous line!
```

we obtain the following output:

```
Line 1            "What's it all about,          Alfie?"

Word 0 is         "What's"
Word 1 is         "it"
```

```
Word 2 is        "all"
Word 3 is        "about,"
Word 4 is        "Alfie?"
------------------------------------------------
Line 2           "Catch 22        "

Word 0 is        "Catch"
Word 1 is        "22"
------------------------------------------------
Line 3           ""


------------------------------------------------
Line 4           "Nothing on the previous line!"

Word 0 is        "Nothing"
Word 1 is        "on"
Word 2 is        "the"
Word 3 is        "previous"
Word 4 is        "line!"
------------------------------------------------
```

14.8 Visualization

The visualization of mathematical objects often provides important insights. Computers are particularly adept at producing precise beautiful images.

Unfortunately, drawing pictures on the computer's screen is often closely tied to the computer's operating system; such a program created for the Windows operating system is unlikely to work on an X-windows (UNIX) or Macintosh computer. In addition, it can be frustrating and time consuming to understand the intricacies of opening windows, drawing graphic objects, translating between mathematical coordinates and screen coordinates, printing, and so forth.

Still, visualization is too important to dismiss and so we offer guidance on how to use C++ to draw pictures in a platform-independent manner.

One strategy, which we consider only briefly, is to write a C++ program whose textual output is used as input to another system with built-in graphics capabilities. For example, the C++ program could write a file in which each line contains a pair of real values. This file would then be loaded into a system such as MATLAB or *Mathematica* and plotted.

Alternatively—and this is the technique we explore—we can write a C++ program whose output is a well-established graphics file such as GIF, JPEG, or EPS. This file can then either be displayed on the computer's screen (using an appropriate viewing program) or incorporated into a word-processing document (such as MS Word or LaTeX).

14.8.1 Introducing and installing the `plotutils` package

Let's start with an assumption. We do not want to learn how graphics files (such as GIF, JPG, or EPS) work; we just want to think about drawing in mathematical terms. Line segments should be specified by their end points and circles by their centers and radii. Someone else should work out how these various graphics formats work and provide us with an easy way to draw. Fortunately, someone else has done exactly that.

The GNU `plotutils` package is free software that provides device-independent drawing tools with output to a variety of popular graphics file formats. The package can be downloaded from the following Web site,

```
http://www.gnu.org/software/plotutils/
```

The package arrives as a single compressed "tar" file[5] in which all the `plotutils` files are held. Installation is mildly complicated. Installation instructions can be found in the files `INSTALL` and `INSTALL.pkg`, but these can scare the uninitiated. Follow this outline.

- *Unpack.* It is necessary to extract the various source files for the `plotutils` files from the downloaded file. On some computers, this may be as simple as double-clicking the downloaded file. On a UNIX system (or in Cygwin on Windows), you can unpack with the following command,

  ```
  tar xfz plotutils-2.4.1.tar.gz
  ```

  (The precise file name may depend on the version you downloaded; be sure to fetch the latest.)

  You should now have a folder on your hard drive named `plotutils-2.4.1` (or something similar).

- *Configure.* The next step is to run the `configure` program found in the folder you have just unpacked. This is done from a command line such as this:

  ```
  ./configure --enable-libplotter --prefix=directory
  ```

  Here, `directory` is a directory where you want the package to be installed. This may be your home directory, a subdirectory of your home directory, or a public directory such as `/usr/local` or `/usr/public`. Please note that the installation procedure (described below) adds files to various subdirectories of these directories; if those subdirectories do not already exist, the install process creates them for you. These subdirectories are named `bin`, `lib`, `include`, and so on.

  The option `--enable-libpotter` enables the C++ version of `plotutils`. (Without this option, only the C version is built.)

---

[5]The word *tar* is an acronym for *tape archive*; this file format provides a mechanism for packaging many files together and need not be associated with storing data on a tape.

- *Build.* The package is built using a single command: `make`. Even on a fast computer, this process takes a while.

- *Create the documentation.* The `plotutils` package comes with a reference manual (and other documentation). To build this, type `make dvi`. This creates the file `plotutils.pdf` in the `info` subdirectory. This can be read (and printed) using Adobe Acrobat Reader or another PDF viewer.

- *Install.* Thus far, all the work of configuring and building the package takes place in the directory (folder) where you unpacked `plotutils`. The final step is to transfer the pieces to the directory you specified in the configuration step. This is quick and easy; type the command `make install`.

  This copies header files (such as `libplotter.h`) to the `include` subdirectory of the directory you specified in the configuration process, library files to `lib`, executable programs to `bin`, and so on.

If you have a Linux computer, it is possible that `plotutils` is already installed. If not, a precompiled version can be found by visiting this Web site,

`http://rpmfind.net/linux/RPM/index.html`

and searching for `plotutils`. The package is then installed using Linux's `rpm` command (for which you must have superuser privileges).

Installing `plotutils` is nontrivial, but not insurmountable. If possible, seek a friendly computer expert to assist you through this process. Rest assured that installing `plotutils` is the most difficult step in using it to draw pictures.

14.8.2 Drawing with `plotutils`—a first example

We begin by creating a program (see Program 14.11) to draw the symbol $\otimes$. It is easy to describe this symbol mathematically. It consists of two line segments and a circle. To be specific, the first line segment joins the points $(-1,-1)$ to $(1,1)$ and the second line segment joins $(-1,1)$ with $(1,-1)$. The circle is centered at the origin with radius $\sqrt{2}$. Once we set up `plotutils`, drawing the picture is just that easy.

In broad strokes, the steps to produce the image are these.

- Select the appropriate subclass of `Plotter`,

- Declare the `Plotter` object with the appropriate arguments,

- "Open" the `Plotter` and perform other initialization including establishing a coordinate system and pen attributes,

- Draw the image, and

- Close the `Plotter`.

Finally, we need to compile and run the program.

Choosing the appropriate plotter type

The `plotutils` package is capable of producing various types of graphics files including `gif`, `eps`, and several others. Fortunately, the methods for creating images for these various formats are all identical. If we write code to produce one type of file and subsequently decide we prefer another type, it is easy to change the code to implement the new choice.

Each drawing is produced by an object that is a subclass of the type `Plotter`. To create a `gif` image, we declare an object of type `GIFPlotter`; for an `eps` image, we declare a `PSPlotter` object. For other graphics formats, see the file `plotutils.pdf` included in the `info` subdirectory.

At this point you may be wondering, which type of file should I choose? Here's some guidance. Graphic image files can be divided into two broad categories: bit images and vector graphics. Bit image files are (essentially) a matrix of pixels; they are useful for representing photographs and are manipulated using "paint" style programs. On the other hand, vector graphic files represent their drawings using mathematical primitives such as line segments, ellipses, Bezier curves, and so forth. They are edited using "draw" style programs. For our first project, drawing $\otimes$, we use `PSPlotter` to produce an `eps` file. If were interested in drawing a picture of a Julia set (by plotting individual points), then we would use a `GIFPlotter`.

These various constructors are defined in the header file `plotter.h`, so the directive `#include "plotter.h"` is needed.

Arguments to the constructor

Regardless of which subtype of the class `Plotter` we choose, the declaration of the `Plotter` object is the same. The constructor requires four arguments like this: *PlotterType(input, output, error, parameters)* where

- `input` is an object of type `ifstream` such as `cin`. At present, this input stream is not used, so the simplest choice is to set this equal to `cin`.

- `output` is an object of type `ofstream` such as `cout`. When the `Plotter` object writes its results, it sends it to this argument. Therefore, it is usually not a good idea to set this equal to `cout`. Rather, it is better to create a separate object of type `ofstream` to write to a file on disk. Thus, we need the directive `#include <fstream>` and to declare an output object like this:

  ```
  ofstream pout("filename");
  ```

  and use `pout` for the second argument to the constructor.

- `error` is also an object of type `fstream` to which error messages are written. If something goes awry during the drawing process, the `Plotter` object reports the problem through this output stream. The simplest choice is to set this to `cerr`, the standard error output stream.

- `params` is an object of type `PlotterParams`. The `plotutils` documentation explains various options that can be embedded in this object and then passed to the `Plotter`. However, the simplest thing to do is to specify no special options. Just declare an object of type `PlotParams` and pass that as the fourth argument to the constructor.

Thus, a `Plotter` object is created with code that looks like this:

```
ofstream pout("picture.eps");
PlotterParams params;
PSPlotter P(cin, pout, cerr, params);
```

With this in place, the plotter object `P` is created. See Program 14.11 lines 11–13.

Other initialization

After a `Plotter` object is constructed, there are a few additional steps to take before we can begin drawing. The first of these is mandatory: the `Plotter` needs to be "opened." This is analogous to opening a file for writing. To open a `Plotter` we use its `openpl()` method. This method returns a negative value if anything goes amiss. See lines 16–19 of Program 14.11.

The next step is to establish a coordinate system. This is done by specifying the coordinates of the lower left and upper right corners of a square[6] in the plane in which you wish to draw. For our project, the region extending from $(-\frac{3}{2}, -\frac{3}{2})$ to $(\frac{3}{2}, \frac{3}{2})$ is sufficient because it encompasses both line segments and the circle centered at the origin of radius $\sqrt{2}$.

The coordinate system is established with the `space` or `fspace` methods. Both of these take four arguments corresponding to the x and y coordinates of the lower left corner followed by the coordinates of the upper right corner. The only difference between these two methods is that `space` takes integer arguments whereas `fspace` takes real (the `f` is for `float`) arguments. Thus, on line 22 of Program 14.11 we have

```
P.fspace(-1.5, -1.5, 1.5, 1.5);
```

but we could have used

```
P.space(-2, -2, 2, 2);
```

instead.

We are nearly ready to draw; all we need to do now is select our pen (this happens on lines 25 and 26 of Program 14.11). First we select the line width of the pen with the `flinewidth` method. The argument produces a line thickness relative to the overall size of your drawing area (established with the `space`/`fspace` method). The color is chosen with either the `pencolorname` or `pencolor` method. The `pencolorname` method takes a `char*` string argument giving the English name[7]

---

[6]It is possible to specify a nonsquare rectangular region, but this results in a distorted image. The drawing region you specified is mapped to a square "viewport". If your region is not square, then it is rescaled horizontally or vertically to make it square. This causes circles to become ellipses, and so on.

[7]Of course, not all color names are recognized by this method, but the basic ones are.

of the color you want whereas the `pencolor` takes three `int` arguments that specify the red, blue, and green components of the color. These arguments run from 0 to 65,535 ($2^{16} - 1$). Invoking `P.pencolor(0,0,0)` loads the pen with black ink and `P.pencolor(0xffff,0xffff,0xffff)` gives white (this is an instance where base 16 is more convenient than decimal).

There are other pen attributes we could set to produce dotted lines. The documentation included with `plotutils` explains these. Later (Program 14.14) we use `filltype` and `fillcolorname` to draw circles with an opaque interior; the circle we draw in Program 14.11 has a transparent interior so we do not deal with this issue here.

Draw the picture and finalize

The $\otimes$ picture is drawn with three simple statements. The two line segments (lines 29–30) are drawn with the `line` method:

```
P.line(-1,-1,1,1);
P.line(-1,1,1,-1);
```

(There is an analogous `fline` method that takes real-valued arguments.)

The circle is drawn with the `fcircle` method (line 33):

```
P.fircle(0.,0.,sqrt(2.));
```

(There is an analogous `circle` method that takes integer-valued arguments.)

There are many other drawing methods in the `plotutils` package for creating rectangles, ellipses, circular and elliptical arcs, and Bezier curves. These can be combined into compound paths (e.g., polygons) which, if closed, can contain a fill color. One can also write text (in a variety of fonts) and plot individual points (or place marker symbols at specified coordinates). See the `plotutils` documentation for more information.

After all drawing is complete, we close the `Plotter` with the `closepl()` method (see line 36). This method returns a negative value if anything goes awry with this step.

Compiling and running the program

The `plotutils` package contains two important pieces: a *header* file `plotter.h` and a *library* file `libplotter.a` (or something similar).

We know that a `#include "plotter.h"` directive is needed but if the file is not in a standard location, then we need to tell the compiler where to find this file. We also need to tell the compiler to use the `plotter` library and (if necessary) where this file is located.

Suppose the program is named `xo.cc`. If `plotter.h` and the `plotter` library are in standard locations, the compiler command is as simple as this:

```
g++ xo.cc -lplotter
```

However, if the compiler cannot find these files, it will complain. For the g++ compiler (and many others), these are specified using the -I (for headers) and -L (for libraries) options. For example, suppose these files are located on your computer as follows.

Header file	/home/zelda/include/plotter.h
Library file	/home/zelda/lib/libplotter.a

To compile, use the following command.

```
g++ xo.cc -I/home/zelda/include -L/home/zelda/lib -lplotter
```

(See Appendix A.1.4 for more detail about the -I, -L, and -l options.)

When you run the program (named a.out or something similar) it produces the file xo.eps. You can now use this file in a LATEX document or for whatever purpose you wish. The output of the program is shown in Figure 14.2.

The full code for the program xo.cc follows.

Program 14.11: A program to draw the symbol ⊗.

```
 1  #include <iostream>
 2  #include <fstream>
 3  #include "plotter.h"
 4  using namespace std;
 5
 6  const char* IMAGE_FILE = "xo.eps";
 7
 8  int main(int argc, char** argv) {
 9
10    // Construct the plotter
11    ofstream pout(IMAGE_FILE);
12    PlotterParams params;
13    PSPlotter P(cin, pout, cerr, params);
14
15    // Open the plotter
16    if (P.openpl() < 0) {
17      cerr << "Unable to open plotter" << endl;
18      return 1;
19    }
20
21    // Set up my coordinate system
22    P.fspace(-1.5, -1.5, 1.5, 1.5);
23
24    // Set drawing parameters
25    P.flinewidth(0.05);
26    P.pencolorname("black");
27
28    // Draw the X
29    P.line(-1,-1,1,1);
30    P.line(-1,1,1,-1);
31
32    // Draw the circle
33    P.fcircle(0.,0.,sqrt(2.));
34
35    // Close the image
```

```
36    if (P.closepl() < 0) {
37       cerr << "Unable to close plotter" << endl;
38       return 1;
39    }
40
41    return 0;
42 }
```

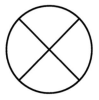

Figure 14.2:   The symbol ⊗ drawn by Program 14.11.

Although we can use C++ to draw pictures such as ⊗, mouse-driven drawing programs are more convenient for producing such simple diagrams. The advantage in using C++ to draw accrues when we want to visualize a mathematical object that requires nontrivial computation to produce.

We demonstrate this now with three examples.

14.8.3 Pascal's triangle modulo two

Pascal's triangle is a table of the binomial coefficients; the nth row (beginning with $n = 0$) contains the values $\binom{n}{0}$, $\binom{n}{1}$, ..., $\binom{n}{n}$. There are myriad interesting properties of binomial coefficients and Pascal's triangle. One of the more striking is to look at Pascal's triangle with entries taken modulo 2. That is, we plot a black point for each odd entry of Pascal's triangle (and leave even entries blank).

Because the image is an array of dots (pixels), we use a GIFPlotter object to do the drawing. Although we could create an array to hold all the entries in Pascal's triangle, we opt instead to create two arrays to hold individual rows (the current row built from the previous row).

Here is the code which should be self explanatory.

Program 14.12: A program to visualize Pascal's triangle modulo 2.

```
1  #include <iostream>
2  #include <fstream>
3  #include "plotter.h"
4  using namespace std;
5
```

```
6    const char* IMAGE_FILE = "pascal.gif";
7    const int   NROWS = 2048;
8
9    int main() {
10
11     // Place to hold row of Pascal's triangle and a copy to generate the
12     // next row.
13     int row[NROWS+1];
14     int prev_row[NROWS+1];
15     for (int i=0; i<=NROWS; i++) {
16       row[i] = 0;
17       prev_row[i] = 0;
18     }
19
20     // Construct the plotter
21     PlotterParams params;
22     ofstream pout(IMAGE_FILE);
23     GIFPlotter P(cin, pout, cerr, params);
24
25     // Open the plotter
26     if (P.openpl() < 0) {
27       cerr << "Unable to open plotter" << endl;
28       return 1;
29     }
30
31     // Set up my coordinate system
32     P.space(-1,-1,NROWS+1,NROWS+1);
33
34     // Set drawing parameters
35     P.pencolorname("black");
36
37     // Generate rows of Pascal's triangle (mod 2)
38     for (int r=0; r<NROWS; r++) {
39       row[0] = 1;
40       for (int j=1; j<r; j++) {
41         row[j] = (prev_row[j-1]+prev_row[j])%2;
42       }
43       row[r] = 1;
44
45       // Plot points for this row
46       for (int j=0; j<=r; j++) {
47         if (row[j] == 1) {
48           P.point(NROWS-r,j);
49         }
50       }
51
52       // copy row to prev_row
53       for (int j=0; j<=r; j++) prev_row[j] = row[j];
54     }
55
56     // Close the image
57     if (P.closepl() < 0) {
58       cerr << "Unable to close plotter" << endl;
59       return 1;
60     }
61
```

```
62    return 0;
63  }
```

The output of this program (Figure 14.3) is the beautiful fractal known as Sierpinski's triangle.

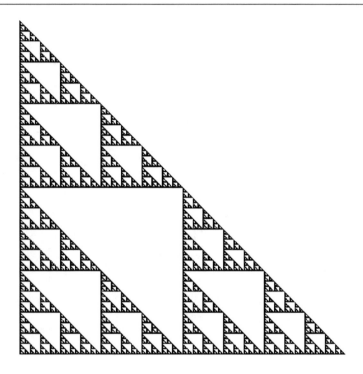

Figure 14.3:   Visualizing Pascal's triangle modulo 2.

14.8.4   Tracing the motion of a point moving randomly in a triangle

The next example visualizes the following process. Three fixed points (named *A*, *B*, and *C*) are located at the vertices of an equilateral triangle. A moving point *X* begins at one of these locations and then moves to a new location at random. At each step, the new location is the midpoint of one of the segments *AX*, *BX*, *CX*—each with probability $\frac{1}{3}$. Imagine that each time we find a new location for *X*, we plot a point in the plane. After many iterations, what do we see?

Program 14.13 draws a picture by precisely this mechanism. Because the image is a collection of pixels, it makes sense to use a `GIFPlotter` to produce a `gif` image file.

The code is reasonably straightforward. One feature we want to emphasize is the procedure `draw_point` on lines 11–13. This procedure takes two arguments: a `Plotter` and a `Point` (see Chapter 6). This procedure is invoked inside `main` on line 78 where it is passed a `GIFPlotter` named `P` (and a `Point` named `X`). At first glance this may appear to be a programming error: the procedure requires a `Plotter` but is invoked with a `GIFPlotter`. This mismatch of types is not a problem because a `GIFPlotter` is a subclass of `Plotter`.

Program 14.13: A program to plot points in a triangle by a random method.

```
1   #include <iostream>
2   #include <fstream>
3   #include "plotter.h"
4   #include "Point.h"
5   #include "uniform.h"
6   using namespace std;
7
8   const char* IMAGE_FILE = "dance.gif";
9
10  // Draw a point on a plotter.
11  void draw_point(Plotter& P, const Point& X) {
12    P.fpoint(X.getX(), X.getY());
13  }
14
15  int main(int argc, char** argv) {
16    // check that user specified number of points
17    if (argc < 2) {
18      cerr << "Usage: " << argv[0] << " npoints" << endl;
19      return 1;
20    }
21
22    int NPOINTS;
23
24    // convert 2nd arg to an integer; make sure its positive
25    NPOINTS = atoi(argv[1]);
26    if (NPOINTS < 1) {
27      cerr << "Usage: " << argv[0] << " npoints" << endl;
28      cerr << "where npoints is a positive integer" << endl;
29      return 1;
30    }
31
32    // Set the corners of the triangle
33    Point A(0.,0.);
34    Point B(1.,0.);
35    Point C(0.5, sqrt(3.)/2.);
36
37    Point X; // the dancing point
38
39    // Construct the plotter
40    PlotterParams params;
41    ofstream pout(IMAGE_FILE);
42    GIFPlotter P(cin, pout, cerr, params);
43
```

```
44   // Open the plotter
45   if (P.openpl() < 0) {
46     cerr << "Unable to open plotter" << endl;
47     return 1;
48   }
49
50   // Set up my coordinate system
51   P.fspace(0.,0.,1.,1.);
52
53   // Set drawing parameters
54   P.pencolorname("black");
55
56   // Draw the corners
57   draw_point(P,A);
58   draw_point(P,B);
59   draw_point(P,C);
60
61   // Start X at A
62   X = A;
63
64   for (int k=0; k<NPOINTS; k++) {
65     switch(unif(3)) {
66     case 1:
67       X = midpoint(A,X);
68       break;
69     case 2:
70       X = midpoint(B,X);
71       break;
72     case 3:
73       X = midpoint(C,X);
74       break;
75     default:
76       cerr << "This can't happen" << endl;
77     }
78     draw_point(P,X);
79   }
80
81   // Close the image
82   if (P.closepl() < 0) {
83     cerr << "Unable to close plotter" << endl;
84     return 1;
85   }
86
87   return 0;
88 }
```

When the program is run we again see Sierpinski's triangle! See Figure 14.4.

14.8.5 Drawing Paley graphs

Our final drawing example is a program to draw Paley graphs. In this context, a *graph* is a pair (V,E) where V is a finite set of vertices and E is a set of unordered pairs of distinct vertices; elements of E are called *edges*. The edge $\{u,v\}$ is said to *join* its end points, u and v; we call such vertices *adjacent*. Note that *is adjacent*

Figure 14.4: An image based on a random process in a triangle.

to is a symmetric relation because the pairs in E are unordered. Furthermore, self-loops are not allowed (edges that join a vertex to itself). Such graphs are often called *simple* graphs.

It is useful to visualize graphs by drawings. The vertices are drawn as dots and edges as line segments (or curves) joining their end point's dots. In our program, we draw the vertices as small circles.

The *Paley graphs* form an interesting class of graphs. Let $n \geq 3$ be an integer. The vertices of G_n are the integers $\{0, 1, 2, \dots, n\}$. Two vertices, u and v, of G_n are adjacent (joined by an edge) if and only if $u - v$ is a perfect square (quadratic residue) in $\mathbb{Z}_n$. Because the *is adjacent to* relation is symmetric, we require that -1 be a perfect square in $\mathbb{Z}_n$; otherwise G_n is undefined.

For example, let $n = 5$. The vertices of G_5 are $\{0, 1, 2, 3, 4\}$. The (nonzero) quadratic residues in $\mathbb{Z}_5$ are 1 and 4. Therefore, the edges of G_5 are $\{0, 1\}$, $\{1, 2\}$, $\{2, 3\}$, $\{3, 4\}$, and $\{0, 5\}$. In other words, G_5 is a 5-cycle. However, G_7 is undefined because $-1 \equiv 6$ is not a perfect square in $\mathbb{Z}_7$.

Our program to draw Paley graphs reads the parameter n from the command line (see lines 11–25 of Program 14.14) and then stores all squares in $\mathbb{Z}_n$ in a `set` object named `squares` (lines 31–35). After checking that -1 is a quadratic residue (lines 40–43), we create a table of x and y values for the vertex locations. We evenly place the vertices on a circle of radius n centered at the origin.

Next we declare and initialize a `PSPlotter` object (lines 57–71) that sends its output to a file named `paley.eps` (lines 8, 58, and 59). We choose a vector graphic format (`eps`) because our drawing consists of mathematically defined curves and not a field of points. We are ready to begin drawing.

We first draw the edges (lines 74–83) as line segments. Then we modify the pen so that enclosed regions (such as circles) are filled with white ink. This is accomplished in two steps: a call to the `fillcolorname` method (line 86) to set the fill color to white and then `P.filltype(1);` (line 87) to activate region filling. The vertices are now drawn as circles of radius $\frac{1}{2}$ (lines 88–90).

Note that if we had not activated region filling, then the circles would have been transparent and we would see the full lengths of the line segments meeting at the centers of the circles; it is more esthetically pleasing to cover that region. Furthermore, it is important that we draw the edges before the vertices. If we were to reverse the order, then the edges would be drawn on top of the circles, and we would see the portion of the line segment that ought to be hidden.

The rest of the program releases the memory allocated for the arrays holding the coordinates of the vertices and closes the `PSPlotter` (line 93 to the end).

Here is the program.

Program 14.14: A program to draw Paley graphs.

```
1  #include <iostream>
2  #include <fstream>
3  #include <set>
4  #include "plotter.h"
5  #include "Mod.h"
6  using namespace std;
7
8  const char* IMAGE_FILE = "paley.eps";
9
10 int main(int argc, char** argv) {
11   // check that user specified number of points
12   if (argc < 2) {
13     cerr << "Usage: " << argv[0] << " n" << endl;
14     return 1;
15   }
16
17   int n;  // number of vertices in the graph and the modulus
18
19   // convert 2nd arg to an integer; make sure it's positive
20   n = atoi(argv[1]);
21   if (n < 3) {
22     cerr << "Usage: " << argv[0] << " n" << endl;
23     cerr << "where n is an integer greater than 2" << endl;
24     return 1;
25   }
26
27   // We'll only be working in Z_n
28   Mod::set_default_modulus(n) ;
29
30   // Find all perfect squares in Z_n
31   set<Mod> squares;
32   for(int k=0; k<n; k++) {
```

```
33      Mod S(k*k);
34      squares.insert(S);
35    }
36
37    // Check if -1 is a perfect square
38    Mod neg1(-1);
39
40    if (squares.find(neg1) == squares.end()) {
41      cerr << "-1 is not a perfect square in Z_" << n << endl;
42      return 1;
43    }
44
45    // Create tables of the x and y coordinates of the vertex centers
46    float* x = new float[n];
47    float* y = new float[n];
48
49    double theta = 2*M_PI/n;
50
51    for (int i=0; i<n; i++) {
52      x[i] = n * sin(i*theta);
53      y[i] = n * cos(i*theta);
54    }
55
56    // Construct the plotter
57    PlotterParams params;
58    ofstream pout(IMAGE_FILE);
59    PSPlotter P(cin, pout, cerr, params);
60
61    // Open the plotter
62    if (P.openpl() < 0) {
63      cerr << "Unable to open plotter" << endl;
64      return 1;
65    }
66    // Set drawing parameters
67    P.flinewidth(0.05);
68    P.pencolorname("black");
69
70    // Set up my coordinate system
71    P.space(-n-1, -n-1, n+1, n+1);
72
73    // Draw the edges
74    for (int i=0; i<n-1; i++) {
75      for (int j=i+1; j<n; j++) {
76        // if i-j is a perfect square ...
77        Mod diff(i-j);
78        if (squares.find(diff) != squares.end()) {
79          // ... then we draw the edge
80          P.fline(x[i],y[i],x[j],y[j]);
81        }
82      }
83    }
84
85    // Draw the vertices
86    P.fillcolorname("white");
87    P.filltype(1);
88    for (int i=0; i<n; i++) {
```

```
89        P.fcircle(x[i], y[i], 0.5);
90      }
91
92      // Release allocated arrays for x and y
93      delete[] x;
94      delete[] y;
95
96      // Close the image
97      if (P.closepl() < 0) {
98        cerr << "Unable to close plotter" << endl;
99        return 1;
100     }
101
102     return 0;
103   }
```

When the program is run with $n = 17$, the picture in Figure 14.5 is produced. The Paley graph G_{17} is interesting because it does *not* contain four vertices that are pairwise adjacent (a clique of size 4) and it does *not* contain four vertices that are pairwise nonadjacent (an independent set of size 4). All graphs on 18 vertices (or more) must contain either a clique or an independent set of size 4.

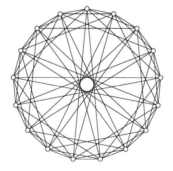

Figure 14.5:   The Paley graph on 17 vertices.

14.9   Exercises

14.1 Suppose a `string` variable contains the decimal representation of an integer. For example, `string s = "7152";`. Explain how to convert the `string` to an `int`.

14.2 A `string` variable named `file_name` contains the name of a file. We want to read data from that file using an `ifstream` object named `fin`. Show how `fin` should be declared.

14.3 A programmer wants to write the 26 lowercase letters a to z on the computer's screen. The following code does not work.

```
for (int   k=0; k<26; k++) {
  cout << 'a' + k;
}
```

What is wrong and how can this code be repaired?

14.4 A free group is the set of all formal products one can create using a set of generators $\{a,b,c,\ldots\}$ and their inverses. If a generator and its inverse are adjacent in the product, then they can be removed. The empty product serves as the identity element. For example,

$$\left(abc^{-1}b\right) \cdot \left(b^{-1}cba\right)$$

simplifies to *abba* or *ab²a*.

Create a class named `Free` that represents elements in a free group. Use a `string` variable to hold the group element using lowercase letters for the generators and the corresponding uppercase letters for their inverses (i.e., A for a^{-1}).

Include an `operator*` for the group product and an `inverse()` method. Also provide a `<<` operator for printing `Free` objects to the screen.

There are several procedures declared in the `cctype` header that you can use for this exercise. These can be used to examine and modify the cases of single letter characters. See Appendix C.6.3 (starting on page 411) for a description of `isupper`, `toupper`, and so on. Read the introductory material carefully as `toupper` does not behave precisely as you might expect.

14.5 Write a program that reads a text file and reports the frequency of each of the 26 letters. Run the program on various texts and see if you can distinguish languages.

You can find books in plain text format at http://www.gutenberg.org.

14.6 *Polygrams.* In Exercise 14.5 we analyzed writing by looking at the frequency of single letters. In this exercise we consider a more sensitive analysis. For a positive integer n, an *n-gram* is a sequence of n letters. For example, the word *banana* contains the 2-grams *ba*, *an*, and *na*.

Create a program to count the *n*-grams found in a text and report the ten most frequent *n*-grams found for each of $n = 2$, 3, and 4.

14.7 *Koch curve.* The Koch curve is a fractal formed by the following iterative process. Begin with a horizontal line segment, say from $(0,0)$ to $(1,0)$. Draw an equilateral triangle using the middle third of this segment as one side, and then erase the side that was part of the original segment. The new path consists of four segments:

$$(0,0) \to \left(\frac{1}{3},0\right) \to \left(\frac{1}{2},\frac{\sqrt{3}}{6}\right) \to \left(\frac{2}{3},0\right) \to (1,0)$$

Now repeat this basic step again on each of the four segments to create a polygonal path with 16 segments. The first two steps of this process are shown in the figure.

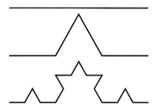

The Koch curve is the result of carrying out this process *ad infinitum*.

Write a program to draw (an approximate version of) the Koch curve to a user-specified number of iterations of the basic step.

14.8 *Mandelbrot set.* Let $c \in \mathbb{C}$ and define $f_c(z) = z^2 + c$. Consider the sequence of iterates

$$z_0 = 0, \qquad z_{k+1} = f_c(z_k).$$

For some values of c this sequence tends to infinity, $|z_k| \to \infty$ as $k \to \infty$, but for other values of c this sequence stays bounded.

The *Mandelbrot set* M is the set of all complex c such that the iterates of f_c (starting at 0) stay bounded. Create a program to draw a picture of the Mandelbrot set.

One can show that if for some k we have $|z_k| > 2$, then the iterates z_n tend to infinity. So, to test if $c \in M$, one can compute (say) the first 500 iterates z_k testing if any have magnitude greater than 2. If so, we know that $c \notin M$. However, if for all $k \le 500$ we have $|z_{500}| < 2$, one might guess that $c \in M$. This guess might be incorrect, but this approach yields a reasonable approximation.

Furthermore, based on this criterion, any complex number c with $|c| > 2$ cannot be an element of M, so one need not check a complex number $a + bi$ with $|a| > 2$ or $|b| > 2$.

Chapter 15

Odds and Ends

It is not the purpose of this book to give an exhaustive coverage of every feature of C++. We have, by design, omitted those parts of the language that (in our judgment) are not useful for mathematical work. In this final chapter we present an assortment of additional topics that may also be of interest and utility in your work.

15.1 The `switch` statement

C++ provides a variety of control structures such as `if`, `for`, `while`, and `do`. Here we discuss an additional control structure: the `switch` statement.

The `switch` statement is used to control execution depending on the value held in an integer variable. Suppose that x holds an integer value. If x equals 1, we perform operation I, if it equals 2 we perform operation II, if it equals 3 or 4 we perform operation III, and if it holds any other value, we do operation IV. Here is how we would structure this situation using `if` statements.

```
if (x == 1) {
  // Do operation I
}
else {
  if (x == 2) {
    // Do operation II
  }
  else {
    if ((x == 3) || (x == 4)) {
      // Do operation III
    }
    else {
      // Do operation IV
    }
  }
}
```

As you can see, this is difficult to read (and awkward to type). With `switch`, the code is much clearer:

```
switch(x) {
 case 1:
   // Do operation I
```

```
    break;
  case 2:
    // Do operation II
    break;
  case 3:
  case 4:
    // Do operation III
    break;
  default:
    // Do operation IV
}
```

The outer structure of the statement is switch(x) {...} where x is an integer type. Within the curly braces is a sequence of case statements. Each case is followed by a specific integer value. When a switch statement is executed, the computer looks for a case that matches the value in x. If it finds one, it executes whatever code it finds following the case statement until it finds a break; statement. If no case statement matches the value of x, the computer looks for a default statement and executes the code that follows. (If there is no default statement, then execution skips to the statements following the closing curly brace of the switch.)

Notice that two case statements can precede a section of code (for the values 3 and 4).

C++ does not require a break; statement between two sections of a switch statement. For example, if we were to delete the break; at the end of case 2, then (when x equals 2), first operation II would execute and then operation III would execute. This is known as *falling through* cases. In general, it is much better to balance every case with a matching break unless (as in the example) two cases execute the exact same code. The default case (because it appears at the end of the switch) does not require a break.

There is no way to specify a range of values in a case statement. So, for example, if you want different behaviors depending on whether x is negative, zero, or positive, you should use if statements.

The switch statement is useful for creating a menu of options for the user. A prompt is typed on the screen listing various options; the user specifies the desired option by typing a number or letter. The following program illustrates how this is done; in it, the switch statement plays a central role. The user specifies the desired option by typing a letter. We allow uppercase and lowercase letters and so two case statements are given for each option.

Program 15.1: A program to illustrate the switch statement.

```
1  #include <iostream>
2  #include <string>
3  using namespace std;
4  int main() {
5
6    bool again = true;
7    while(again) {
8      cout << endl;
```

```
9      cout << "What would you like to do?" << endl;
10     cout << "   (a) Prove Fermat's little theorem" << endl;
11     cout << "   (b) Prove Fermat's last theorem" << endl;
12     cout << "   (c) List some Fermat primes" << endl;
13     cout << "   (d) End this program" << endl;
14     cout << "Type the character and hit return: ";
15
16     string response;
17     cin >> response;
18
19     if (response.empty()) continue;
20
21     char option = response[0];
22
23     switch(option) {
24     case 'a':
25     case 'A':
26       cout << endl
27            << "Apply Lagrange's theorem to the group Z_p." << endl;
28       break;
29
30     case 'b':
31     case 'B':
32       cout << endl
33            << "I have a great proof for this but, alas, it will"
34            << endl << "not fit on this computer screen." << endl;
35       break;
36
37     case 'c':
38     case 'C':
39       cout << endl << "Here are the only ones I know:" << endl
40            << "3 5 17 257 25537." << endl;
41       break;
42
43     case 'd':
44     case 'D':
45       cout << endl << "Goodbye." << endl;
46       again = false;
47       break;
48
49     default:
50       cout << endl << "Sorry. Your response was not recognized."
51            << endl << "Please try again." << endl;
52     }
53   }
54   return 0;
55 }
```

Here is a sample run of the program.

```
What would you like to do?
   (a) Prove Fermat's little theorem
   (b) Prove Fermat's last theorem
   (c) List some Fermat primes
   (d) End this program
Type the character and hit return: a
```

```
Apply Lagrange's theorem to the group Z_p.

What would you like to do?
  (a) Prove Fermat's little theorem
  (b) Prove Fermat's last theorem
  (c) List some Fermat primes
  (d) End this program
Type the character and hit return:      b

I have a great proof for this but, alas, it will
not fit on this computer screen.

What would you like to do?
  (a) Prove Fermat's little theorem
  (b) Prove Fermat's last theorem
  (c) List some Fermat primes
  (d) End this program
Type the character and hit return: C

Here are the only ones I know:
3 5 17 257 25537.

What would you like to do?
  (a) Prove Fermat's little theorem
  (b) Prove Fermat's last theorem
  (c) List some Fermat primes
  (d) End this program
Type the character and hit return: x

Sorry. Your response was not recognized.
Please try again.

What would you like to do?
  (a) Prove Fermat's little theorem
  (b) Prove Fermat's last theorem
  (c) List some Fermat primes
  (d) End this program
Type the character and hit return: D

Goodbye.
```

15.2 Labels and the `goto` statement

In addition to `switch` and all the various other control structures, C++ also provides the simple `goto` statement. For good reasons, use of the `goto` statement is highly discouraged. It can lead to programs that are impossibly difficult to understand. There is, however, one situation where a `goto` is handy: breaking out of nested loops. Here is how `goto` works.

Any statement in C++ may be prefixed by a label. A statement is labeled by a name; a label name is formed according to the same rules as variables are named. To label a statement, use the label name followed by a colon, like this:

```
increment: x += 10;
```

This names the statement x += 10; as increment. Now, anywhere in the same procedure as this labeled statement, we may have the statement goto increment;. Whenever the goto statement is invoked, the program immediately jumps to the statement labeled increment and proceeds from there.

If possible, use one of C++'s other control structures rather than resorting to the use of goto. However, if you need to break from a nested loop, a goto is a clean solution. Here's the issue: if you want to exit a for loop before the end condition has been reached, you can execute a break statement. For example,

```
int k;
for (k=0; k<100; k++) {
  if (x[k] < 0) break;
  // otherwise do something
}
```

This loop stops running if any element of the array x is negative. The break statement forces the for loop in which it is contained to stop looping. Therefore, if the break is enclosed in, say, a double for loop, only the inner loop is interrupted. How can we stop the execution of the outer loop? Unfortunately, break won't help us.

A solution is to label the first statement after the double for loop and use a goto. Here is an example.

```
#include <iostream>
using namespace std;

int main() {
  int i,j;

  for (i=0; i<10; i++) {
    for (j=0; j<10; j++) {
      if (i+j == 15) goto alpha;
    }
  }

 alpha:
  cout << "i = " << i << " and j = " << j << endl;
  return 0;
}
```

The two for loops execute until i+j reaches the value 15. At that point, the loops are interrupted and execution proceeds at the statement labeled alpha. The output is as follows.

```
i = 6 and j = 9
```

In practice, this situation rarely occurs. You are unlikely to ever need a goto statement in your programs.

15.3   Exception handling

In the course of a computation unexpected situations can arise. For example, the computer might attempt to find the inverse of a noninvertible object. If the computer is asked to divide by zero, it signals the invalid result in one of two ways: the division `1./0.` results in the special value `inf` (which stands for ∞) and the division `0./0.` results in the special value `nan` (which stands for *not a number*). Similarly, in Chapter 9 we designed the `Mod` class so that if the `inverse` method is invoked on a noninvertible `Mod` object, the return value is an invalid `Mod`.

To check that a `Mod` computation took place without error, the `is_invalid` method can be applied to the results. To check if a floating point calculation resulted in `inf` or `nan`, use the C++ procedures `isinf()` or `isnan()`.[1]

In some cases, it is difficult to communicate an abnormal event through a `return` value. In this case, C++'s exception system is useful.

Although you might elect not to use the exception system, it is worth knowing its basics because packages that you download from the Web might use this feature.

15.3.1   The basics of `try`, `throw`, and `catch`

C++ provides an alternative method for detecting and handling such situations. When a procedure detects an erroneous situation, it can *throw an exception*. When the exception is thrown, the normal flow of the program is interrupted and control is passed to code that *catches* the exception. The code that might throw an exception is contained inside a *try* block. The exception-handling mechanism involves the three keywords `try`, `throw`, and `catch` and is structured like this:

```
// start of the procedure
try {
  // some calculations
  if (something_bad_happens) throw object;
  // more calculations
}
catch (object_type var) {
  // deal with bad situation
}
// rest of the procedure
```

To illustrate how this works, we present the following short program that asks the user for two numbers and returns their quotient.

---

[1] In order to use these procedures, you need the directive `#include <cmath>` at the beginning of your program. Also, at the time of this writing, there is a bug in the Mac OS X version of C++ of `cmath`. Examine the file `/usr/include/gcc/darwin/default/c++/cmath`. Comment out the lines that read `#undef isinf` and `#undef isnan`.

Program 15.2: A program to divide two numbers and illustrate basic exception handling.

```cpp
#include <iostream>

using namespace std;

int main() {
  double x,y;
  try {
    cout << "Enter numerator: ";
    cin >> x;
    cout << "Enter denominator: ";
    cin >> y;

    if (y==0.) throw x;

    cout << x << " divided by " << y << " is " << x/y << endl;
  }
  catch (double top) {
    cerr << "Unable to divide " << top << " by zero" << endl;
  }

  cout << "Thank you for dividing." << endl;

  return 0;
}
```

After some initialization, we encounter the `try` block (lines 7–16). Within this block, we prompt the user for two real numbers. Then, just before we divide, we test if the denominator is zero. At this point there are two possible ways the program might proceed.

- If the denominator is zero, the exception is thrown (line 13). In this case, we throw the value of the numerator. The remaining statements inside the `try` block are skipped; line 15 is not executed.

  At this point, the computer skips to the end of the `try` block and searches for a `catch` statement whose type agrees with the type of the object that was thrown. Because `throw x;` throws a `double` value, execution proceeds to the `catch` statement on line 17. The variable `top` named in the `catch` statement is set equal to the value thrown (the value held in x).

  Now the statements embedded inside the `catch` block execute. In this example, there is only one statement (line 18) that prints an error message on the screen.

  Once the `catch` block is finished, execution continues at line 21.

- Alternatively, if the denominator is not zero, then the `throw` statement on line 13 does not execute. Instead, the program continues with the remaining statements inside the `try` block. In this example, there is only one more statement (line 15).

Because no exception was thrown, the catch block is skipped, and execution proceeds to the statements following the try block—that is, to line 21.

Here are two runs of the program to illustrate how this works.

```
Enter numerator: 17
Enter denominator: 4
17 divided by 4 is 4.25
Thank you for dividing.
```

```
Enter numerator: -9.8
Enter denominator: 0
Unable to divide -9.8 by zero
Thank you for dividing.
```

For this example, it would be simpler to use an if/then/else construction. The exception-throwing mechanism is particularly useful when the exception is thrown by one procedure and caught by another. We illustrate this idea in the following program.

Program 15.3: A program to illustrate catching exceptions thrown by other procedures.

```cpp
1   #include <iostream>
2
3   using namespace std;
4
5   double quotient(double p, double q) {
6     if (q==0.) throw p;
7     return p/q;
8   }
9
10  int main() {
11    double x,y;
12    try {
13      cout << "Enter numerator: ";
14      cin >> x;
15      cout << "Enter denominator: ";
16      cin >> y;
17
18      double q = quotient(x,y);
19
20      cout << x << " divided by " << y << " is " << q << endl;
21    }
22    catch (double top) {
23      cerr << "Unable to divide " << top << " by zero" << endl;
24      exit(1);
25    }
26
27    cout << "Thank you for dividing." << endl;
28
29    return 0;
30  }
```

This program defines a procedure named quotient (see lines 5–8) that takes two double arguments and returns their quotient. If this procedure is invoked with q equal to zero, then the quotient is undefined. In this case, the procedure throws an exception. Notice that there is no try/catch block in the quotient procedure; it is the responsibility of the procedure that invokes quotient (in this example, main) to handle the exception.

Inside main, the call to quotient is embedded in a try block (lines 12–21). If quotient runs normally (i.e., its second argument is not zero), then line 18 executes normally. Execution then proceeds to line 20, then the catch block (lines 22–25) is skipped, and the program continues at line 27.

However, if (at line 18), the second argument sent to quotient is zero, then quotient throws an exception. The value of the numerator is thrown. The rest of the try block is skipped; that is, line 20 is not executed. At this point the computer searches for a catch statement whose type matches the type of the thrown value. Because a double value is thrown by quotient, it is caught at line 22. Now the catch block executes with top set equal to the thrown value. An error message is printed (line 23) followed by a call to a system procedure named exit. This causes the program to stop executing immediately. Consequently, the statements following the catch block—statements that would typically execute were it not for the call to exit—are not executed.

If possible, the catch block should repair any damage caused by the exceptional situation so the program can continue running. However, some errors are sufficiently serious that continued execution does not make sense. In that case, a call to exit causes the program to stop running. Note that exit can be called inside any procedure (not just in main).

The exit procedure takes an integer valued argument; this value is passed back to the operating system. If your C++ program is invoked by a script, the script can use that value to detect that something unusual occurred during the execution of the C++ program. By convention, a return value of 0 signals normal execution. That is why we end our main programs with the statement return 0;. However, if the program is forced to stop executing inside a procedure other than main, then exit can be used to return a value to the operating system.

Here are two runs of Program 15.3.

```
Enter numerator: 10
Enter denominator: 4
10 divided by 4 is 2.5
Thank you for dividing.
```

```
Enter numerator: -10
Enter denominator: 0
Unable to divide -10 by zero
```

15.3.2   Other features of the exception-handling system

Multiple catches

A `try` block must be followed by a `catch` block. However, there may be more than one `catch` block, provided each catches a different type. The general structure looks like this:

```
try {
  // code that might generate exceptions
}
catch (type_1 x1) {
  // handle this exception
}
catch (type_2 x2) {
  // handle this exception
}
...
catch (type_n xn) {
  // handle this exception
}
// rest of the code
```

If an exception is thrown in the `try` block, it is caught by the `catch` block that matches the type of the object that was thrown. That, and only that, `catch` block is executed; the other `catch` blocks are skipped. Of course, if no exception is thrown, none of the `catch` blocks executes.

Uncaught exceptions and a catch-all block

It is important that there be a `catch` block for every type of exception that might be thrown by a `try` block. Otherwise, the uncaught exception causes the program to terminate by calling the system procedure `abort`. The `abort` procedure is a more drastic version of `exit`. `abort` takes no arguments and prints a message such as `Abort trap` on the screen.

It is conceivable that you might not know every type of exception your program might produce. For example, you may be using a package you downloaded and you might not be familiar with the various exceptions that package's procedures can produce. In such a case, you can create a catch-all block that catches any exception. Here's how: after the various known exceptions, create a block that looks like this,

```
catch(...) {
  // code to handle an exception of an unknown type
}
```

It is difficult for a catch-all block to handle errors because any information contained in the thrown object is lost; unlike a typical `catch` block, there is no value sent to a catch-all block.

Objects to throw

Any C++ object may be thrown by a `throw` statement. If you were to create your own arc sine function, it would be natural to throw an exception if the argument to the function were outside the interval $[-1, 1]$. In this case, it would be sensible to throw the argument of the function so the calling procedure knows that it sent an illegal value.

Alternatively, we can throw an object specifically designed to convey detailed information on the error. For example, we can create a class named `TrigException` like this:

```
class TrigException {
public:
  double value;
  string message;
};
```

Then our arc sine function would have the following form.

```
double arc_sin(double x) {
  if ((x<-1.) || (x>1.)) {
    TrigException err;
    err.value = x;
    err.message = "Argument to arc_sin not in [-1,1]";
    throw err;
  }
  // calculate the arc sine of x, etc.
}
```

A procedure that calls `arc_sin`, or other trigonometric functions of our own creation, can be structured like this:

```
try {
  // various trig calculations
}
catch (TrigException E) {
  cerr << "Trouble occurred during the calculation" << endl;
  cerr << E.message << endl;
  cerr << "The faulty argument was " << E.value << endl;
  exit(1);
}
```

Packages you download from the Web might contain their own exception types. For example, a geometry package might include a procedure to find the point of intersection of two lines. What should such a procedure do when presented with a pair of parallel lines? A sensible answer is to throw an exception.

Rethrowing exceptions

It is possible for a procedure to throw an exception, catch it, partially handle the exception, and then rethrow the exception to be handled by another procedure. It is unlikely that you will need this feature for mathematical work. Nevertheless, we describe how this is done.

A `catch` block that both handles an exception and also passes that exception on is structured like this:

```
catch(type x) {
  // handle the exception
```

```
  throw;
}
```

When this `catch` is invoked, the various statements in the `catch` block are executed. Finally, the statement `throw;` is reached; this causes the original exception that was caught to be thrown again.

The keyword `throw` in a procedure declaration

When you write a procedure that throws exceptions, it is advisable to announce explicitly the types of exceptions that might be thrown. This is done immediately after the argument list as in the following example.

```
double quotient(double p, double q) throw(double) {
  if (q==0.) throw p;
  return p/q;
}
```

If a procedure throws more than one type of exception, we list the various types within the parentheses like this:

```
return_type proc_name(type1 p1, type2 p2, ..., typeN pN)
    throw(xtype1, xtype2, ..., xtypeM) {
      // the procedure
}
```

The addition of a `throw` list to the introduction of a procedure makes it clear to the user of the procedure what kinds of exceptions the procedure might throw. Users of the procedure need only inspect the first line to deduce this information and do not need to hunt through the code for the various places an exception is generated.

Exceptions that are not listed on the list of `throw` types cannot escape from the procedure. Either they must be caught inside the procedure or they trigger a runtime error.

15.4 Friends

Private data members of a class are accessible to the class's methods; other procedures cannot inspect or modify these values. This is the essence of data hiding and it protects objects from being corrupted. When you use a class, you do not interact directly with its data, but only use its public methods.

However, when you create a class, it may be useful to create some procedures that are permitted to access the class's private elements. For example, recall the `Point` class of Chapter 6. In addition to the various methods (member procedures of the class), we also defined the procedures `dist` and `midpoint`. Neither of these is a member of the class `Point`, and so they need to use `getX` and `getY` to learn their arguments' coordinates. C++ provides a mechanism by which `midpoint` and `dist`

can bypass data hiding; we do this by declaring these procedures *friends* of the class
`Point`.

The original header file for the `Point` class, `Point.h`, is given in Program 6.1.
Here we present an alternative version in which `dist` and `midpoint` are declared to
be friends of the `Point` class.

Program 15.4: A new `Point.h` header with `friend` procedures.

```
1   #ifndef POINT_H
2   #define POINT_H
3   #include <iostream>
4   using namespace std;
5
6   class Point {
7
8   private:
9     double x;
10    double y;
11
12  public:
13    Point();
14    Point(double xx, double yy);
15    double getX() const;
16    double getY() const;
17    void setX(double xx);
18    void setY(double yy);
19    double getR() const;
20    void setR(double r);
21    double getA() const;
22    void setA(double theta);
23    void rotate(double theta);
24    bool operator==(const Point& Q) const;
25    bool operator!=(const Point& Q) const;
26
27    friend double dist(Point P, Point Q);
28    friend Point midpoint(Point P, Point Q);
29
30  };
31
32  ostream& operator<<(ostream& os, const Point& P);
33
34  #endif
```

Notice that `dist` and `double` are now declared inside the `Point` class declaration, and their declarations begin with the keyword `friend`. The `friend` keyword
signals that they are not `Point` class members (i.e., methods), but rather privileged
procedures that are permitted to access the private elements of `Point`.

With this header in place, the definitions of `dist` and `midpoint` (in the file
`Point.cc`) look like this:

```
double dist(Point P, Point Q) {
    double dx = P.x - Q.x;
    double dy = P.y - Q.y;
    return sqrt(dx*dx + dy*dy);
```

```
}

Point midpoint(Point P, Point Q) {
  double xx = ( P.x + Q.x ) / 2;
  double yy = ( P.y + Q.y ) / 2;
  return Point(xx,yy);
}
```

Efficiency

By coding the `midpoint` and `dist` procedures as friends of the `Point` class, we can bypass the calls to `getX` and `getY`. Consequently, these procedures should be faster than the standard (nonfriend) versions.

Another way to improve performance is to code these procedures to use call by reference (instead of call by value). In that case, the `midpoint` procedure would begin like this:

```
Point midpoint(const Point& P, const Point& Q)
```

To send a `Point` object by value requires the transmission of two `double` values; this comprises more bytes than call by reference where only a reference to the parameters needs to be sent to the procedure.

Let us experiment with the effects of these differences. Let us write a program that generates a long list of points and then finds the midpoints for all pairs of points on the list. In this experiment, we use four different versions of `midpoint`:

- A standard (nonfriend) version with call by value,

- A standard version with call by reference,

- A friend version with call by value, and

- A friend version with call by reference.

The results[2] of this experiment are summarized in the following chart:

Time in seconds (no optimization)

	Standard version	Friend version
Call by value	214	166
Call by reference	191	144

From these results, we observe that the `friend` version outperforms the corresponding standard version; this is thanks to the friend's ability to access data elements directly. We also observe that the call-by-reference versions outperform the corresponding call-by-value versions; this is thanks to the reduced number of bytes that need to be sent to the `midpoint` procedure.

---

[2]This experiment was performed on an 800 MHz G4 Mac OS X system using version 3.3 of the Gnu compiler. The list of points contained 20,000 entries and so 400 million calls to `midpoint` were generated.

However, these results were obtained when none of the optimization options for the compiler was engaged. Modern compilers can deduce when a call to `getX` may be streamlined away and the private value can be safely sent directly into the procedure. Hence, when the experiment is repeated with all optimization options engaged, we get the following results.

Time in seconds (full optimization)

	Standard version	Friend version
Call by value	66	72
Call by reference	45	44

From this second experiment we observe that there is no advantage (in this instance) to setting up the procedures as friends, but we still achieve better performance using call by reference.

15.5 Other ways to create types

15.5.1 Structures

C++ is an extension to the C language. The centerpiece of this extension is the concept of a class. An ancestor to the class concept is known as a *structure*. Like classes, structures are assemblies of data, and can be used as data types. However, structures consist only of data (no associated methods) and one cannot extend structures using inheritance.

Structures are declared in a manner that is similar to how we declare classes. This is how we declare a structure type that represents an ordered pair of real numbers:

```
struct RealPair {
  double x;
  double y;
};
```

With this in place, we may define variables of type `RealPair`, like this:

```
RealPair p;
```

Now, to access the data held in p, we use the familiar dot notation: `P.x` and `P.y`.

You never need to use structures in your programming. The exact same behavior can be achieved using classes. Here is how `RealPair` would be declared.

```
class RealPair {
public:
  double x;
  double y;
};
```

15.5.2 Enumerations

An *enumeration* is a variation on the integer data types. Suppose our program deals with different sorts of infinity. In the context of real numbers, we might have separate $+\infty$ and $-\infty$, but in the context of complex numbers we have a single "complex infinity."

In C++ we can create a type to represent these three options. The syntax looks like this:

```
enum infinity { minusInfinity, plusInfinity, complexInfinity };
```

With this in place, `infinity` becomes a type and variables may be declared to be of type `infinity`:

```
infinity X;
```

Now `X` may be assigned one of the three values declared in the enumeration; for example, `X = plusInfinity;`.

Behind the scenes, the enumeration values (listed between the curly braces) are given integer values. Therefore, enumeration types may be used in `switch` statements:

```
switch(X) {
case minusInfinity:
  // action
  break;
case plusInfinity:
  // action
  break;
case complexInfinity:
  // action
  break;
}
```

It is not necessary to use enumeration types. The same effect can be achieved with constant integer values:

```
const int minusInfinity = 0;
const int plusInfinity = 1;
const int complexInfinity = 2;
```

Using enumerations does have some advantages. If you decide to add an addition kind of infinity, then you just add that new value to the enumeration list. If a variable is declared to be an enumeration type, then the compiler will complain if you attempt to assign a value to that variable that isn't one of the allowed enumeration values. This can help prevent errors.

15.5.3 Unions

A *union* is a data structure that allows different types of data to be held in the same location in the computer's memory. It is a trick that is designed to save memory. You should not use these things. We mention how they work just for your amusement.

A union is a type (like classes, structures, and enumerations). To declare a union that may hold an integer or a double value, you write this:

```
union number {
  double x;
  long   i;
};
```

Now we can declare a variable to be of type `number`:

```
number Z;
```

The variable z can hold only one data value, but that value may be either a `double` or a `long`. For example, to assign a real value to z we write, for example, `Z.x=0.12;`. Or we can assign an integer value like this: `Z.i=-11;`.

What happens if we assign a value to a union using one of its types and then access its value using another? This is just begging for trouble. The value extracted is bound to be garbage. Consider this program.

```
#include <iostream>
using namespace std;

union number {
  double x;
  long i;
};

main() {
  number Z;
  Z.x = sqrt(2.);
  cout << "Z.x = " << Z.x << endl;
  cout << "Z.i = " << Z.i << endl;
  return 0;
}
```

When run, it produces the following output.

```
Z.x = 1.41421
Z.i = 1073127582
```

15.5.4 Using `typedef`

The keyword `typedef` is used to define synonyms for existing types. For example, rather than using `double` and `long` to declare variables, we mathematicians might prefer to use the single letters R and Z instead. To do this we use the following statements,

```
typedef double R;
typedef long Z;
```

The newly defined names may save you a lot of typing. For example, instead of writing `complex<double>` to declare complex variables, we can use a `typedef`:

```
typedef complex<double> C;
```

Choosing different names for types can improve the readability of your code and save you a bit of typing. In addition, using your own type names may prove invaluable when revising your code. Suppose, for example, you initially used `typedef` to make R a new name for `double`. Subsequently, you find that you don't really need the accuracy of `double` variables and that your arrays are too large for your computer. So you decide to use `float` variables instead. All you need to do is to edit the single line `typedef double R;` to read `typedef float R;` and recompile your code.

15.6 Pointers

Our approach to C++ has scrupulously avoided pointers. We have been able to do this for two reasons. First, call by reference obviates much of the need to use pointers. Second, pointers are often used to create intricate data structures. Fortunately, the Standard Template Library provides a sufficiently rich assortment of ready-made structures, that it is not necessary for us to make new ones. In short, we simply don't need pointers for our work.[3]

However, pointers are used by many C++ programmers, and a package that you might download from the Web may require you to use pointers. We therefore close our exploration of C++ with an overview of pointers.

15.6.1 Pointer basics

An ordinary variable holds a value that represents an integer, a real number, a character, or an object of some class. A *pointer* is a variable whose value is a location in the computer's memory. Given an ordinary variable, one can learn the address of that variable using the address-of operator, `&`. For example, if x is an `int` variable, then `&x` is the location in memory where the value is actually held. This is illustrated by the following program.

```
#include <iostream>
using namespace std;

int main() {
  int x = 12;
  cout << "x   = " << x << endl;
  cout << "&x = " << &x << endl;
  return 0;
}
```

---

[3]The primary exception to this rule is the `this` pointer by which an object refers to itself.

The output of this program looks like this.[4]

```
x  = 12
&x = 0xbffff980
```

By convention, memory addresses are expressed in base 16; the `0x` prefix signals that the value is in hexadecimal.

Pointer values can be saved in variables. In C++, every variable must have a type and we declare pointer variables using an asterisk `*`. If the pointer variable points to a location in memory holding a value of type `base_type`, then the pointer variable is declared to be of type `base_type*`. For example, if `z` is type `double`, then `&z` is type `double*`. This is illustrated in the following program.

```
#include <iostream>
using namespace std;

int main() {
   double z = 1.2;
   double* zp;

   zp = &z;

   cout << "z  = " << z << endl;
   cout << "zp = " << zp << endl;
   return 0;
}
```

```
z  = 1.2
zp = 0xbffff980
```

15.6.2   Dereferencing

Suppose the pointer variable `zp` is type `double*` (i.e., a pointer to a `double` value). We can use `zp` to inspect and to modify the value held in the memory location `zp`. The expression `*zp` stands for the value stored in the location to which `zp` points. So, if we have the statement `cout << *zp;`, the value held at location `zp` is printed. Similarly, the statement `*zp = 3.2;` changes the value held at location `zp` to 3.2, but does not change the pointer `zp` itself. These ideas are illustrated in the following example.

Program 15.5: Illustrating pointer dereferencing.

```
1  #include <iostream>
2  using namespace std;
3
4  int main() {
5     double z = 1.2;
```

---

[4]If you run this program on your computer, the output might be different because the variable x might be stored at a different memory location.

```
6     double* zp;
7     zp = &z;
8
9     cout << "zp  = " << zp << endl;
10    cout << "*zp = " << *zp << endl << endl;
11
12    *zp = M_PI;
13
14    cout << "zp  = " << zp << endl;
15    cout << "z   = " << z << endl;
16
17    return 0;
18  }
```

On lines 9 and 10 we print the pointer itself and the value to which it refers. Then, on line 12, we use the pointer to modify the value pointed to by zp. Because zp points to the location that holds z, the value of z changes. Here is the output of the program.

```
zp  = 0xbffff980
*zp = 1.2

zp  = 0xbffff980
z   = 3.14159
```

The unary[5] operations & (address of) and * (pointer dereference) are inverses of each other. The first converts a variable into a pointer to that variable, and the second converts a pointer to a memory location to the value held at that location.

Program 15.5 contains two hints about the perils of pointers. First, the variable z is modified by an expression that does not use the letter z (line 12). This makes the code more difficult to understand, more likely to contain bugs, and harder to debug. Second, if we neglected to initialize zp (line 7), then we don't know where the pointer zp points. Consider this code.

```
#include <iostream>
using namespace std;

int main() {
  double z = -1.1;
  double* zp;
  *zp = 9.7;

  cout << "z   = " << z << endl;
  cout << "&z  = " << &z << endl;
  cout << "zp  = " << zp << endl;
  cout << "*zp = " << *zp << endl;

  return 0;
}
```

---

[5]The operations & and * have different meanings when used as binary operations: bitwise-and and multiplication.

When run on one computer we have the following result.

```
z    = -1.1
&z   = 0xbffff980
zp   = 0xbffffa60
*zp  = 9.7
```

Note that zp points to some unknown location (it does not point to z) and so the statement *zp = 9.7; changed some unpredictable location in the computer's memory.

When this program is run on another system, the following message is produced: Segmentation fault. This message was spawned by the statement *zp = 9.7; because, on this second computer, zp pointed to a memory location that was "off limits" to the program. (It is also possible to see the message Bus error; this also arises from using bad pointers.) This latter behavior is actually much better than the former because we see right away that something is wrong. In the first instance, no error messages were generated; this type of bug is insidious and difficult to find.

15.6.3 Arrays and pointer arithmetic

As we mentioned in Section 5.2, there is a connection between pointers and arrays. Recall that arrays can be declared in two ways. If, when we are writing our program, we know the size of the array, we declare it like this:

```
int A[10];
```

However, if the size of the array cannot be determined until the program is run, we use new and delete[]:

```
// determine the value of n
int* A;
A = new int[n];
// use the array
delete[] A;
```

In this second case, the declaration of the variable A is indistinguishable from the declaration of a pointer-to-int variable. Indeed, in both cases, the variable A is a pointer. By convention, the name of an array is a pointer to the first element (index 0) of the array. This is illustrated by the following example.

```
#include <iostream>
using namespace std;

int main() {
  int A[10];
  for (int j=0; j<10; j++) A[j] = 10*j+5;

  cout << "The array is: ";
  for (int j=0; j<10; j++) cout << A[j] << " ";
  cout << endl;

  cout << "A[0] = " << A[0] << endl;
  cout << "A    = " << A << endl;
  cout << "*A   = " << *A << endl;
```

```
    return 0;
}
```

```
The array is: 5 15 25 35 45 55 65 75 85 95
A[0] = 5
A    = 0xbffff950
*A   = 5
```

The expressions A[0] and *A have exactly the same meaning. Both stand for the first element of the array and may be used interchangeably. Thus, if we want to change the first element of A to, say, 28, we may use either of the statements *A = 28; or A[0] = 28;. The second, however, is easier to understand and therefore preferable.

Conversely, it is possible to use the subscript notation [] for pointers that were not created to be arrays.

```cpp
#include <iostream>
using namespace std;

int main() {
   int  z = 23;
   int* A = &z;

   cout << A[0] << endl;
   cout << A[1] << endl;

   return 0;
}
```

```
23
-1073743488
```

We see that A[0] yields the value 23. This follows from the fact that A[0] is identical to *A. However, for this example, the notation *A is preferable because A does not refer to an array.

Which leads to the question: In this context, what is A[1]? To answer, we need to understand pointer arithmetic.

Consider this program.

```cpp
#include <iostream>
using namespace std;

int main() {

   int  z = 23;
   int* A = &z;

   cout << A << endl;
   cout << A+1 << endl;

   return 0;
}
```

The first line of output shows that A equals 0xbffff980. It would therefore make sense that the second output statement would produce 0xbffff981; that, however, is incorrect. Here is the output.

```
0xbffff980
0xbffff984
```

We see that A+1 is 4 bigger than A. The reason adding one increases an int * pointer by 4 is that (at least on my computer) the size of an int is 4 bytes. That is, if A points to an int variable in the computer's memory, then A+1 points to the next (possible) int.

In general, one may add an integer to (or subtract an integer from) any pointer. If p is a pointer to an object of type T, then adding *n* to p causes the pointer to change by $n \times$ sizeof(T) bytes. Likewise, p++ increases p by sizeof(T) bytes.

It does not make sense to add, to multiply, or to divide a pair of pointers. However, it does make sense to subtract a pair of pointers of the same type. The result is an integer value defined as the difference in bytes divided by the size of the kind of object to which the pointers point. The following code

```
int   a = 34;
int   b = 49;
int* A = &a;
int* B = &b;

cout << B-A << endl;
```

prints 1 on the screen.

We now return to our discussion of arrays. An array A of objects of type T places the objects contiguously in memory. The location of A[1] is sizeof(T) bytes after the location of A[0]. Therefore, because A points to the first element of the array (i.e., to A[0]), A+1 points to the next element of the array (i.e., to A[1]). In general A+k points to A[k]. Ergo,

A[k] is exactly the same as *(A+k).

15.6.4 new **and** delete **revisited**

In Section 5.5 we introduced the new and delete[] operators. These were used to allocate and to release space for arrays. Here we discuss a slightly different use of new and delete.

Let T be a class. The usual way to declare a variable of type T is with a statement like this:

```
T x;
```

We find such declarations inside the body of a procedure. Such variables are *local* to the procedure. Once the procedure terminates, the variable is lost. Variables declared in this usual manner are stored in a portion of the computer's memory called *the stack*.

An alternative method for setting up a variable is to use `new`. If T is a type, we have the following statements,

```
T* xp;
xp = new T;
```

The `new` statement allocates `sizeof(T)` bytes of memory for a new object of type T. The zero-argument constructor (if any) is invoked when initializing the new object. Then, a pointer to that new object is assigned to `xp`. We may supply arguments to the type T constructor using this syntax:

```
xp = new T(arg1, arg2, ..., argN);
```

When we use `new` (as opposed to the ordinary method of declaring variables) the memory for the object does not reside on the stack. Instead, it resides in a different portion of the computer's memory called the *heap*. The practical difference is that the object defined using `new` persists even after the procedure in which it was created ends.

Once the program is finished using an object created with `new`, the memory allocated to that object should be released. This is done with a `delete` statement:

```
delete xp;
```

Notice that the syntax here is slightly different from the situation in which we are freeing an array; in the latter case we use `delete[]`. The square brackets indicate that an entire of array of objects is being freed.

Once an object is created using `new`, it can be used just as any other object. However, we must remember that the variable `xp` is *not* of type T, but rather is a pointer to an object of type T. Therefore, if T has a method named, say, `reset`, it is an error to invoke this method with the expression `xp.reset()`. Instead, we should write `(*xp).reset()`. Likewise, to access a data member of this object (say, it has a data element named a), we do not use the expression `xp.a`; rather, we write `(*xp).a`.

There is an alternative way to write `(*xp).reset()` and `(*xp).a`. The C++ operator `->` is defined as follows.

- `xp->reset()` means `(*xp).reset()`, and

- `xp->a` means `(*xp).a`.

The combination of a hyphen and a greater-than symbol is meant to look like an arrow pointing to the part of `*xp` we want.

15.6.5 Why use pointers?

There are a few situations in which pointers are useful, but in nearly all cases, one can do fine without them. Before call by reference was introduced into C++, procedures written in C that needed to modify their arguments used pointers instead of the values of the variables. For example, a procedure to swap the values of two `double` variables would be written like this:

```
void swap(double* a, double* b) {
  double tmp = *a;
  *a = *b;
  *b = tmp;
}
```

To use this `swap` procedure on variables x and y of type `double`, we would write `swap(&x,&y);`. The ampersands are necessary because this version of `swap` requires pointer-to-`double` arguments.

Similarly, suppose T is a class and that objects of type T are large (say, hundreds of bytes). Then each time we pass an object of type T to a procedure, the object is duplicated. Where possible, we can improve efficiency by passing a reference to the object instead of using call by value. Alternatively, we can declare the procedure's argument to be of type pointer-to-T. Passing a pointer is efficient, but we need to remember to pass the address of the object, and not the object itself. This is summarized in the following chart:

Declaration	Invocation	Efficiency
`void proc(T x) {...}`	`proc(x);`	slow
`void proc(const T& x) { ... }`	`proc(x);`	fast
`void proc(T* x) { ... }`	`proc(&x);`	fast

There is a natural way to avoid using `new` to create new instances of objects in the heap. Recall that primary purpose for doing this is to avoid passing large objects to and from a procedure. For example, a graph theory package would naturally include a procedure for partitioning the vertex set into connected components. This hypothetical procedure, named `components`, would take a single argument (of type `Graph`) and return a partition. The standard way to define such a procedure is like this:

```
Partition components(Graph g) {
  Partition p;
  // ...
  return p;
}
```

Then, when we have the statement `P = components(G);` the computer would first make a copy of G (to be held in g inside the procedure). Then the `Partition` computed by the procedure (p) is copied back to the calling procedure to be saved in P. If the graph is large, this is highly inefficient.

Some programmers would solve this problem by using pointers. In that paradigm, the procedure would look like this:

```
Partition* components(Graph* gp) {
  Partition* pp;
  pp = new Partition;
  // ...
  return pp;
}
```

To invoke this procedure, we use a statement like this: `P_ptr=components(&G);`.
Now `G` does not have to be copied; a pointer to `G` is all that needs to be sent to the
procedure and a pointer to the newly calculated partition is all that needs to be sent
back. It is then incumbent on the programmer to remember to `delete pp;` or else
suffer a memory leak.

There is a better, third alternative. We can use call-by-reference parameters to
prevent repeated copying, and avoid pointers altogether. In this case, we define the
procedure like this:

```
void components(const Graph& g, Partition& p) {
  // calculate p from g
}
```

In this case, we call the procedure with a statement of the form `components(G,P);`.
The procedure would begin by erasing any data that happen to be in `P` and then
overwriting with the new partition.

Thankfully, there are hardly any situations in C++ that require the use of a pointer.
You may be required to use pointers when using other people's packages. An instance of this is the Standard Template Library's `sort` procedure (see Section 7.4).

Every method in a class may use a pointer named `this`. The `this` pointer refers
to the object that invoked the method; that is, if we call `X.method()`, then `method`
may use a pointer named `this` that points to `X`. This is handy if `method` modifies
`X` and then wants to return a copy of `X`. This behavior is desirable when defining
operators such as `+=` for a class. See Section 6.8.

15.7   Exercises

15.1 Create a `Card` class to represent cards from a standard 52-card deck. The class
should have the following features.

- A zero-argument constructor that creates a default card (say, the ace of
spades) and a two-argument constructor that sets the the value and suit
of the card. The user should be able to create cards such as this:

```
Card boss(ACE,SPADES);
Card weak(2, CLUBS);
```

- Comparison operators `<`, `==`, and `!=`.

- An `operator<<` for writing cards to the screen in English words, such as
`ace of spades` or `four of diamonds`.

15.2 What is the difference between the following two statements?

```
T* xp = new T(10);
T* xp = new T[10];
```

15.3 In Exercise 5.11 you were asked to create a program that repeatedly asked for large blocks of memory by using `new` without balancing calls to `delete[]`. When `new` is unable to allocate space (because memory has been exhausted), it throws an exception of type `std::bad_alloc`. Revise your program so that it exits gracefully (instead of crashing) when memory is exhausted.

15.4 C++ uses punctuation symbols for a variety of purposes. Many of these symbols serve multiple purposes in the language. For example, `*` is used for multiplication and for declaring arrays (and for a few other purposes). In addition, these symbols can be doubled (e.g., `&` versus `&&`) or combined with others (e.g., `!=`) to form additional meanings.

Find as many meanings for these symbols (either singly or in combination) as you can:

`+ - * / = < > ! & | ~ :`

Don't bother listing all of the combined arithmetic/assign operators such as `+=`. Classes may overload many of these symbols; list only the standard overloads.

15.5 Create a class named `Constructible` to represent numbers that can be determined using the classical construction tools: straight edge and compass. Specifically, define *constructible numbers* recursively by declaring that all integers are constructible; the sum, difference, product, and quotient[6] of constructible numbers are constructible; and the square root of a constructible number is constructible. The complex numbers that can be built this way correspond to points in the plane that can be constructed using a straight edge, a compass, and a given unit length.

The objects in this class should represent their numbers exactly. That is, the number $2 + \sqrt{5 - \sqrt{2}}$ should be held as such, and not as a decimal (`double`) approximation.

This class should include the following methods.

- A zero-argument constructor to create the constructible number zero.
- A one-argument constructor that takes a `long` integer argument.
- A copy constructor.
- A destructor.
- An assignment operator.
- The binary operators `+` `-` `*` `/` (either two `Constructible` arguments or a `Constructible` and a `long` integer) and unary `-`.
- A `sqrt()` method that returns the square root of the constructible number on which it was invoked.

---

[6]Of course, division by zero is forbidden.

- A `value()` method that returns a `complex<double>` value giving the approximate decimal value of the constructible number.

- An `operator<<` procedure that writes the number to an output stream in a suitable format such as TEX

```
2 + \sqrt{5 - \sqrt{2}}
```

or *Mathematica*

```
2 + Sqrt[5 - Sqrt[2]]
```

or some sensible method of your choosing.

15.6 *Tautology Checker.* Create a program to check if Boolean expressions are tautologies. A *Boolean expression* is an algebraic expression involving variables $(a, b, c, \ldots)$ and logical operations (and, or, not, implies, iff, etc.). A *tautology* is a Boolean expression that evaluates to TRUE for all possible truth values of its variables. For example, $((x \to y) \land x) \to y$ is a tautology.

The program should be invoked either (a) with a command-line argument specifying a file that contains the Boolean expression to be tested, or (b) with no command-line argument, in which case the user is prompted to type in the Boolean expression.

The variable names may be any single lowercase letter (a to z). Use the following for operation symbols.

Symbol	Meaning
+	or $\lor$
*	and $\land$
−	not $\sim$
>	implies $\to$
<	implied by $\leftarrow$
=	equivalent $\leftrightarrow$
0	TRUE
1	FALSE

Use reverse Polish notation (RPN) for the expressions (these are easier to process than ordinary algebraic expressions). For example, the expression $((x \to y) \land x) \to y$ would be entered as `x y > x * y >`. Spaces are optional, so this may also be entered as `xy>x*y>`.

It may be convenient for the user to enter an expression over several lines. The user should indicate that the expression is finished by typing a period.

Part IV

Appendices

Appendix A

Your C++ Computing Environment

The methods by which you type, compile, and run programs varies between different computers with different operating systems.

There are two broad approaches to this edit/compile/run process.

- You use separate tools in terminal windows. That is, you type commands in a window to invoke an editor for creating and modifying your C++ files, type another command to compile your code, and another command to run the program.

  The code in this book was developed using the emacs editor and compiled using the g++ compiler. These tools are available for many computer systems (Windows, Macintosh, UNIX) and available for free from the Free Software Foundation (www.gnu.org).

- You use an integrated development environment (IDE). This is a software application that provides a built-in text editor (for creating and modifying your code), a compiler, a debugger, and a means to run your program. Such environments include Microsoft's Visual Studio on Windows computers, Apple's Xcode on Macintosh OS X, KDevelop on Linux, and Metrowerk's Code Warrior which runs on several platforms.

Which approach you prefer is a matter of taste. The installation of these tools ranges from simple to complex. You may need some assistance from a friendly computer science colleague in getting started. If you can type in the code in Program 1.1, compile it, and run, then you are well under way!

In this Appendix we provide specific advice to help you get started programming.

A.1 Programming with a command window and a text editor

In this approach your work is performed using two windows: a text editor and command shell. The specific tools we discuss come from the Gnu/UNIX world, but are available for Windows, Macintosh, Linux, and many other systems. It is possible that these tools are already present on your computer (especially if you are using Linux or some variation of UNIX).

A.1.1   What you need and how to get it (for free)

To program in this manner you require the following.

- A shell command window (running `sh` or one of its variants such as `bash` or `csh`).

- A text editor (the `emacs` or `Xemacs` editor, or some other editor designed for programming).

- A C++ compiler (such as `g++` from Gnu).

- The `make` program to automate the compiling process (optional).

- The Doxygen program for creating Web-based documentation for your own program (optional). See Appendix B.

If you are using a UNIX computer, these tools are likely to be already installed. For Windows and Macintosh users, you have the following options.

Cygwin on Windows

For Windows, we recommend that you install Cygwin (available for free from `www.cygwin.com`). This is a system that provides UNIX-like tools including the pieces you need: a `bash` terminal window, the `g++` compiler, `emacs` and `Xemacs` text editors, and the `make` program.

Follow the download and installation directions at the Cygwin Web site (see "Setting Up Cygwin").

Next, run the `setup.exe` program which presents you with a long list (separated into categories) of packages to install. Select the packages you want to install. We recommend the following.

```
bash
doxygen
emacs
emacs-X11
gcc
gcc-g++
make
xemacs
xorg-x11-xwin
xterm
```

The Cygwin `setup` program may install other packages that you did not select because it understands how the packages you want depend on other packages. For example, selecting the `xemacs` package triggers the inclusion of other packages needed to provide the X-window environment.

If all has gone well, new entries will be present in your Start menu, including an entry to start a `bash` shell:

```
Start > Programs > Cygwin > Cygwin Bash Shell
```

Macintosh tools

Macintosh OS X comes with many of the tools you need. For example, the Terminal application launches a `bash` shell.

If X11 has not been installed (see the Utilities folder inside the Applications folder), then install it using the disks (CD/DVD-ROM) that came with your computer.

You need the Developer Tools. Check if there is a folder named Developer at the top level of your hard drive (and be sure that inside that folder there is a subfolder named Applications containing the Xcode application). Alternatively, open a Terminal window and type the command: `g++ --version`. If the computer complains `g++: command not found` then you need to install the Developer Tools, but if it responds with the version of `g++` installed, then you know that the compiler is already installed on your computer. The `make` program should also be installed (try `make --version`).

If the Developer Tools are not on your computer, they are either available on the disks that came with your computer or you can download them (for free) from Apple's Web site (`developer.apple.com`). Get a free membership to their Developer Connection, navigate to Developer Tools, and download Xcode.

Next you need a text editor such as `emacs`. You can find Macintosh-style versions of `emacs` for free on the Web; see:

`http://home.att.ne.jp/alpha/z123/emacs-mac-e.html`

Finally, install the Doxygen application, available for free from `doxygen.org`. This is a standard Macintosh application that you install simply by dragging its icon to your Applications folder. (See Appendix B.)

A.1.2 Editing program files

To create the files for your C++ project you need a text editor such as `emacs`. Do not use a word processor, such as Microsoft Word, for this purpose.

A good programming editor makes your life easier. As you type your program, the editor automatically indents lines of code the proper amount so you can see the structure. It also highlights keywords in color. If, for some reason, the editor does not indent your typing the distance you expect or a keyword does not appear in color, then you have a quick visual clue that you mistyped something. Smart text editors (such as `emacs`) detect what sort of file you are editing and place you into an appropriate mode[1] for that sort of file.

Begin by creating a folder (directory) where you want to save your program. If you wish to use code you have already created for another project (or code from this book on the accompanying CD), you can copy the files you want into this directory.

---

[1] In `emacs`, editing a `.h` file places you in an editing mode meant for C, and not for C++. To switch to C++ mode, type: `M-x c++-mode` where `M-x` means "meta-x".

A typical C++ program has several files: .h header files and .cc code files. Save them all in the working directory you created.

The emacs editor takes some practice because its conventions are different from other (non-UNIX) programs. The mouse-driven menus in emacs (or Xemacs) make getting started easier, but it is worth learning the keyboard shortcuts for common tasks including creating, opening, and saving files. I recommend you work through the tutorial that comes built in to emacs (available from the **Help** menu).

A.1.3   Compiling and running your program

By now you have created and saved your program in its own directory (folder) on your computer and you are ready to see if it works. Open a shell window[2] and navigate to your working directory. The shell command you need to change directories is cd (for change directory). Typing cd (and then pressing return) places you in your "home" directory. To move to a subdirectory type cd subdir where "subdir" is the name of the subdirectory. To move up a level in the hierarchy, type cd .. or to move to a directory directly, type the full path, like this:

```
cd /home/barney/programming/twin-primes
```

If you lose track of where you are in your computer's file structure, the command pwd (for print working directory) reports your current location.

One more useful shell command is ls. The ls command lists all the files in the current directory.

Single file programs

If your program consists of just a single file (e.g., poem.cc) then compile your program with this command:

```
g++ poem.cc
```

After a brief pause, one of two things happens.

- With luck, the computer responds by typing the shell prompt (typically a single dollar sign $). This means that the program compiled without errors and created a ready-to-run program. Depending on your system, the program is named a.out (most UNIX systems) or a.exe (on Windows/Cygwin).

- Or, more likely, the compiler types warnings and error messages. Unfortunately, these can be difficult to understand. The error messages usually report line numbers where the compiler got into trouble. However, that is not necessarily the line number where the programming error resides. You will quickly learn that an error reported for line n is often caused by a missing semicolon on line $n-1$. Another common error is to forget the using namespace std; statement.

---

[2]Use a bash or similar shell window. If you are using Windows, do not open a DOS command window; instead, use the bash command terminal supplied by Cygwin.

When you get error messages, go back to your editor and figure out what's wrong. Make some changes and try compiling again.

When compiling is successful, the program is ready to run. In the shell window type

```
./a.out
```

(or `./a.exe` for Windows/Cygwin). The program runs and the output appears on the screen.

If nothing appears to be happening and the shell prompt has not returned, your program may be stuck in an infinite loop. To force the program to stop, type a Control-C (while holding down the Control key on your keyboard, press the C key).

You might not want to name your program `a.out` (or `a.exe`). You may either compile as described above and change the name of the executable, or you can compile with the following command,

```
g++ poem.cc -o poem
```

The `-o poem` option causes the compiler to write the executable program to a file named `poem`. (Use `poem.exe` for Windows.)

Multiple file programs

In most cases, your program spans multiple files. Typically, each class has a `.h` file and a `.cc` file. The `main()` procedure is in yet another file. For example, suppose your project consists of the following files: `gcd.h`, `gcd.cc`, `Mod.h`, `Mod.cc`, and `main.cc`.

The simplest way to compile your program is with this shell command:

```
g++ gcd.cc Mod.cc main.cc
```

This compiles the program in these three source files and writes the executable to `a.out` (or `a.exe`). Alternatively, the shell command `g++ *.cc` compiles all files that end with `.cc` in the current directory.

The `.h` files do not need to be mentioned in the compile command; they are automatically absorbed into the appropriate `.cc` files by the `#include` directives.

If your project has just a few `.cc` files, then compiling them all with a single command works well. Imagine, however, that your project has ten different `.cc` files. You compile with the convenient `g++ *.cc` command, and the compiler complains about an error in one of the files. You fix the offending `.cc` file and compile again. The annoyance with this approach is that C++ compiling can be time consuming. So, rather than compiling all the `.cc` files at once, it is possible to compile them one at a time (fixing errors in each as you go), and then combine the pieces into a single executable program. Here is how you do this.

Suppose the `.cc` files are named `one.cc`, `two.cc`, ..., `nine.cc`, and finally `main.cc` (which is the only source file that contains a `main()` procedure). To compile just the file `one.cc` type this:

```
g++ -c one.cc
```

The -c option tells the compiler that this is just a piece of the whole program, so don't try to build a full executable a.out. Instead, the compiler creates a file named one.o. If there is a problem while compiling one.cc, we fix the error and compile this part again. This is much faster than recompiling all the files every time you modify just one of them.

Of course, you also need to compile two.cc with the command g++ -c two.cc, and so on for all of the .cc files.

Finally, you have collected several .o files that the computer needs to stitch together to make the executable a.out (or a.exe). This phase is known as *linking* the object (.o) files. To link the .o files together, type this:

```
g++ *.o
```

Alternatively, you may name all of the .o files on the command line instead of using the wildcard *.o. You may also use the -o option to write the executable file to a different file name.

If your project has just a few .cc files, then this technique is quite manageable. If you have many .cc files, then it is easy to lose track of which .o files are up to date with their corresponding .cc files. Fortunately, there is a good way to automate this procedure using the make program. We describe this tool in a moment (see Appendix A.1.5) but first we introduce additional options you can give to the compiler.

A.1.4 Compiler options

In the previous section we introduced two options that can be used with the g++ compiler. The -o option is used to specify the name of the final executable program and the -c option is used to compile a .cc file just to a .o file and not to try to link the object files into an executable program.

Here are some other options you can give to the g++ compiler.

- -Wall: This tells the compiler to issue all warnings. There are C++ statements that are not exactly errors, but are not fully legitimate C++. The compiler can proceed, but something is not quite right. With this option specified, all possible warnings are issued.

  It's a good idea to use this option and fix your code so that all warnings disappear.

- -ansi and -pedantic: These options disable nonstandard C++ features. Compiler manufacturers often include extensions to the standard C++ language. If you plan to use your program only on your own computer, then there is no serious problem in taking advantage of these nonstandard options. However, if you try to compile your program on another computer with a different C++ compiler, these special features might not be available. By including these options you ensure the portability of your code.

- -O, -O2, and -O3: These options engage the compiler's optimizer. (Note: The O is the uppercase letter oh, and not the numeral zero.) Without these options, the computer compiles your code as quickly as possible, but does not produce the best possible (i.e., fastest) executable code. There are several "tricks" the compiler can apply during the compilation process and these options request greater and greater levels of optimization. The basic optimization is -O (which is equivalent to -O1). Each of -O2 and -O3 employ additional optimization techniques.

- -g: This causes the compiler to produce a program that can be run using a debugging tool (gdb). The gdb debugger is complicated to use and beyond the scope of this book. Some advice on debugging your code is presented in Appendix A.3.

- -p and -pg: This causes the compiler to insert extra code in your program so that it can be analyzed with a profiling tool (prof and gprof, respectively). These tools measure how much time your program spends in each of its procedures and how many times each procedure is called. If you need to make your program run faster, then this information tells you where to concentrate your efforts.

- -I *directory*: This option tells the compiler where to look for nonstandard header files. If you have downloaded and installed some special C++ package, you may need a #include directive for its header files. These header files might not be installed in a location that the C++ compiler knows to check. One solution to this problem is to copy the header files into your project's directory. A better alternative is to tell the compiler where to find the special header files and for that you use the -I option. (See the next bullet for more details.)

- -L and -l options: Packages that you download from the Internet may contain *libraries*—these are similar to .o files in that they contain compiled programs that can be linked into your final program. For example, the Gnu Multiprecision Package (GMP) is distributed with .h header files and libraries. One uses the -I option (described above) to specify the location of the .h files and the -L and -l options to specify the location and name of the library.

  For example, on my computer the GMP header files are located in the directory /sw/include so if I have #include <gmpxx.h> in my program, I need the -I /sw/include option when I compile my programs.

  Furthermore, because the GMP procedures are embedded in a library, I need to tell the computer where the library is located (because on my computer, it's in a nonstandard location) and the name of the library I need. The -L option specifies the location of the library. For my situation, I would type -L/sw/lib to specify the location. The names of the GMP libraries are gmp and gmpxx (we need both when working in C++). To tell g++ to use these libraries, we use the options -lgmp -lgmpxx.

More details: The $-I$ option is information for the *compiler* whereas the $-L$ and $-l$ options are used by the *linker*. So if I were to compile the various source files of my program separately, the commands I use would look like this:

```
g++ -c alpha.cc -I /sw/include
g++ -c beta.cc -I /sw/include
g++ -c gamma.cc -I /sw/include
g++ -c main.cc -I /sw/include
g++ *.o -L/sw/lib -lgmp -lgmpxx -o my_program
```

While writing and debugging your program, I recommend you use the $-$Wall, $-$ansi, and $-$pedantic options. Once the program is working properly, do a final compilation with $-$O3.

A.1.5 Introduction to make

A program with just a few .cc files is easy to compile with a single command, either g++ *.cc or

```
g++ alpha.cc beta.cc gamma.cc main.cc -o myprog
```

Alternatively, each piece can be individually compiled and the .o files linked with these commands:

```
g++ -c alpha.cc
g++ -c beta.cc
g++ -c gamma.cc
g++ -c main.cc
g++ alpha.o beta.o gamma.o main.o -o myprog
```

However, if the project involves more .cc files, it is annoying to need to compile each .cc file separately, and difficult to remember which .cc files have been successfully compiled into the corresponding .o files. One is tempted to type g++ *.cc and endure the slow compilation process.

Happily there is an alternative: The program make automates the entire process. To use make you create a file in your project directory named Makefile and then type make at the shell command prompt.

A basic Makefile for compiling a program based on four .cc files (alpha.cc, beta.cc, gamma.cc, and main.cc) is presented in Program A.1.

Program A.1: A basic Makefile.

```
1  OBJS = alpha.o beta.o gamma.o main.o
2  CXXFLAGS = -Wall -pedantic -ansi
3
4  myprog: $(OBJS)
5          g++ $(OBJS) -o myprog
6
7  clean:
8          rm -f *.o
```

Line 1 defines the object (.o) files that need to be created. Line 2 sets the options that g++ is to use when compiling the .cc files.

Line 4 indicates that the program myprog depends on the files listed in OBJS. The syntax $(OBJS) tells make that the variable OBJS is to be replaced by its value (i.e., the list of .o files on line 1). The program myprog is called a *target* and line 4 means that before myprog can be made, each of the files after the colon (i.e., the four .o files) must be made.

Line 4 tells make on which files myprog depends, whereas line 5 tells make how to create the program myprog from the list of .o files. Notice that line 5 is indented; what you can't see is that this indentation is caused by entering a TAB character into the file (and not by typing several spaces). The TAB is mandatory. Line 5 is equivalent to this:

```
g++ alpha.o beta.o gamma.o main.o -o myprog
```

In other words, line 5 tells make how to link the object files into the executable program named myprog.

(Ignore lines 7 and 8 for now.)

Interestingly, there is no indication in the Makefile of how to create the .o files from the .cc files. That's because make is smart enough to figure that out. It knows that if there is a file named alpha.cc in the working directory, then it is compiled into alpha.o using the command g++ -c alpha.cc.

However, by setting the special make variable CXXFLAGS on line 2, make adds -Wall, -pedantic, and -ansi.

With this Makefile in the same directory as alpha.cc, beta.cc, gamma.cc, and main.cc, we type make at the shell prompt. Here is what we see.

```
$ make
g++ -Wall -pedantic -ansi    -c -o alpha.o alpha.cc
g++ -Wall -pedantic -ansi    -c -o beta.o beta.cc
g++ -Wall -pedantic -ansi    -c -o gamma.o gamma.cc
g++ -Wall -pedantic -ansi    -c -o main.o main.cc
g++ alpha.o beta.o gamma.o main.o -o myprog
```

Now suppose we want to make a change to, say, beta.cc. If we type make again, this is what happens:

```
g++ -Wall -pedantic -ansi    -c -o beta.o beta.cc
g++ alpha.o beta.o gamma.o main.o -o myprog
```

Notice that only beta.cc is recompiled and then myprog is relinked. The other .o files do not need to be rebuilt, and they weren't.

If your project has many .cc files, they might not fit conveniently on a single line in the Makefile. You can use a backslash character \ to show that the definition of OBJS continues onto the next line, like this:

```
OBJS = one.o two.o three.o four.o five.o six.o \
       seven.o eight.o nine.o main.o
```

Now let's return to lines 7 and 8. Line 7 defines a target named clean with no dependencies and line 8 tells make what to do when we want to make clean. Line 8 causes the shell command `rm` (for re̲mo̲ve) to delete all files that end with `.o`, that is, the object files. Once the program is functioning properly, we do not need the `.o` files and they can be removed. To cause this to happen, we simply type `make clean`.

(Note: The standard way to run the `make` program is to type `make` *target-name* at the command prompt. However, if *target-name* is omitted, the first target found in the `Makefile` is built. Hence typing `make` or typing `make myprog` are the same for this `Makefile`. However, to cause `make` to run the commands for `clean`, we need to type `make clean`.)

There are two other reasons why we might want to remove all the `.o` files.

- Once the program is working properly, we may wish to recompile with `-O3` added to line 2 of the `Makefile`. To force `make` to compile everything again, we type `make clean` and then `make`.

- This simple `Makefile` cannot determine that `.cc` files may depend on `.h` files. A change to a `.h` file should be followed by a recompilation of all `.cc` files that `#include` it. It is possible to construct a more elaborate `Makefile` that deals with this issue, but a simpler solution is this: if you change header files, do a `make clean` followed by a `make`.

A.2   Programming with an integrated development environment

An integrated development environment (IDE for short) is a computer application that provides all the tools you need to write programs. It includes a text editor (for creating and modifying your `.h` and `.cc` files), a compiler, a debugger, and many other tools. They are extremely powerful and can be used to work on massive software engineering projects involving teams of programmers. These tools provide a dizzying array of menus and control buttons that can intimidate a novice programmer.

Fortunately, you can begin by using only a small fraction of the IDE's capabilities. As your familiarity with the tool grows, you can gradually add more of its features to your repertoire.

In this section we give a brief introduction to Microsoft Visual Studio (for Windows) and Xcode (for Macintosh OS X, available for free from Apple).

Please note that software manufacturers regularly release new versions of their products. The new versions may have different features and the locations of menus and other controls may vary from version to version.

Common features

Visual Studio and Xcode share some basic organizing principles. In both you create *projects*. A project houses all the pieces you need for your program such as

.cc and .h files.[3] You either create these component files using the IDE's built-in editor or you add files that have already been created elsewhere (via an **Add** menu item). Once the files are written and incorporated into the project, one *builds* the project; this means compiling the various source files and creating an executable program. Next, we *run* the program (within the IDE environment) and observe the action in a terminal window. If any errors crop up during the build phase, the IDE provides a convenient way to jump to the problematic portion of the source file.

Both IDEs provide a debugger. You set *break points* in the program; the break points cause the program to pause at the statement where they are set. While the program is suspended, you can inspect the values held in variables. You can then start the program running again. If a subsequent break point is reached (or the same break point is met again), you have another chance to inspect the contents of variables.

Both IDEs have a concept of a *build configuration*. The different configurations set different options for how the project is to be built. By default, both provide configurations named *Debug* and *Release*. Typically, one works in the *Debug* configuration until the program is working as expected. Then one switches to *Release* for a final build.

A.2.1 Visual C++ for Windows

Visual Studio is a integrated development environment sold by Microsoft. Install this on your computer using the disks provided.

Starting a new project

Launching Visual Studio[4] brings you to a start page with a button labeled **New Project** that you should click.

A "New Project" window opens with a list of project types (on the left) and templates (on the right). Follow these steps.

1. Select **Visual C++ Projects** from the left and then **Win32 Console Project** from the right. Type in a name and a location for your project in the lower portion of the box: the *location* is the full path name of a folder on your hard drive (e.g., C:\research\math-programs\). The *name* is the name you want to give to your project (e.g., twin-primes). Fill these in and click **OK**.

2. A dialogue box appears; in the left of this box click on **Application Settings**. On the right, check **Empty Project** under **Additional Options**. Click **Finish**.

---

[3]More complicated projects can include images, sounds, video, and the like.
[4]You can launch Visual Studio from your Start menu.

Adding files to the project

You are now ready to create your program files. From the **File** menu[5] select **Add New Item**. A dialogue box opens from which you select either **C++ File** or **Header File** (and you can safely ignore all the other options). Choose the type of file and type in its name in the text box at the bottom of the window. Note that Visual Studio expects the name of your C++ source file to end with `.cpp` (and not `.cc` as we have been using in this book). Feel free to use whichever naming convention you prefer. Click **OK**.

You can now start typing your program. Create additional source files by using **Add New Item** until you have created all the files you need.

If there is a file already on your computer that you want to add to this project (perhaps a file you created in a different project), use the **Add Existing Item** menu[6] option.

If you want to edit a file that is already in your project, double-click on its icon in the **Solution Explorer** on the right side of your screen.

Compiling

Once your source files are collected in your project, it's time to compile. Under the **Build** menu, select **Build Solution**. This engages the compiler. As the compiler runs, it reports its progress in a window at the bottom of your screen.

When the compiler finishes you see, with luck, a message that the build succeeded. More likely, however, there are errors in your program in which case a "Build Error Task List" appears in the bottom of your screen. Read the first error carefully and then double-click on it. Visual Studio opens the source file containing the error and takes you directly to the line at which it ran into trouble (which might not be the faulty line in your program).

After you fix the first error on the Task List, you have two choices. You can proceed to fixing the second error or you can try to compile again. Sometimes fixing the first error also fixes several subsequent errors (e.g., if you forgot to declare a variable). In this case, rebuilding removes several errors from the Task List and helps you move efficiently to the next trouble spot.

Iterate this process of building the project and fixing errors until no errors or warnings remain.

Running and debugging your program

Once the program compiles without errors, it's time to start it running. To do this, go to the **Debug** menu and choose **Start Without Debugging**. When the program starts running, a terminal window opens showing the program's output and provides a place for you to type any input the program requires. When the program finishes, the

---

[5]Or from the **Project** menu.

[6]Available either in the **File** or **Project** menu.

message Press any key to continue is typed in the terminal window—press a key and the window closes.[7]

If your program is not behaving as you expected, then you need to find the error in your program. This is more difficult than fixing problems that arose during the compilation process. Fortunately, the debugger in Visual C++ is extremely helpful (also see the advice in Section A.3 later in this appendix). Open any of your C++ code files. To the immediate left of your program is a gray strip; click next to a statement in your program and a large red dot appears in this strip. You have just set a break point in your program. Place these break points wherever you want your program to pause. Now select **Start** from the **Debug** menu. The program runs until it reaches the first break point and then pauses. In the lower portion of the screen you can find a list of variables and their values. You can check these to be sure they are what you expect. When you are ready to continue, select **Continue** from the **Debug** menu. The program resumes, pausing again at the next break point.

When running under the debugger, the terminal window closes immediately after the program ends (so you can't see what was written to that window). To prevent this from happening either set a break point at the return 0; statement in main() or add these lines to the end of your main():

```
cout << "Program finished" << endl;
cin.get();
```

Now when you run the terminal window does not disappear until you type a return after seeing your Program finished message.

Final version

Once your program compiles and runs without errors it is time to perform one final build. When Visual Studio starts a new project, the project is in a *Debug* config- uration. Now that our program is working, we switch to a *Release* configuration by pulling down the **Build** menu and selecting **Configuration Manager**. This opens a dialogue box whose topmost control is a pull-down menu of configurations. Switch this from "Debug" to "Release" and click the **Close** button.

Having switched the configuration to *Release*, build the project again. Visual Stu- dio uses different compiler options for different configurations and, by default, the *Release* configuration has compiler optimizations enabled so the executable program runs faster.

(Note: You can modify the compiler options for the *Release* configuration. Pull down the **Project** menu and select the **Properties** item for your project (the last item in this menu). Select the **C/C++** tab on the left and then the **Optimization** subtab. Now, on the right, you find the various options that can be set to adjust the optimization. In general, choose optimizations that favor higher speed as opposed to smaller size.)

---

[7]Note: If you choose **Start** from the **Debug** menu, then the terminal window closes immediately after the program finishes running.

With the final version built, you do not have to use Visual Studio to run your program. Using the Windows, navigate to the directory that houses your project (e.g., `C:\research\math-programs\twin-primes`). Within this folder, open a folder named `Release` where you should find an executable file named (based on our illustration) `twin-primes.exe`. This file can be copied to any convenient location on your hard drive. You can run this program either by double-clicking it, or by opening a DOS command window, going to the appropriate director, and typing its name.

A.2.2   Xcode for Macintosh OS X

Starting a new project

Launch Xcode[8] and select **New Project** from the **File** menu. Scroll until you reach the heading **Command Line Utility** under which you select **C++ Tool**; click **Next**. Type in a name for your project and select a directory (folder) in which this project is to reside. Click **Finish**.

A project window opens. Notice that a `main.cpp` file is already present in the project. You may either edit this file (double-click its name to get started) or you can delete it: select the file (single-click on its name) and press the Delete key on your keyboard. Choose **Delete References & Files** and `main.cpp` is deleted from the project and from your hard drive.

Adding files to the project

You are now ready to create your program files. From the **File** menu select **New Empty File**. This opens a new window in which you can type your program (either a header `.h` file or a code `.cc` file).

Start typing and be sure to save your work. The first time you save you are prompted for a name and location for your file. Save your work inside the project folder. Next, the computer asks if you wish to add this file to your project; click **Add to Project** and then **Add**.

If there is a file already on your computer that you want to add to this project (perhaps a file you created in a different project), select[9] **Add to Project** from the **Project** menu. Navigate until you find the file you want and then click **Add**. A second panel opens at the top of which is an option to **Copy items into destination group's folder (if needed).** Click the check box next to this to indicate that you want a copy of the file made inside your project; then click **Add**.

To edit a file that is already in your project, simply double-click on its name in the project window.

---

[8]Launch Xcode by double-clicking on its icon. The Xcode application is not in the main Applications folder. Rather, use the Finder to navigate to the top level of your hard disk, open the Developer folder and then the Applications folder inside there.

[9]If the **Add to Project** option is dimmed, click once on the name of your project just below the heading "Groups & Files" in the left portion of your main project window.

Compiling

Once your source files are collected in your project, it's time to compile. Click on the **Build** icon in the project window (or select **Build** from the **Build** menu) and the compiler gets to work.

When the compiler finishes you see, with luck, the message "Build succeeded" at the bottom left of the project window. More likely, however, is that there are errors to fix. The number of errors found in each file is reported in the main project window. You should also see a small red stop sign (octagon) with a white X at the lower right corner of the project window; click on this to bring up a list of errors. As you click on each error, the lower portion of the window jumps to the trouble spot in the relevant file.

After you repair the first error, you have two choices. You may proceed to fixing the second error or you can try to compile again. Sometimes fixing one mistake resolves subsequent errors; in this case, rebuilding removes several items at once and allows you to move efficiently to the next trouble spot.

Iterate this process of building and fixing errors until no errors or warnings remain.

Running and debugging your program

Once the program compiles without errors, it's time to run it. Simply click on the **Build and Go** button on the project window (or select **Build and Run** from the **Build** menu) and the program starts executing. A window opens to show you the output of the program and provide a place for you to type any input the program requires.

If your program is not behaving as you expected, then you need to find the error(s) in your program. This is more difficult than fixing problems that arose during the compilation process. Fortunately, the debugger in Xcode is extremely helpful (also see the advice in Section A.3 later in this appendix). Open any of your C++ code files (by double-clicking its name in the project window). If you click in the left margin next to any of the statements in the code a small black arrow appears. This indicates that you have set a break point into the program. Place these break points wherever you want the program to pause. To remove a break point, simply click on it.

Once the break points are set, choose **Build and Debug** from the **Build** menu. The program starts running and then stops at the first break point it encounters. A list of variables with their current values is displayed. If some of the variables are classes, you can inspect the data members (including private members) for those objects.

Continue executing the program by pressing the **Continue** button on the debugger window. The program now resumes execution until it reaches another break point or the end of `main()`.

Final version

Once your program is functioning properly it is time to perform one final build. In the **Project** menu, find the **Set Active Build Configuration** submenu, and select

Release.

Next select **Edit Project Settings** from the **Project** menu. A window appears with four panel selectors at the top; choose **Build** from these. Reading from the top of the page, there is a drop-down menu for "Configuration"—choose "Release." Next there is a drop-down menu labeled "Collection"—choose "Code Generation."

You should see a list of setting/value pairs. Find the "Optimization Level" setting and, using the up/down arrows to the right, set the optimization level to "Fastest [–O3]."

Now close the Info window and click the **Build** button on the main project window. This creates a "release" version of your program.

Once this final version is built, you do not have to run it in Xcode. Instead, you can find the executable program inside the project directory. (Open the project directory in the Finder, double-click the folder named "Build" and then the folder named "Release.") You may now either double-click the executable file you find there, or copy it to another location on your hard drive. Alternatively, you can open a Terminal window and change to the directory containing the executable program. For example, if the project (and the program) is named say, `poetry`, you launch the program by typing `./poetry` once you are inside the appropriate directory.

A.3   General advice on debugging

Finding mistakes in computer programs can be a frustrating painful experience. Here we present some advice for how to find errors in programs and some common pitfalls.

- **Keep procedures short**. It can be difficult to understand what's happening in a long procedure. Break long procedures into a few shorter ones. Then you can test each individual piece to make sure it behaves as expected.

- **Name and document your procedures with care**. Suppose you have a procedure declared like this:

```
double my_proc(double x, double y, long j);
```

  This may make perfect sense to you today, but when you return to your work after the weekend, you may be confused by what `my_proc` does and what its arguments represent.

  Liberally comment your code and use names that reflect the meaning of the symbols.

- **Print, print, print!** Sprinkle your code with output statements to show the contents of variables at each point of the program. Here's a handy way to do this. In each procedure, define a `bool` variable named `DEBUG` and wrap your debugging code in an `if` statement, like this:

```
int count_instances(const Group& G) {
  const bool DEBUG = true;
  ...
  if (DEBUG) {
    cerr << "i = " << i << " and j = " << j << endl;
  }
  ...
}
```

Once this procedure is working, you can edit the first line of the procedure to read

```
const DEBUG = false;
```

thereby switching off all the debugging code you entered.

Alternatively, if you are working with an IDE, you can set break points and inspect variables as the program runs.

- **Be meticulous in your use of** const. Suppose we have a procedure declared like this:

```
int count_generators(const Group& G);
```

If this procedure invokes a Group method on G that has not been flagged const, then—even if that method does not modify G—the compiler reports an error.

- **Make sure all variables are declared**. It is easy to forget to declare a variable or to misspell its name.

- **Watch for memory leaks.** If your class uses a destructor to free memory, you can insert debugging code into the class's destructor so you can observe when (if) the destructor is doing its job.

- **Turn on all warnings**. There are legal C++ statements that, almost surely, are not what you intend. The C++ compiler may help you find such errors. Here is a classic example.

```
#include <iostream>
using namespace std;

int main() {
  long a = 10;
  long b = 23;
  if (a=b) {
    cout << "They are equal" << endl;
  }
  return 0;
}
```

This program compiles without error and the output is the mystifying

```
They are equal
```

where clearly a and b do not hold equal values.

However, by turning on warnings (e.g., with -Wall in g++) we get the following warning message from the compiler concerning line 7 of the program:

```
suggest parentheses around assignment used as truth value
```

This is not an especially helpful message, but at least it does point us to line 7 where, in fact, the problem lies. The a=b ought to be a==b.

- **Remember** the using namespace std; statement. If the compiler complains that it doesn't know what cout is, then you probably forgot the using statement.

- **Do not name your program** test. This is a subtle problem that can be truly maddening. Suppose you name the following program test.cc and compile it as an executable named test:

```cpp
#include <iostream>
using namespace std;
int main() {
  cout << "Does it work?" << endl;
  return 0;
}
```

The program compiles witout error and now you attempt to run it by typing test at the command prompt. What happens? Apparently nothing. The message Does it work? never appears on your screen.

There is *nothing* wrong with the program. The problem is that UNIX computers already contain a standard program named test. When you type test at the command prompt, that version of test is run (not yours). To make sure the test in the current directory is the one that is executed, type ./test at the command prompt.

A way to avoid this problem is to name your program try or attempt.

Appendix B

Documentation with Doxygen

Document your code

Write a description for every class, method, and procedure you create. Explain what the input parameters represent (and what the valid ranges for these variables are) and what the output means. Include descriptions of what the variables inside the procedure mean.

This vital step in programming takes a bit of effort, but ultimately saves you hours. Over time, you will develop a personal library of classes and procedures. At some point you will want to reuse a class or procedure you have already created. If you don't document, you will spend nearly as much time trying to figure out what the procedure is for as it would take to rewrite it!

The most likely reader of your comments is you. However, a colleague might learn that you have created a C++ class that he or she would like to use in another project. Your comments will make your colleague's work much simpler (and your colleague won't be repeatedly visiting your office asking you what the procedures do and what the parameters represent).

The simplest way to document your code is to include C++ comments in the .h and .cc files. However, there is a variety of tools that—with scant extra effort—make your documentation much easier to use (and beautiful to behold as well). These tools scan your C++ files for specially structured comments that are used to create Web pages (or LaTeX documents) that are easy to use.

In this appendix we describe one such tool: Doxygen (available for free download from www.doxygen.org). There you can find the source code and, more conveniently, already compiled, ready-to-use versions for Windows, Mac OS X, and Linux.

Doxygen is rich in features. The description we present here is just a primer, but is sufficient to get you started.

B.1 Doxygen comments

Doxygen reads your C++ source files (.cc and .h) and looks for specially structured comments. A C++ single-line comment begins with a double slash // and a multiline C++ comment is enclosed between /* and */. Doxygen reads what you write in those comments. To tell Doxygen to read a particular comment begin either with a triple slash /// (for single-line comment) or with the sequence /** (for a multiple-line comment).

B.1.1 Documenting files

Each file (either .cc or .h) should begin with a brief comment explaining the file's general purpose. Structure the comment like this:

```
/**
 * @file Polygon.h
 * @brief Declaration of the Polygon class
 */
```

The tag @file gives the name of the file and the @brief tag gives a short description of what the file is for. Notice the double asterisk on the first line; Doxygen looks for this to process this comment.

Note: The initial asterisks on lines 2 and 3 are optional and do not affect Doxygen's output. These stars are useful to the human reader making this portion of the file stand out clearly as a comment.

B.1.2 Documenting procedures

Procedures are declared in a .h file and the full code is given in a .cc file. To document a procedure for processing by Doxygen, we place special comments *before* the procedure declaration in the .h file.

We generally write two comments. The first is a short, single-line overview of what the procedure does. The second is a long, multiple-line description of the procedure, its input arguments, and its return value.

The brief comment begins with a triple slash ///. We follow this with a few words that describe the procedure.

The long comment begins with /** giving a more elaborate description of the procedure. This includes special tags to document the input parameters and the return value. Here is an example.

Program B.1: Documenting a procedure for Doxygen.

```
 1  /// Number of real roots of a quadratic.
 2  /**
 3   * This procedure determines the number of real roots of a quadratic
 4   * polynomial by examining its discriminant.
 5   *
 6   * @param a coefficient of x-squared
 7   * @param b coefficient of x
 8   * @param c constant term
 9   * @return the number of real roots of ax^2 + bx + c
10   */
11  int nroots(double a, double b, double c);
```

Program B.1 is an excerpt from a header file (say, `nroots.h`). The brief description (line 1) tells us the procedure's purpose.

The expanded comment (starting at line 2) provides more detail including a description of the technique used. Following the description are tags that describe the input parameters (lines 6–8) and the return value (line 9). The input parameters are documented like this:

```
@param param_name role of this input parameter
```

The return value is documented with an `@return` tag that describes the value returned by this procedure.

Note that the initial asterisks on lines 3–9 are optional and are ignored by Doxygen. They serve the human reader by showing that this stretch of the file is a comment.

Finally, the declaration of the `nroots` procedure is given on line 11.

The code for this procedure is in a separate `.cc` file. For the sake of completeness, we give it here.

```
#include "nroots.h"

int nroots(double a, double b, double c) {
  double d; // The discriminant of the quadratic

  d = b*b - 4.*a*c;

  // Check sign of the discriminant and return number of roots

  if (d > 0.) {
    return 2;
  }
  if (d < 0.) {
    return 0;
  }
  return 1;
}
```

Note that there are no Doxygen comments here, however, there are ordinary comments to help the human reader understand the code.

B.1.3 Documenting classes, data, and methods

Classes and their members are documented for Doxygen in a manner similar to ordinary procedures. The class and its members receive a brief description and then a detailed description. Methods for the class have their parameters and return values (if any) documented as well. In addition, the data members should be documented.

To illustrate these, we present a class named `Polygon`. The full `.h` file is presented in Program B.2. Notice that the file, the class, the class's constructors and methods, and the class's data elements all receive comments to be processed by Doxygen. For the data members, a brief description is sufficient, so we omit the detailed elaboration.

Notice that constructors and `void` methods (such as `move_vertex`) do not have an `@return` tag because these do not return a value. The method `perimeter` takes no arguments, so no `@param` tag is given.

Program B.2: Documenting a class and its members for Doxygen.

```
1   /**
2    * @file Polygon.h
3    * @brief Declaration of the Polygon class
4    */
5
6   #ifndef _POLYGON_
7   #define _POLYGON_
8
9   #include <vector>
10  #include "Point.h"
11
12  /// Planar polygons
13  /**
14   * The Polygon class represents a closed n-gon in the plane. The
15   * vertices of the Polygon are held as a vector of Point objects.
16   */
17  class Polygon {
18  private:
19    /// Hold the vertices of the Polygon
20    vector<Point> vertices;
21
22    /// The number of points in the Polygon
23    int np;
24
25  public:
26    /// Default constructor
27    /**
28     * This default constructor creates an empty Polygon, i.e., a
29     * Polygon without any points.
30     */
31    Polygon();
32
33    /// Create a basic n-gon
34    /**
35     * This creates a polygon with a specified number of points. The
36     * points are placed evenly around the unit circle.
37     * @param n the number of points.
38     */
39    Polygon(int n);
40
41    /// Copy constructor
42    /**
43     * Creates an exact copy of another Polygon.
44     * @param P the Polygon we're copying
45     */
46    Polygon(const Polygon& P) {
47      np = P.np;
48      vertices = P.vertices;
49    }
50
```

```
51    /// Move a vertex to a new location
52    /**
53     * Note: If an invalid vertex index is given, no change is made to
54     * this Polygon.
55     *
56     * @param j index of the vertex we want to relocate
57     * @param P new location for vertex j
58     */
59    void move_vertex(int j, Point P);
60
61    /// Perimeter of this Polygon
62    /**
63     * If the Polygon has fewer than two vertices, its perimeter is 0.
64     * If it consists of exactly two vertices, the perimeter is twice
65     * the distance between those points.
66     *
67     * @return the perimeter of this polygon
68     */
69    double perimeter() const;
70  };
71
72  #endif
```

For the sake of completeness, here is the `.cc` file that accompanies `Polygon.h`.

```
/**
 * @file Polygon.cc
 * @brief Code for the Polygon class methods
 */

#include "Polygon.h"

Polygon::Polygon() {
  np = 0;
  vertices.clear();
}

Polygon::Polygon(int n) {
  // If n is negative, set n to be zero.
  if (n < 0) n = 0;

  // If is zero, just clear the vertices data structure
  if (n==0) {
    np = 0;
    vertices.clear();
    return;
  }

  // At this point, n is positive

  np = n;  // set number of points to n
  vertices.resize(n); // allocate enough room

  for (int k=0; k<n; k++) {
    double theta = (2*M_PI*k)/n;
    vertices[k] = Point(cos(theta),sin(theta));
  }
```

```
}

void Polygon::move_vertex(int j, Point P) {
  // If the index is invalid, no action is taken
  if ( (j<0) || (j >= np) ) return;
  vertices[j] = P;
}


double Polygon::perimeter() const {
  // The perimeter is zero unless we have at least two points
  if (np <= 1) return 0.;

  double ans = 0.; // place to accumulate the distances

  // add up the distances between successive vertices
  for (int k=0; k<np-1; k++) {
    ans += dist(vertices[k],vertices[k+1]);
  }

  // Add in the distance between the first and last points
  ans += dist(vertices[0], vertices[np-1]);

  return ans;
}
```

The only Doxygen comments in `Polygon.cc` are to document the file. Other, ordinary comments are included to clarify the code to a human reader.

B.2   Using Doxygen

If Doxygen is not already installed on your computer, download it from its Web site (`www.doxygen.org`) and install it according the instructions found there.

You should also collect all the files for your project into a single folder (directory) on your computer. You probably have done this already.

Now launch the Doxygen tool. You are presented with a graphical user interface ("GUI") window that specifies the easy steps for you to follow. See Figure B.1.

B.2.1 Configuring Doxygen

The first step is to configure Doxygen. In the configuration process you set various options (and I'll give you recommendations on what to select) and then save the configuration as a file named `Doxyfile`.

To begin the configuration process, click the **Wizard...** button. A new window opens with four tabs. See Figure B.2. The four tabs are named **Project**, **Mode**, **Output**, and **Diagrams**. Here's how to complete each of these pages.

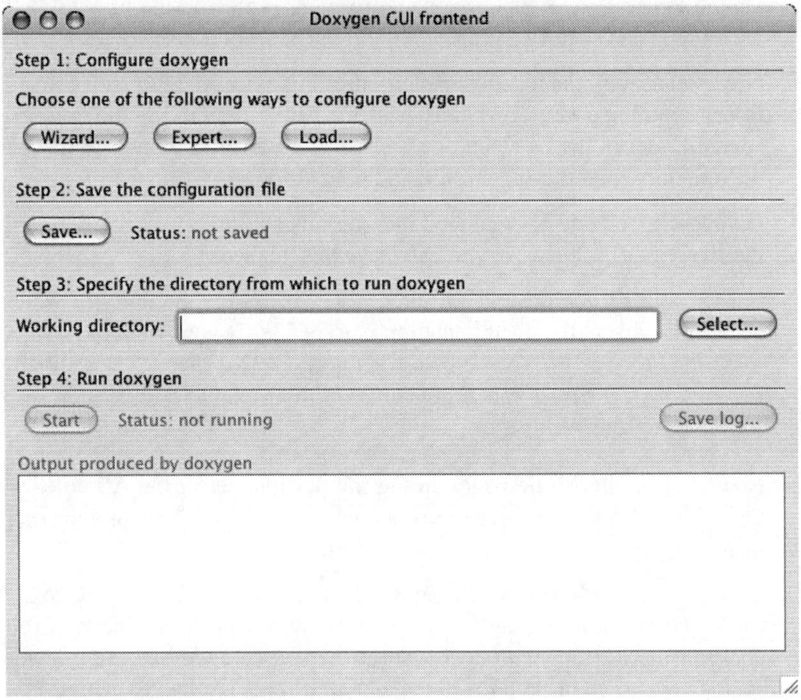

Figure B.1:   Doxygen GUI window.

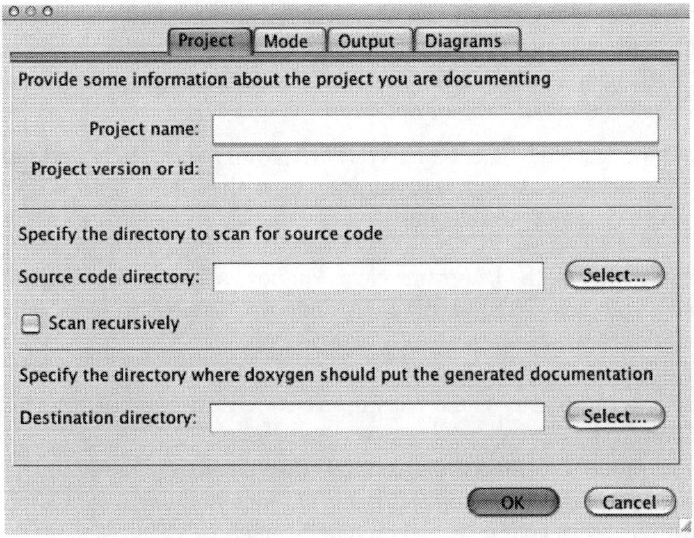

Figure B.2:   Doxygen configuration panel.

- On the **Project** page, type in a name for your project in the first field. If you desire, you can specify a project version in the second field.

  Next you specify the location of the source code (i.e., the full path name of the folder containing your .cc and .h files). Either type in the full path to your source folder in the text field, or use the first **Select...** button to navigate to the directory (and the Wizard fills in the field for you).

  If your source code is organized into various subfolders, check the **Scan recursively** box. Otherwise, you can leave this unchecked.

  Finally, specify the destination directory. This is the directory in which the beautifully formatted documentation is to reside. If you wish, this can be the same directory as the source code. As needed, Doxygen creates subdirectories in which it places the documentation. For example, it creates an html subfolder for the Web pages.

- Now click to the **Mode** page. In the top portion, select the **All entities** button. This tells Doxygen to create documentation for all parts of your program including private data members, and so on.

  Check the **Include cross-referenced source code in the output** box. This causes Doxygen to create Web pages that show your source code. (You can click on the documentation to look at the source code and click inside the source code to go back to the documentation. This is a handy feature.)

  Finally, click the **Optimize for C++ output** button.

- Now click to the **Output** page. Check the **HTML** box. Under that box either select **plain HTML** or **with frames and navigation tree**. If you are only building documentation for one procedure or a small class, then the **plain HTML** option is sufficient. However, if the project has more than one source file, select the **with frames** option.

  For now, leave the other options (**LaTeX**, **Man pages**, **Rich Text Format**, and **XML**) unchecked. Later, if you wish to explore these other output formats, you can change your selection.

- Finally, click to the **Diagrams** page. For now, select the **No diagrams** option. Later, you can explore the different kinds of diagrams that Doxygen can create for you.

You are now finished using the Wizard, so click **OK** to return to the main Doxygen GUI window.

You now proceed to **Step 2** on the GUI: save the configuration file. When you click **Save...** you will be prompted to save the configuration in a directory. Choose the same directory as your code and use the file name Doxyfile.

Doxygen is now configured for your project. All of Doxygen's settings are saved in the file named Doxyfile in the same directory as your source code.

If, after you have saved the `Doxyfile`, you wish to modify your settings, select **Load...** in the Doxygen GUI to select the `Doxyfile` you wish to modify, and then use **Wizard...** to choose the new options. If you want finer control, you may modify the settings using the **Expert...** button.

You are now ready to run Doxygen to generate the Web pages.

B.2.2 Running Doxygen

Continuing in the Doxygen GUI, **Step 3** asks for a directory in which Doxygen is to run. Use the **Select...** button to navigate to the same directory that holds the code and holds the `Doxyfile` configuration file.

Finally, press the **Start** button in **Step 4**. If all is well, a subfolder is created named `html` into which the Web pages for your documentation are deposited. Open this folder and double-click on `index.html`. You are greeted by a main page from which you can explore all the documented files, classes, procedures (functions), methods, and so forth.

If there are mistakes in the formatting of your Doxygen comments, error messages appear in the box labeled **Output produced by doxygen** in the main GUI window.

B.2.3 More features

In this brief introduction to Doxygen we presented just a few of the documentation tags available. Reviewing tersely, we have seen:

- `@file` and `@brief` document the overall purpose of a file, and

- `@param` and `@return` to document the inputs and outputs of procedures and methods.

Doxygen provides many other tags and we mention just a few of them here.

- *Cross references*. Doxygen provides a tag named `@see` that you can use to insert a cross-reference to another procedure or method.

- *Sample code*. It is useful to include a snippet of C++ code to illustrate how a procedure is used. To do this, enclose the code fragment between `@code` and `@endcode`, like this:

```
/// Find the median
/**
 * This procedure finds the median value in an array of numbers.
 * @code
 * double* values;
 * // allocate and populate the values array
 * m = median(values, n);
 * @endcode
 * @param vals array of real numbers
 * @param nels number of elements in the array
 * @return the median value in the array
```

```
*/
double median(double* vals, int nels);
```

- *Mathematical notation.* It is possible to include mathematical formulas into
  the documentation using TEX. This requires that TEX be installed on your
  system. In TEX, mathematical notation is enclosed between single dollar signs.
  For Doxygen, enclose your notation between @f$ tags, like this:

```
/// Approximate the Riemann zeta function
/**
 * This procedure calculates an approximation to the Riemann
 * zeta function by expanding @f$\zeta(s)=\sum_{k=1}^n 1/k^s@f$.
 * @param s argument to the zeta function
 * @param n number of terms in the series
 * @return an approximation to @f$\zeta(s)@f$
 */
double zeta(double s, int n);
```

  It is also possible to typeset displayed formulas enclosed in the tags @f[and
  @f].

- *Bugs.* Sometimes programs don't work properly and give incorrect results.
  Until you have your code working properly, you can flag what is wrong with a
  @bug tag.

```
/// Group element powers
/**
 * Given a Group element g and an integer n, calculate g^n.
 * @bug This isn't working for negative exponents.
 *
 * @param g an element of the group
 * @param n the power to which we wish to raise g
 * @return g multiplied by itself n times
 */
```

- *Author, date, and version.* With the tags @author, @date, and @version you
  can flag who did what and when on a large-scale project.

Many more Doxygen tags can be found in the Doxygen documentation (available
online).

In addition to Doxygen tags, it is possible to insert HTML code in your comments.
For example, suppose you implemented an algorithm you found on a Web page in
your code and you wish to give proper attribution to the source you consulted. You
can include a link to that Web page like this:

```
/**
 * Description of this procedure...
 * <p>
 * The algorithm we use was invented by Ann Expert and is described
 * <a href="http://www.math.major-univ.edu/~expert/algo.html">on
 * this Web page</a>.
 */
```

Appendix C

C++ Reference

This appendix gives a brief review of the C++ language as developed in this book. After you have read this book, you can use this appendix as a refresher course.

C.1 Variables and types

All variables in C++ must be declared before they are used. The declaration specified the type of data the variable holds. The type of a variable may be one of the fundamental types or a class.

C.1.1 Fundamental types

- Boolean: `bool`

- Character: `char`

- Integer: `short, int, long, long long`[1]

- Real: `float, double, long double`

The integer types may be preceded with the keyword `unsigned` to shift their range to nonnegative values. The `long long` and `long double` types might not exist on all systems.

C.1.2 Standard classes/templates

C++ provides a large collection of standard classes. Many of these require a `#include` directive to load the appropriate header file.

For example, to work with complex numbers, we include the `<complex>` header. A typical complex number is declared `complex<double> z;` and a Gaussian integer is declared `complex<long> w;`.

---

[1] Use `__int 64` in Visual Studio.

Character strings are handled most conveniently using the standard `string` class (see Chapter 14); use the header `#include <string>`.

There is a variety of standard classes for input/output (see Chapter 14) and to serve as containers (see Chapter 8).

C.1.3   Declaring variables

The simplest declaration statement specifies the type of a variable: `int j;`.

Two or more variables of the same type may be declared in a single statement: `long a,b;`.

A variable may be given an initial value when it is declared: `double x = 3.5;`.

A variable declared using a template class should include the type parameters between < and > symbols, like this: `complex<double> z;`.

Class variables often take arguments (sent to the constructor methods) when declared: `Mod a(3,10);`.

There are three natural locations for a variable declaration.

- Inside the body of a procedure (such as `main` or a class method). For example,

```
int main() {
  double x;
  ...
}
```

- Inside the declaration of a procedure or method. For example,

```
long gcd(long a, long b) {
  ...
}
```

- Inside a looping control structure. For example,

```
for (int k=0; k<10; k++) {
  cout << k << endl;
}
```

C.1.4   Static variables and scope

Variables are created when they are declared and are lost when the section of the program in which they are declared finishes. Two different procedures (say, `alpha` and `beta`) may both declare a variable named `x`, but `alpha`'s `x` and `beta`'s `x` have nothing to do with each other. All variables are local to the procedure in which they are declared. (It is possible, but ill advised, to create global variables in C++. We don't discuss how.)

When a procedure finishes, the values held in its variables are lost unless the variables are declared `static`. In this case, the value held in the variable is retained between calls to the procedure.

C.1.5 Constants and the keyword `const`

If a variable is declared with a given value and the program never changes that value, then the variable should be declared `const`. For example, in a program that uses the Golden Mean, we would have this: `const phi = (1.+sqrt(5.))/2.;`.

Arguments to procedures that are not modified by the procedure should be declared `const`. For example, suppose we create a procedure to calculate the sum of the elements in a set of integers. Because such a procedure does not modify the set, we declare it such as this:

```
long sum(const set<long>& S) {
  ...
}
```

A third use of the keyword `const` is to certify that a method does not change the object on which it is invoked. For example, if we are creating a `LineSegment` class, a method that reports the length of the segment would not modify the segment. In the header file, say `LineSegment.h`, we would find this:

```
class LineSegment{
private:
  // variables to specify the segment
public:
  // constructors, etc.

  double length() const;
};
```

and in the program file, say `LineSegment.cc`, we find this:

```
double LineSegment::length() const {
  // calculate and return the length
}
```

C.1.6 Arrays

If the size of an array can be determined before the program is run, it may be declared like this: `int alpha[20];`.

However, if the size of an array is unknown until the program is running, use a dynamically sized array like this:

```
int n;
cout << "Enter array size: ";
cin >> n;
int* alpha;
alpha = new int[n];
...
delete[] alpha;
```

Remember that every array allocated with `new` should be released with a `delete[]`.

C.2   Operations

Here we describe the behavior of C++'s fundamental operations. Remember, however, that classes can override these meanings (e.g., the << operator does input when used with `cin` but its fundamental use is to shift bits).

C.2.1   Assignment

The statement a=b; takes the value held in the variable b and places that value in the variable a. Assignment statements can be combined. The statement a=b=c; is equivalent to a=(b=c); and the effect is to take the value in c and store that value in both b and a.

C.2.2   Arithmetic

The basic arithmetic operators are +, -, *, and /. For integer types, the mod operator is %. There is no exponentiation operator, but the `pow` and `exp` procedures fill this gap.

The - operator can be used as a unary operator; this negates its argument. For example, if a holds the value 3, then b=-a; assigns to b the value -3 (leaving a unchanged).

The arithmetic operators can be combined with assignment. For example, a+=b; is equivalent to a=a+b;.

For integer types, the operators ++ and -- cause the variable to which they are applied to increase (respectively, decrease) by one. There is a difference between ++a and a++. Do not write code that exploits this subtlety.

C.2.3   Comparison operators

Numerical types can be compared for equality, inequality, and order using these operators:

```
==    !=    <    <=    >    >=
```

The result of these operators is a value of type `bool`.

C.2.4   Logical operators

C++ provides the following operators for `bool` values,

```
&&    ||    !
```

These are logical *and*, *or*, and *not*.

C.2.5 Bit operators

The individual bits in an integer variable can be manipulated using the following operators,

     & | ˜   ˆ   <<   >>

These are (bitwise) *and, or, not, exclusive or, left shift,* and *right shift*.

C.2.6 Potpourri

Square brackets are the subscript operator, used to access elements of an array. The first element of an array has index 0.

Parentheses are used for grouping, but also indicate the invocation of a procedure or method: `y = alpha(x,n);`.

Dot (`.`) is the member-of operator used to specify a data member of method of a class: `x.size()` or `thing.first`.

The question mark/colon trigraph is an abbreviated *if/then* statement. The statement `a=q?b:c;` assigns the value of `b` to `a` if `q` is true; otherwise (`q` is false) it assigns the value of `c` to `a`.

The keyword `sizeof` is an operator that returns the number of bytes required to hold a type; for example, `sizeof(double)`.

The comma is used for separating procedure arguments and when declaring more than one variable of a given type (e.g., `double x,y;`). It also has the obscure purpose of combining two (or more) expressions into a single statement. For example, in a `for` loop, if we want to initialize two counters, we may have a code such as this:

```
int a, b;
for (a=N,b=0; a-b>0; a--,b++) {
  // stuff to do
}
```

This initializes `a` with the value held in `N` and `b` with 0.

The following operators are for working with pointers.

Ampersand is the address-of operator. If `x` is an `int` variable, then `&x` returns a pointer to `x` of type `int *`. The ampersand is also used to specify call-by-reference semantics in procedures.

Asterisk is the pointer dereferencing operator. If `p` is a pointer to an integer value, then `*p` is the value held in the location pointed to by `p`.

The arrow operator, `->`, combines the action of `.` and `*`. For example, if `p` is a pointer to an object, then `p->x` is equivalent to `(*p).x` (the data element `x` of the object `p` points to) and `p->alpha()` is equivalent to `(*p).alpha()` (invoke method `alpha` on the object pointed to by `p`).

C.3   Control statements

Statements in C++ must be terminated by a semicolon. Collections of statements can be grouped together to form a compound statement by enclosing those statements in curly braces. All statements enclosed in the curly braces must end in a semicolon, but the compound does not require a semicolon after the closing brace. (Exception: When defining a class, the closing semicolon must be followed by a semicolon.)

Normally, statements are executed in the order encountered. However, various control structures may be used to modify this behavior.

C.3.1 `if-else`

The simplest form of this structure is this:

```
if (expression) {
  statements;
}
```

Here, `expression` is evaluated. If its value is `true`, then the `statements` are executed; if `false`, then the `statements` are skipped.

We also have the following richer form,

```
if (expression) {
  statements1;
}
else {
  statements2;
}
```

Again, `expression` is evaluated. If it yields `true`, then `statements1` are executed and `statements2` are skipped. Otherwise (`expression` evaluates to `false`) the opposite occurs: `statements1` are skipped and `statements2` are executed.

The `?:` operator is a compact version of the `if-else` structure.

C.3.2 Looping: `for`, `while`, and `do`

The `for` statement has the following format,

```
for (start_statement; test_expression; advance_statement) {
  work_statements;
}
```

This statement results in the following actions. First, the `start_statement` is executed. Then `test_expression` is evaluated; if `false` then the loop exits and control passes to the statement following the close brace. Otherwise (`test_expression` is `true`), the `work_statements` are executed, then the `advance_statement` is executed, and finally the `test_expression` is evaluated. If `false`, the loop exits and if `true`, the entire process repeats.

The `while` statement is structured as follows,

```
while (test_expression) {
  work_statements;
}
```

The `test_expression` is evaluated first and, if `false`, the `work_statements` are skipped and execution passes to the next statement after the close brace. Otherwise (`test_expression` is `true`), the `work_statements` are executed, then `test_expression` is re-evaluated and the process repeats.

The `do` statement is structured as follows,

```
do {
  work_statements;
} while (test_expression);
```

Here, the `work_statements` are executed first, and then the `test_expression` is evaluated. If `true`, the process repeats; if `false`, the loop terminates.

All three looping constructs (`for`, `while`, and `do`) support the use of the statements `break;` and `continue;`. A `break;` statement causes the loop to exit immediately. A `continue;` statement causes the loop to skip the remaining work statements and attempt the next loop (starting with `test_expression`).

C.3.3 `switch`

A `switch` statement controls execution depending on the value of an expression. The format is this:

```
switch (expression) {
case val1:
  statements1;
  break;

case val2:
  statements2;
  break;

...

default:
  default_statements;
}
```

Here, `expression` is evaluated to yield an integer result. If there is a `case` statement whose label matches the value of `expression`, control passes to that point and the statements following the `case` label are executed until a `break` statement is reached. If no matching label can be found, then statements following the `default` label are executed. Groups of statements may be preceded with more than one label.

The labels `val1`, `val2`, and so on, must be specific numbers (not variables).

C.3.4 `goto`

C++ contains a `goto` statement whose syntax is `goto label;`. The causes execution to pass to a statement that has been flagged with the name `label`. In general, the use of `goto` is discouraged as it can lead to unintelligible programs. However, one use is for breaking out of a double loop:

```
for (int a=0; a<N; a++) {
  for (int b=0; b<N; b++) {
    if (beta(a,b) < 0) goto aftermath;
    // other stuff
  }
}
aftermath:  cout << "All finished" << endl;
```

C.3.5 Exceptions

The keywords `try`, `throw`, and `catch` are used to implement C++'s exception-handling mechanism. Typical code looks like this:

```
try {
  statements;
  if (something_bad) throw x;
  more_statements;
}
catch(type z) {
  recovery_statements;
}
```

If `something_bad` is `false`, execution continues with `more_statements` and the `recovery_statements` are skipped. However, if `something_bad` evaluates to `true`, then `more_statements` are skipped and the value `x` is "thrown". Assuming that `x` is of type `type`, the exception is "caught" by the `catch` statement and `recovery_statements` are executed.

The statements inside the `try` block need not have an explicit `throw` statement; the procedures invoked inside this block may throw exceptions. See Section 15.3.

C.4 Procedures

Procedures (often called functions in the programming community) are subprograms designed to do a particular job.

Procedures are declared by specifying a return type (if none, write `void`), followed by the procedure's name, followed by a list of arguments (with their types). The value returned by the procedure is given by a `return` statement.

Two procedures may have the same name provided they have different number and/or types of arguments.

C.4.1   File organization

In general, it is best to separate the *declaration* of a procedure from its *definition*. The declaration is placed in a header file (suffix .h) and the full definition is placed in a code file (suffix .cc).

For example, suppose we wish to declare a procedure named nroots that returns the number of real roots of a quadratic polynomial $ax^2 + bx + c$. The header file would contain the following single line,

```
int nroots(double a, double b, double c);
```

The .cc file would contain the full specification:

```
int nroots(double a, double b, double c) {
  double d = b*b - 4.*a*c;
  if (d < 0.) return 0;
  if (d > 0.) return 2;
  return 1;
}
```

C.4.2   Call by value versus call by reference

By default, C++ procedures use call-by-value semantics. That is, when a procedure (such as nroots) is invoked, the values of the arguments in the calling procedure are copied to the local variables in the procedure. Although procedures may modify the copies of the arguments, the original values (in the parent procedure) are unaffected.

However, variables can be designated to use call-by-reference semantics. This is indicated by inserting an ampersand between the type and the argument. In this case, a procedure can change a value from its calling procedure.

Here is an example:

```
void alpha(int x) { x++; }
void beta(int &x) { x++; }
int main() {
  int a = 5;
  int b = 5;
  alpha(a);
  beta(b);
  cout << a << endl;
  cout << b << endl;
  return 0;
}
```

The procedure alpha increases a copy of a, so main's a is unaffected. However, the procedure beta increases the variable b itself, so its value becomes 6. The output of this program is this:

```
5
6
```

Call by reference is useful if the objects passed to a procedure are large. Passing a reference is faster than making a copy of the object.

C.4.3   Array (and pointer) arguments

When an array is passed to a procedure, C++ does not make a copy of the array; instead it sends a pointer to the first element of the array.

For example, suppose we write a procedure to sum the elements in an array of `double` values. Here is the code.

```
double sum(const double* array, long nels) {
  double ans = 0.;
  for (long j=0; j<nels; j++) ans += array[j];
  return ans;
}
```

Here, `nels` specifies the number of elements in the array. No duplication of the array is made. Instead, a pointer to the first element is passed. One implication of this is that a procedure can modify the entries in an array argument. Because the `sum` example we presented here does not, in fact, modify the elements of the array, we certify that with the keyword `const`.

More generally, a pointer can be passed to a procedure. In this case, the value pointed to by the pointer can be modified by the procedure. However, it is simpler to use reference arguments.

C.4.4   Default values for arguments

Arguments to procedures may be given default values. If an argument is given a default value, then all arguments to its right must also be given default values. For example:

```
void example(int a, double x = 3.5, int n = 0) {
  ...
}
```

Calling `example(4)` is tantamount to `example(4,3.5,0)`.

C.4.5   Templates

Earlier we considered a procedure to sum the elements in an array. This procedure works only for floating point (`double`) arrays. The identical code (but with different types) would be used for a procedure to sum an integer array. Rather than write different versions for each type of array, we can write a procedure template such as this:

```
template <class T>
T sum(const T* array, long nels) {
  T ans = T(0);
  for (long j=0; j<nels; j++) ans += array[j];
  return ans;
}
```

In this example, the symbol `T` acts as a "type variable"—it may stand for any type.

C.4.6 `inline` **procedures**

A procedure may be declared as `inline`; this causes the compiler to generate a different style of object code that uses more memory but runs faster. In general, it is not necessary to use this keyword because modern compilers automatically inline procedures when they determine it is advantageous to do so.

C.5 Classes

New data types are created in C++ through the use of classes and class templates. Various ready-to-use classes are provided with C++ such as `string`, `vector`, and `complex`. Other classes are available for download from the Web and from commercial software vendors. Finally, programmers can create their own classes.

C.5.1 Overview and file organization

Suppose we wish to create a class named `MyClass`. The specification for the class is broken across two files: `MyClass.h` contains a declaration of the class and `MyClass.cc` contains code for the class's methods.

The code in `MyClass.h` typically looks like this:

```
class MyClass {
private:
    // private data (and methods)
public:
  MyClass();  // basic constructor
  // other constructors

  int method1(double x, double y);
  // other methods
};
```

In the file `MyClass.cc` we give the code for the constructors and other methods for the class, like this:

```
#include "MyClass.h"

MyClass::MyClass() {
  // code for the constructor
}

int MyClass::method1(double x, double y) {
  // code for this method
}
```

Alternatively, code for constructors and methods may be given inline in the `.h` file. This is advisable when the code is only a few lines long.

Data and methods listed in the `private` section are accessible only to the methods inside the class (but see Appendix C.5.7). Data and methods listed in the `public` section are accessible to all parts of the program.

It is wise to designate all data in a class private and to provide get/set methods to inspect/manipulate the data.

C.5.2   Constructors and destructors

A constructor is a class method that is invoked when an object of the class is created (e.g., when declared). Constructors do not have a return type, but may have arguments.

Be sure to have a zero-argument constructor for your class. This constructor is invoked when a variable is declared in the simple form: `MyClass X;`. The object X is then initialized using the zero-argument constructor.

Classes may have constructors with arguments. For example, if `MyClass` has a constructor with a single integer argument, then the declaration `MyClass X(17);` invokes that constructor. Such a constructor is also invoked in all the following situations.

```
MyClass X;
X = MyClass(17);

MyClass Y = 17;

MyClass Z;
Z = 17;
```

Some constructors allocate storage (with `new`). When such an object goes out of scope, the allocated memory needs to be recovered (or the program suffers a memory leak). To accomplish this, a destructor needs to be specified. In the `.h` file, a public method is declared like this:

```
class MyClass {
private:
  int* BigTable;   // table of numbers
...
public:
...
  ~MyClass();      // destructor declaration
...
};
```

and in the `.cc` file:

```
MyClass::~MyClass() {
  delete[] BigTable;   // or other clean-up code
}
```

Alternatively, with a short destructor, the code may be written inline in the `.h` file.

C.5.3 Operators

The usual C++ operators (such as + and *) apply to the built-in data types (int, double, and so on). These operators may also be used for classes by creating operator methods and procedures.

Suppose we wish to define + for objects of type MyClass; that is, if A and B are type MyClass, then we want to ascribe a meaning to A+B. Typically, we would declare a method within the body of the MyClass declaration (in the .h file) like this:

```
class MyClass {
private:
  ...
public:
  ...

  MyClass operator+(const MyClass& Z) const;
};
```

and in the .cc file give the code:

```
#include "MyClass.h"

MyClass MyClass::operator+(const MyClass& Z) const {
  ...
}
```

Then, when the compiler encounters the expression A+B, it applies the operator+ method for object A with object B passed (by reference) to Z.

(Note the double appearance of the keyword const. The first const certifies that this method does not modify the argument Z and the second certifies that this method does not modify the object on which it is invoked.)

Alternatively, we could implement A+B with a procedure that is not a class method. In a .h file we declare the procedure like this:

```
MyClass operator+(const MyClass& U, const MyClass& V);
```

and in the corresponding .cc file we give the code:

```
MyClass operator+(const MyClass& U, const MyClass& V) {
  ...
}
```

Unary operators (e.g., for negation) are declared as class methods like this:

```
class MyClass {
  ...
  MyClass operator-() const;
};
```

or as procedures like this:

```
MyClass operator-(const MyClass& U);
```

Binary operators may be used to combine objects of different types. If the left operand of the operator is of type MyClass, then the operator may be defined as a method of MyClass. For example, for the operator MyClass+int, use this:

```
class MyClass {
  ...
  MyClass operator+(int j) const;
};
```

However, for int+MyClass, a procedure needs to be declared like this:

```
MyClass operator+(int j, const MyClass& Z);
```

The increment ++ and decrement -- operators have two forms: ++A and A++. We recommend defining only the prefix form. This is done with an class method like this:

```
class MyClass {
  ...
  MyClass operator++();
};
```

(Note: Typically ++A is used to increase the value of A by one, so this method modifies A. That is why we do not include const after the parentheses.)

It is possible to declare a postfix form of these operators. To do this, we give a "dummy" argument of type int like this:

```
class MyClass {
  ...
  MyClass operator--(int x);
};
```

C.5.4 Copy and assign

If objects A and B are of type MyClass, then the expression A=B has a default meaning that can (and sometimes should) be overridden.

The default meaning of A=B is to copy the data fields of B into the corresponding data fields of A. That is, if class MyClass has data fields x, y, and z, then A=B; has the effect of performing the three assignments

```
A.x = B.x;   A.y = B.y;   A.z = B.z;
```

Finally, the new value of A is returned.

This behavior is appropriate in many cases, especially if the data fields are the basic types. However, if one of these fields, say x, is an array, then the action A.x = B.x; does not copy the array B.x into A.x (as, presumably, we would want). Rather, it causes A.x to point to the same location in memory as B.x, so subsequent modifications to array elements in B.x are also applied to the array A.x because these arrays are now housed in the same memory.

To achieve the desired behavior, we need to write a new assignment operator. The effect of this operator is to copy the data from B to A. In addition, an assignment operator should return the value of A (after it is updated).

To do this, we declare an operator= method inside the class definition like this:

```
class MyClass {
  ...
```

```
  MyClass operator=(const MyClass& Z);
};
```

In the .cc file, the code looks like this:

```
MyClass MyClass::operator(const MyClass& Z) {
  // set the fields x, y, and z
  // so they are duplicates of Z.x, Z.y, Z.z
  return *this;
}
```

The final `return *this;` statement causes a copy of the object to be the return value of this method; see Appendix C.5.6.

Just as C++ provides a default assignment operator, it also provides a default copy constructor. The default behavior of the declaration `MyClass A(B);` (where B is a previously declared object of `MyClass`) is to do a field-by-field copy of B's data into A. As in the case of assignment, this default behavior may be unacceptable. In such cases (e.g., when `MyClass` data includes an array), we must write our own copy constructor.

To declare a copy constructor, we have the following in the .h file,

```
class MyClass {
  ...
  MyClass(MyClass& Z);
};
```

and in the .cc file:

```
MyClass::MyClass(MyClass& Z) {
  // set the fields x, y, and z
  // so they are duplicates of Z.x, Z.y,, Z.z
}
```

C.5.5 `static` data and methods

In a typical class, each object of the class has its own values for each data field. Sometimes it is desirable to have a value that is shared among all objects in the class. For example, if we wish to keep track of how many objects of type `MyClass` are currently in existence, we can define a variable `object_count` that is shared among all objects of type `MyClass`. Constructors would increment this variable and destructors would decrement it. To distinguish variables that are shared among all objects from ordinary data members that are particular to each instance of the class, we use the keyword `static`. For example, in the .h file we have

```
class MyClass {
private:
  ...
  static int object_count;
public:
  MyClass()  { ...; object_count++; }
  ~MyClass() { ...; object_count--; }
  ...
};
```

and in the .cc file we have

```
MyClass::object_count = 0;
```

One may also have `static` methods. These are class methods that do not apply to any particular object, but to the class as a whole. They can only access `static` data in the class because the other data are particular to each object of the class.

In the example above, we would want a method that reports how many objects of type `MyClass` currently exist. To that end, we add to the `public` section of `MyClass` the following method,

```
static int get_object_count() { return object_count; }
```

Because this method applies to the class `MyClass` and not to any particular object of that type, it is inappropriate to invoke it as `A.get_object_count()` (where `A` is of type `MyClass`). Rather, we write `MyClass::get_object_count()`.

C.5.6 `this`

Class methods may access all data fields of the object on which they are invoked. On occasion, it is useful for a method to refer to the entire object on which it was invoked. For example, when we define `++A` (by declaring an `operator++`), the conventional return value of this operator is the new value assigned to the object.

To do this, C++ provides a pointer named `this`. The `this` pointer refers to the object on which it is invoked. Suppose `MyClass` has a method named `work`. If we use the pointer `this` inside of `work`, then `this` is a pointer to the object `A` and `*this` is the object `A` itself.

So, when we define `operator++()` for `MyClass`, the final statement of that method would be `return *this;`.

Likewise, when we define an assignment operator, the return value of that operator should be the new value of the left-hand argument. Ending with `return *this;` accomplishes this task.

C.5.7 Friends

Private data elements of a class can be accessed by class methods, but not by other procedures. However, it is possible to grant a procedure special access to private data elements by declaring that procedure to be a `friend` of the class.

To do this, we need to declare the `friend` procedure inside the class declaration like this:

```
class MyClass {
private:
  ...
public:
  ...
  friend int buddy(MyClass Z);
};
```

In the .cc file we would see this:

```
int buddy() {
  // calculations using Z's fields
  return answer;
}
```

Note that `buddy` is not a method of `MyClass`, so in the `.cc` file we do not use the prefix `MyClass::`.

It is generally safer not to allow any functions to have access to the data elements of a class. The use of `friend` procedures subverts the goal of data hiding. In other words: Don't use this feature of C++.

Furthermore, it is possible to have a procedure that is a `friend` of two different classes, or for one class to be a `friend` of another. These advanced topics are beyond the scope of this text.

C.5.8 Class templates

C++ implements complex numbers using a class template. For complex numbers with `double` real and imaginary parts, we use the declaration `complex<double>` whereas for Gaussian integers we use `complex<int>`.

We create our own class templates like this:

```
template <class T>
class MyClass {
private:
  T x;   // data element of type T
  ...
public:
  ...
};
```

Here, the `T` acts as a "type variable." Now we can declare objects to be of type `MyClass<int>` or even `MyClass< complex<double> >`. (Note the extra space; we need to avoid typing `<<` or `>>`.)

Templates must be placed in their entirety in the `.h` file and all their methods defined inline.

It is possible to define a template with more than one argument, like this:

```
template <class U, class V>
class MyClass {
private:
  U x;
  V y;
  ...
};
```

C.5.9 Inheritance

A key feature of object-oriented programming is the ability to add features to an existing class to make a new class. For example, we may have a class `Graph` to represent simple graphs (no loops or multiple edges). In some cases, we may wish to consider graphs embedded in the plane (with adjacent vertices joined by line segments), which would be an object of type EmbeddedGraph.

Because there are many instances in which we work with abstract graphs (without embedding) we first create the Graph class. Once this is in place, we build the EmbeddedGraph class. Rather than start from scratch, we extend the Graph class like this:

```
class EmbeddedGraph : public Graph {
private:
  // additional data members to hold the x,y coordinates
  // of the vertices
public:
  ...
};
```

It is useful for the constructors for EmbeddedGraph to use the constructors of Graph as first step. For example, suppose we have a constructor Graph(int n) that creates a graph with *n* vertices. We would want EmbeddedGraph(int n) to create the graph, but also set up a default embedding. To do this, we declare the new constructor like this:

```
class EmbeddedGraph: public Graph {
  ...
public:
  EmbeddedGraph(int n) : Graph(n) {
    // set up the embedding
  }
};
```

Data in Graph that are declared in the private section are not available to the added methods of EmbeddedGraph. If we want to grant access to EmbeddedGraph to these data elements, we move them from private to protected.

```
class Graph {
private:
  // data and methods that EmbeddedGraph never needs to access
protected:
  // data and methods that EmbeddedGraph may access
public:
  // public methods, accessible to all parts of the program
};
```

C.6   Standard functions

C.6.1   Mathematical functions

The following mathematical functions are available in C++. Many of these are declared in the header cmath and you may need a #include <cmath> declaration to use them.

- abs: absolute value of an int. The absolute value function comes in the following flavors.

  - int abs(int x)
  - long labs(long x)
  - long long llabs(long long x)
  - double fabs(double x)
  - float fabsf(float x)
  - long double fabsl(long double x)

  Of these, abs and fabs are the forms most commonly used.

- acos: arc cosine. Usage: double acos(double x). Of course, this requires $-1 \leq x \leq 1$.

- asin: arc sine. Usage: double asin(double x). Of course, this requires $-1 \leq x \leq 1$.

- atan and atan2: arc tangent.

  The first is double atan(double x) and gives the arc tangent of x in the interval $(-\pi/2, \pi/2)$.

  The second is double atan(double x, double y) and gives the angle of the vector from the origin to the point (x, y); the result is between $-\pi$ and π.

- Bessel functions. The following are available.

  - double j0(double x): first kind, order 0.
  - double j1(double x): first kind, order 1.
  - double jn(int n, double x): first kind, order n.
  - double y0(double x): second kind, order 0.
  - double y1(double x): second kind, order 1.
  - double yn(int n, double x): second kind, order n.

  There are also the variants j0f through ynf that replace double with float as well as j0l through ynl that use long doubles.

- cbrt: cube root. Usage: double cbrt(double x). Returns $\sqrt[3]{x}$. Might not be available on all systems.

- ceil: ceiling function. Usage: double ceil(double x). This returns $\lceil x \rceil$.

  There are also the following variants.

  - float ceilf(float x)

     – `long double ceill(long double x)`

- `cos`: cosine. Usage: `double cos(double x)`. Gives $\cos x$.

- `cosh`: hyperbolic cosine. Usage: `double cosh(double x)`. Gives $\cosh x$.

- `erf`: error function. Usage: `double erf(double x)`. Gives

$$\mathrm{erf}\, x = \frac{2}{\sqrt{\pi}} \int_0^x e^{-t^2}\, dt.$$

There is the related function `double erfc(double x)` that gives $1 - \mathrm{erf}\, x$.

- `exp`: e^x. Usage: `double exp(double x)`.

There is a related function `double expm1(double x)` that returns $e^x - 1$.

See also `pow`.

- `floor`: floor function. Usage: `double floor(double x)`. This returns $\lfloor x \rfloor$.

There are also the following variants.

     – `float floorf(float x)`

     – `long double floorl(long double x)`

- `fmod`: mod for reals. Usage: `double fmod(double x, double y)`. This returns $x - ny$ where $n = \lfloor x/y \rfloor$.

- Gamma function. The function `gamma` is implemented differently on different computers. In some cases it gives $\Gamma(x)$ but in others it gives $\log \Gamma(x)$.

The related `gammaf` and `gammal` work with `float` and `long double` values, respectively.

Free of ambiguity, `lgamma` gives $\log \Gamma(x)$ and `tgamma` gives $\Gamma(x)$.

Some systems have additional variations of `lgamma`. On a UNIX system, type `man gamma` or `man lgamma` for more information.

- `hypot`: hypotenuse. Usage: `double hypot(double x, double y)`. Returns $\sqrt{x^2 + y^2}$.

- `log`: natural logarithm. Usage: `double log(double x)`. Returns $\log_e x$.

The related `double log1p(double x)` that returns $\log(x+1)$.

Also `double log10(double x)` returns $\log_{10} x$.

- `pow`: exponentiation. Usage: `double pow(double x, double y)`. Returns x^y. See also `exp`.

- `sin`: sine. Usage: `double sin(double x)`. Returns $\sin x$.

- `sinh`: hyperbolic sine. Usage: `double sinh(double x)`. Returns $\sinh x$.

- `sqrt`: square root. Usage: `double sqrt(double x)`. Returns $\sqrt{x}$.

- `tan`: tangent. Usage: `double tan(double x)`. Returns $\tan x$.

- `tanh`: hyperbolic tangent. Usage: `double tanh(double x)`. Returns $\tanh x$.

C.6.2 Mathematical constants

Various constants are defined[2] via the `<cmath>` header. Here is a list. `M_PI`

C++ name	Value
`M_E`	e
`M_LOG2E`	$\log_2 e$
`M_LOG10E`	$\log_{10} e$
`M_LN2`	$\ln 2$
`M_LN10`	$\ln 10$
`M_PI`	π
`M_PI_2`	$\pi/2$
`M_PI_4`	$\pi/4$
`M_1_PI`	$1/\pi$
`M_2_PI`	$2/\pi$
`M_2_SQRTPI`	$2/\sqrt{\pi}$
`M_SQRT2`	$\sqrt{2}$
`M_SQRT1_2`	$1/\sqrt{2}$

C.6.3 Character procedures

The following procedures require `#include <cctype>` and provide convenient tools for manipulating characters. Many of these are defined in terms of the `int` type and not the `char` type. This is not an issue with a function such as `islower`. This procedure checks if its argument represents a lowercase letter (from a to z). It would be logical to expect that the argument to `islower` is a `char` and its return value is a `bool`. However, both values are type `int`. This is inconsequential for this particular function because code such as this works as expected:

```
char ch;
...
if (islower(ch)) {
  ...
}
```

---

[2]Not all compilers define these constants. If your compiler does not, you can define them yourself in a header file named, say, `myconstants.h`. Use statements such as this: `const double M_PI = 3.14159...;`.

Problems arise when using a procedure such as `tolower` that converts uppercase characters A to Z to their lowercase form. It would be logical to expect this procedure's argument and return types to be `char`, but this is not the case. Instead, both are type `int`. The consequence is that the statement `cout<<tolower('G');` does *not* print the character `g` on the screen. Instead, it prints the value 103. Why?

When the procedure `tolower('G')` is encountered the char value `'G'` is automatically converted into an `int` value (because that's what `tolower` expects). Effectively, this becomes `tolower(71)` (because G is represented inside the computer as the number 71). Now `tolower` does its work and converts the code 71 for G to the code for g and returns that value: 103. That is the value that is sent (via the `<<` operator) to `cout`.

The issue becomes: how do we convert the value 103 to the desired character g? Simply wrap `char` around the return value from `tolower`. The successful way to convert G to g is to use this: `char(tolower('G'))`.

One additional warning: Do not use `tolower("G")`. The `"G"` is a character array (type `char*`) and not a single character (type `char`). The C++ compiler is able to convert a `char*` to an `int`, so such a statement might generate only a warning from the compiler. When run, this code would act on the memory address of the character array, and not on the character G as you intended.

Here are the procedures. All of them require `#include <cctype>`.

- `isalnum(ch)`: checks if `ch` is either a letter (upper- or lowercase) or a digit (0 through 9).

- `isalpha(ch)`: checks if `ch` is a letter (upper- or lowercase).

- `isdigit(ch)`: checks if `ch` is a digit (0 through 9).

- `isgraph(ch)`: checks if `ch` is a printable character (such as letters, digits, or punctuation) but not white space (such as a space character, a tab, or a newline).

- `islower(ch)`: checks if `ch` is a lowercase letter (a to z).

- `isprint(ch)`: checks if `ch` is a printable character (including letters, digits, punctuation, and space). A nonprintable character is known as a *control character*. Those can be detected with `iscntrl`.

- `ispunct(ch)`: checks if `ch` is a punctuation character.

- `isspace(ch)`: checks if `ch` is a white space character (such as a space, tab, or newline).

- `isupper(ch)`: checks if `ch` is an uppercase letter (A to Z).

- `isxdigit(ch)`: checks if `ch` is a digit from a hexadecimal number (i.e., one of 0 through 9, a through f, or A through F).

- `tolower(ch)`: converts `ch` to a lowercase letter (if `ch` is an uppercase letter). Otherwise, this returns the value `ch` unchanged.

  In most cases, use `char(lower(ch))`.

- `toupper(ch)`: converts `ch` to an uppercase letter (if `ch` is a lowercase letter). Otherwise, this returns the value `ch` unchanged.

  In most cases, use `char(toupper(ch))`.

C.6.4   Other useful functions

The header `<cstdlib>` contains other functions useful for C++ programming. Here we list the ones most useful for mathematical work. (The inclusion of the header `<cstdlib>` might not be required on your system.)

- `abort`: quickly halt the execution of the program. The syntax is `abort()`. This is an "emergency stop" procedure that brings a program to a screeching halt. A message, such as `Abort trap` is written to the screen.

  There are better ways to stop your program. If possible, arrange your code so the program always manages to find its way to the end of the `main()` procedure (or to a `return` statement in the `main`). Alternatively, use `exit` (described below).

- `atof`, `atoi`, `atol`: convert character arrays to numbers. The syntax for these are:

  - `float atof(char* str)`
  - `int atoi(char* str)`
  - `long atol(char* str)`

  In all cases, `str` is a character array (not a C++ `string`) and the procedure converts the character array into a number. For example, `atoi("23")` returns an `int` value equal to 23.

  Some platforms support an `atoll` procedure that converts its character array to a `long long`.

- `exit`: terminate the program. Usage: `exit(int x)`. The preferred method for ending a program is for the execution to reach a `return` statement in the `main`. Sometimes, this is not practical. In such cases, it is convenient to call `exit` to end the program.

  The argument to `exit` is sent as a "return code" back to the operating system. (The `return` value in `main` serves the same purpose.) By convention, use a return value of 0 if all is well and a nonzero value if something amiss occurred (e.g., your program was unable to open a file it needed).

- `rand`: return pseudo random values. Usage: `int rand()`. This returns a "random" integer value between 0 and `RAND_MAX` (inclusive). See Chapter 4.

- `srand`: initialize the random number generator. Usage: `void seed(int x)`. This resets the pseudo random number generator to a given starting position. A good choice for a seed value is the system time. To do this, use the statement `srand(time(0));` (you may need to include the `ctime` header).

Appendix D

Answers

This appendix provides answers and useful comments for nearly all of the exercises in this book.

1.1 The first `*/` after the word and ends the comment. The additional `*/` on the last line causes the error.

1.2 The first message complains that the `cout` object is deemed undeclared despite the fact that the programmer did not forget `#include <iostream>`. What the programmer did forget is the statement `using namespace std;` without which `cout` is not understood.

   The second error caused a `parse error` on line 5. However, there is nothing wrong with line 5. The error is on line 4 where we forgot to include the semicolon. However, the compiler did not get confused by this omission until it reached line 5.

1.3 No. If `//` or `/*` appear inside quotation marks, they are part of the character sequence and are not interpreted as the start of a comment.

1.4 The sequence `\n` starts a new line. The statements `cout << "\n";` and `cout << endl;` have the same effect.

   The sequence `\t` causes the output to skip to the next tab stop in the output. It is useful for lining up data in columns.

   The sequence `\\` causes a single backslash `\` to be written.

1.5   Don't make such a fuss fuss
      This is more fun than learning C++!

2.1 The number 3 is printed to the screen. When the `float` 3.5 is assigned to an `int` variable, there is a loss of precision; the `int` variable holds just the value 3. Then, when the `int` value *e* is assigned to the `double` variable, the `double` is given the value 3.

   A good compiler should warn that assigning a `float` from an `int` can result in loss of precision.

2.2 The output from the program looks like this:

```
10
12
13
10
13.3333
```

Only cout << (4./3)*10 gives the proper output.

2.3 The output of the program looks like this:

```
-2
2
-2
2
```

The mathematician's mod operation $a \bmod n$ is usually only defined for $n > 0$ and $a \bmod n$ is an element of $\{0, 1, 2, \ldots, n-1\}$. Thus, $-5 \bmod 3 = 1$, but C++ returns -2 for (-5)%3. In addition, C++ allows the modulus, n, to be negative.

2.4 Both '3' and 3 are integer types (the first is a char which is considered to be an integer type). The result of this expression is false because the character '3' does not have the same value as the integer 3. Indeed, on many computers '3' has the value 51 because the character '3' is the 51st in the ASCII character set.

2.5 The first occurrence of the expression a==b evaluates to false because, when this expression is reached, a and b hold different values. When a bool value of false is written to the screen, the number 0 is used.

Next, the expression a=b has the *effect* of copying b's value into a and returns the *value* now held in common in a and b. Hence, 10 is printed.

Finally, when the second occurrence of a==b is evaluated, both a and b hold the value 10 and so the expression evaluates to true which is printed as 1 on the screen.

2.6 The result of your experiment might depend on the computer on which it is run. The following is the typical result.

If the numerator and denominator are both int types, both $\frac{0}{0}$ and $\frac{1}{0}$ evaluate to zero.

However, if the numerator or the denominator (or both) are float then the special values nan and inf are returned for $\frac{0}{0}$ and $\frac{1}{0}$, respectively. Clearly inf stands for *infinity* whereas nan stands for *not a number*.

2.7 400.

2.8 (a) A variable name may not begin with a digit.

(b) The minus sign is an operator and may not be part of a variable's name.

(c) The word `double` is a C++ keyword and may not be used as a variable name.

(d) A variable name may not include a colon.

(e) The ampersand is an operator and may not be part of a variable's name.

(f) This begins with a digit, contains a minus sign, and is a number (equal to 0.01).

3.1 $d = 17, x = 6$, and $y = -1$. Other answers for x and y are possible.

3.2 If this procedure is invoked with n equal to -1 we fall into an infinite recursion.

There is also the issue that if a large value of n is passed to this procedure, the result would be too large to fit in a `long`, overflow would result, and the answer returned would be incorrect.

3.3 In this implementation, we take the input argument to be a real number (type `double`) and the return value to be an `int`. Your choice may vary.

```
int signum(double x) {
    if (x < 0.) return -1;
    if (x > 0.) return 1;
    return 0;
}
```

3.4 Here is an iterative version.

```
long fibonacci(long n) {
    if (n < 0) return -1;
    if (n==0) return 1;
    if (n==1) return 1;
    long a = 1;
    long b = 1;
    long c;

    for(int k=2; k<=n; k++) {
        c = a+b;
        a = b;
        b = c;
    }
    return c;
}
```

This is a recursive version.

```
long fibonacci(long n) {
    if (n < 0) return -1;
    if (n==0) return 1;
    if (n==1) return 1;
    return fibonacci(n-1)+fibonacci(n-2);
}
```

3.5 The iterative version is significantly faster than the recursive version. When the iterative version computes F_n, it does $n-1$ additions. However, when the recursive version computes F_n, it requests the calculation of F_{n-1} and F_{n-2}. The calculation of F_{n-1} also requests the computation of F_{n-2}, and this is wasted effort. The amount of work needed to calculate F_n grows exponentially with n.

When asked to calculate F_n, the recursive procedure presented here is called $2F_n - 1$ times (your program may have different behavior). This is easy to show by induction. Let $c(n)$ denote the number of times fibonacci is called when asked to calculate F_n. For $n = 0, 1$ we have $c(n) = 1$ and for $n > 1$ we note that $c(n) = 1 + c(n-1) + c(n-2)$. Now one checks that $2F_n - 1$ satisfies this recurrence with the given initial conditions.

The values you were requested to find are

$$F_{20} = 10{,}946 \qquad F_{30} = 1{,}346{,}269 \qquad F_{40} = 165{,}580{,}141.$$

3.6 Here is a complete program.

```cpp
#include <iostream>
#include <cmath>
using namespace std;

const float zeta2 = M_PI*M_PI/6.;

float zeta2up(long n) {
  float ans = 0.;
  for (long k=1; k<=n; k++) {
    ans += 1./float(k)/float(k);
  }
  return ans;
}

float zeta2down(long n) {
  float ans = 0.;
  for (long k=n; k>=1;  k--) {
    ans += 1./float(k)/float(k);
  }
  return ans;
}

int main() {
  long n = 1000000; // 10^6
  cout << "zeta2up    = " << zeta2up(n) << endl;
  cout << "zeta2down = " << zeta2down(n) << endl;
  cout << "truth      = " << zeta2 << endl;
  return 0;
}
```

The output looks like this.

```
zeta2up    = 1.64473
zeta2down = 1.64493
truth      = 1.64493
```

Observe that zeta2down, which begins the sum with $1/N^2$, gives the more accurate result. The discrepancy is due to roundoff issues. A float holds only so many digits and (on my computer) the epsilon value for a float is around 10^{-7}. This means that by the time the sum reaches $k = 10^6$, the value held in ans is too large for the addition of 10^{-12} to have any effect. Many terms at the end of the series have little or no effect on the sum; their contribution is lost. However, those tail terms do have a cumulative contribution on the actual sum and are needed to give an accurate result. By summing in the reverse order these small terms do not have their contributions lost.

3.7 The important issue here is to use call by reference for the third and fourth arguments. Here is a complete program and its output.

```
#include <iostream>
#include <cmath>
using namespace std;

void xy2polar(double x, double y, double& r, double& t) {
  r = hypot(x,y);
  t = atan2(x,y);
}

void polar2xy(double r, double t, double& x, double &y) {
  x = r*cos(t);
  y = r*sin(t);
}

int main() {
  double x,y,r,t;

  x = -4.; y = -1.;
  xy2polar(x,y,r,t);
  cout << "The point (" << x << "," << y
       << ") in polar coordinates is ("
       << r << "," << t << ")" << endl;

  r = 2.; t = 2.5;
  polar2xy(r,t,x,y);
  cout << "The point in polar coordinates (" << r
       << "," << t << ") is (" << x << "," << y << ")" << endl;

  return 0;
}
```

```
The point (-4,-1) in polar coordinates is (4.12311,-1.81577)
The point in polar coordinates (2,2.5) is (-1.60229,1.19694)
```

3.8 Here is a program. Change long to long long (or __int64) if that type is available to you.

```
#include <iostream>
using namespace std;
```

```cpp
bool is_zero_one(long n) {
  long ones_digit = n % 10;
  if (ones_digit > 1) return false;
  if (n == 0) return true;
  return is_zero_one(n/10);
}

long find_zero_one_mult(long n) {
  for (long k=1; ; k++) {
    long m = k*n;
    if (m < 0) return -1; // overflow without success
    if (is_zero_one(m)) return m;
  }
  return 0;
}

int main() {
  for(long k=1; k<=100; k++) {
    cout << k << "\t" << find_zero_one_mult(k) << endl;
  }
  return 0;
}
```

The first procedure, `is_zero_one`, checks if its input argument consists entirely of 0s and 1s (in base ten) by a recursive algorithm. The second procedure, `find_zero_one_mult`, does its work by trying all positive multiples of the input argument until it succeeds. However, if the multiplies overflow (detected by testing `if(m<0)`) then we return -1 to signal failure.

Changing `long` to `long long` avoids the overflow problem, but then it takes much longer to find the proper multiple.

This program is not efficient. You may notice the computer pausing while it works to find the least multiple of 9 of the required form: 111111111. The program is not likely to find the least such multiple of 99, because it is

$$\underbrace{111111111111111111}_{18 \text{ digits}}.$$

Instead of testing successive multiples of n until we find the result we want, we can step through decimal numbers composed entirely of 0s and 1s and check if n is a divisor. However, this is trickier to program.

To step through the decimal values 1, 10, 11, 100, 101, 110, and so on, we realize that if we interpret these numbers as binary, then we are simply counting 1, 2, 3, 4, 5, and so on. If we had a procedure to convert binary numbers to the corresponding decimal number with the same digits, the task would be easy.

To this end, we create a procedure named `bin2dec` that performs the mapping

$$n = \sum_{j\geq 0} b_j 2^j \mapsto f(n) = \sum_{j\geq 0} b_j 10^j$$

where the b_j are in $\{0,1\}$. There's a nice recursive way to define this transformation:

$$f(n) = \begin{cases} 10f\left(\frac{n}{2}\right) & \text{if } n \text{ is even, and} \\ 10f\left(\frac{n-1}{2}\right) + 1 & \text{if } n \text{ is odd} \end{cases}$$

with base cases $f(0) = 0$ and $f(1) = 1$. [Note: The recursion can be written with a single algebraic expression like this: $f(n) = 10f(\lfloor n/2 \rfloor) + (n \bmod 2)$.]

With this idea in place, we present a second solution to this problem. This program runs much faster than the first.

```cpp
#include <iostream>
using namespace std;

long long bin2dec(long n) {
  if (n==0) return 0;
  if (n==1) return 1;
  return 10*bin2dec(n/2) + (n%2);
}

long long find_zero_one_mult(long long n) {
  for (long long k=1; ; k++) {
    long long m = bin2dec(k);
    if (m < 0) return -1; // overflow without success
    if (m%n == 0) return m;
  }
  return 0;
}

int main() {
  for(long long k=1; k<=100; k++) {
    cout << k << "\t" << find_zero_one_mult(k) << endl;
  }
  return 0;
}
```

4.1 Here is a program that does the job.

```cpp
#include "uniform.h"
#include <iostream>
#include <cmath>
using namespace std;

int main() {
  double x,y,u,v;
  const long nreps = 100000000;
  double sum;
  seed();
  sum = 0;
  for (long k=0; k<nreps; k++) {
    x = unif();
    y = unif();
    u = unif();
    v = unif();
    sum += sqrt( (x-u)*(x-u) + (y-v)*(y-v) );
```

```
    }

    cout << "The average length of the segment is " << sum/nreps
         << endl;
    return 0;
}
```

This gives the following output.

```
The average length of the segment is 0.521435
```

This naturally leads to the question: What is the analytic answer? My colleague James Fill informs me that it is $\frac{1}{15}\left(2+\sqrt{2}+\sinh^{-1}1\right)$ which is approximately 0.521405.

4.2 Here is a program. Note the inclusion of the `uniform.h` header; to compile this program we need this file (call it `buffon.cc`) and the files `uniform.h` and `uniform.cc`.

```cpp
#include <iostream>
#include <cmath>
#include "uniform.h"
using namespace std;

/**
 * This procedure simulates one drop of Buffon's needle.
 * It returns true if the needle crosses a line and false if not.
 */

bool drop() {
    double x1;      // x-coord of one end point of the needle
    double x2;      // x-coord of the other end point
    double theta;   // orientation of the needle

    x1 = unif();
    theta = 2*M_PI*unif();
    x2 = x1 + cos(theta);

    if (floor(x1) == floor(x2)) return false;

    return true;
}

int main() {
    long nreps;   // number of times to drop the needle
    cout << "Enter number of needle drops -> ";
    cin >> nreps;

    seed();
    long count = 0; // number of successful drops
    for (int k=0; k<nreps; k++) {
        if (drop()) ++count;
    }
```

```
    cout << count << " successes out of " << nreps
         << " drops" << endl;
    cout << "frequency = " << double(count)/double(nreps) << endl;
}
```

Here is a sample run of this program.

```
Enter number of needle drops -> 100000000
63653758 successes out of 100000000 drops
frequency = 0.636538
```

This is reasonably close to the theoretical value, $\frac{2}{\pi} \approx 0.63662$.

4.3 Here is some advice on how to approach this exercise. The procedures to generate points are random in a circle and in a triangle should be of the following form,

```
void point_in_circle(double& x, double& y);
void point_in_triangle(double& x, double& y);
```

Using reference arguments enables these procedures effectively to return two values.

For point_in_circle, use a rejection algorithm. That is, generate a point (x,y) uniformly at random in the square $[-1,1] \times [-1,1]$. If $x^2 + y^2 \leq 1$, then return (x,y); if not, try again. This is a good opportunity to use the do-while control structure.

For point_in_triangle, it does not matter in which triangle the points are generated. (Reason: Whether four points lie at the vertices of a convex quadrilateral is invariant under invertible affine transformations.) For simplicity, use the triangle with vertices located at $(0,0)$, $(1,0)$, and $(0,1)$. A rejection algorithm may be used (generate points in a square until finding one in the triangle). However, it is more efficient to find a point uniformly at random in $[0,1] \times [0,1]$ and if it lies in the wrong half of the square, reflect it across the line through $(0,1)$ and $(1,0)$.

To test if four points determine the corners of a convex quadrilateral, first write a procedure to see if the points (x_1,y_1) and (x_2,y_2) lie on opposite sides of the line through (x_3,y_3) and (x_4,y_4). They are if and only if the two triples of points $(x_1,y_1),(x_2,y_2),(x_3,y_3)$ and $(x_1,y_1),(x_2,y_2),(x_4,y_4)$ have opposite orientation (see the figure).

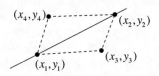

Thus, the points (x_1, y_1) and (x_2, y_2) lie on opposite sides of the line through (x_3, y_3) and (x_4, y_4) if and only if

$$\det \left(\begin{bmatrix} 1 & x_1 & y_1 \\ 1 & x_2 & y_2 \\ 1 & x_3 & y_3 \end{bmatrix} \cdot \begin{bmatrix} 1 & x_1 & y_1 \\ 1 & x_2 & y_2 \\ 1 & x_4 & y_4 \end{bmatrix} \right) < 0.$$

Call this procedure something like opposite_sides and use it to build a procedure that tests if four points determine a convex quadrilateral.

Your experiments should yield the following results. For a triangle, $P(K) = \frac{2}{3}$. For a circle, $P(K) = 1 - \frac{35}{12\pi^2} \approx 0.70448$.

4.4 Despite the presence of trig functions, the first version is somewhat faster than the rejection method on my computer.

4.5 The probability that a point (x, y, z) chosen uniformly at random in $[-1, 1]^3$ lies within distance one of the origin equals the volume of the ball of radius 1 divided by the volume of the cube: $\frac{4}{3}\pi \div 8 = \pi/6 \approx 0.5236$.

In high-dimensional space, the unit ball occupies only a tiny fraction of the cube $[-1, 1]^n$. Therefore, the probability that a given iteration of the rejection algorithm succeeds is small and many iterations are required to generate a point within the unit ball.

An efficient way to generate a point $\mathbf{x} = (x_1, x_2, \ldots, x_n)$ uniformly at random on the unit sphere in $\mathbb{R}^n$ is to assign to each x_i a Gaussian $N(0, 1)$ random value and then to scale by $1/\|\mathbf{x}\|$.

4.6 In order for the procedure to remember the position of the particle from one invocation to the next, use a static variable. Here is the code.

```
#include "uniform.h"

int random_walk() {
  static int position = 0;
  if (unif() > 0.5) {
    ++position;
  }
  else {
    --position;
  }
  return position;
}
```

4.7 Naturally, we need to use a static variable to remember the value between calls. However, a static variable declared in up is not available for use by down, or vice versa. (It is possible to use a global variable, defined outside the scope of any procedure, but this practice is frowned upon.) The solution is to put the static variable in another procedure! Here's a solution.

```
int updown(int change) {
  static int value = 0;
  value += change;
  return value;
}

int up() {
  return updown(1);
}

int down() {
  return updown(-1);
}
```

5.1 $\varphi(100) = 4$, $\varphi(2^9) = 256$, and $\varphi(6!) = 192$.

5.2 The problem is with the declaration `int vals[n];`. Because the value of n cannot be determined when the program is compiled, this statement is illegal. Instead, declare `vals` with the statement `int* vals;` and then allocate space with `vals = new int[n];`. When `vals` is no longer needed, release the allocated memory with `delete[] vals;`.

5.3 $x \equiv 23 \pmod{180}$.

5.5 First, there is nothing wrong with using `bool` types for the array `theSieve`. The decision to use `char` is based on an oddity. Most compilers use one byte to house a `char`, but some use four bytes to house a `bool`. Use the expressions `sizeof(char)` and `sizeof(bool)` to learn the behavior on your computer.

If we are using an array that has only a million entries, then the difference between one and four bytes per cell is not important. But if the array has hundreds of millions (or a billion) entries, then the difference is important as the program may exhaust available memory.

In Chapter 8 we introduce the `vector<bool>` type that provides greater memory efficiency; this uses only one bit for each entry.

5.6 Did you remember to declare your array to hold 21 values?

```
#include <iostream>
using namespace std;

int main() {
  long fib[21];
  fib[0] = 1;
  fib[1] = 1;
  for(int k=2; k<=20; k++) {
    fib[k] = fib[k-1] + fib[k-2];
  }
  for (int k=0; k<=20; k++) {
    cout << k << "\t" << fib[k] << endl;
  }
  return 0;
}
```

5.7 Here is a program.

```cpp
#include <iostream>
using namespace std;

int main() {
  const int STEPS = 15;
  long a[STEPS];
  long b[STEPS];

  a[0] = b[0] = 1;

  for (int k=1; k<STEPS; k++) {
    a[k] = b[k-1];
    b[k] = a[k-1] + 2*b[k-1];
  }

  for (int k=0; k<STEPS; k++) {
    cout << k << "\t" << a[k] << "\t" << b[k] << "\t"
         << double(a[k])/double(b[k]) << endl;
  }
  return 0;
}
```

Notice the use of a const int to declare the array sizes. If we decide to recompile this program to generate arrays of a different size, then we only need to edit this one value.

Here is the output of the program.

0	1	1	1
1	1	3	0.333333
2	3	7	0.428571
3	7	17	0.411765
4	17	41	0.414634
5	41	99	0.414141
6	99	239	0.414226
7	239	577	0.414211
8	577	1393	0.414214
9	1393	3363	0.414213
10	3363	8119	0.414214
11	8119	19601	0.414214
12	19601	47321	0.414214
13	47321	114243	0.414214
14	114243	275807	0.414214

Based on this chart, it is safe to conjecture that $a_n/b_n \to \sqrt{2} - 1$.

Here's one approach to proving this. The recurrence can be written in matrix form as

$$\begin{bmatrix} a_0 \\ b_0 \end{bmatrix} = \begin{bmatrix} 1 \\ 1 \end{bmatrix} \quad \text{and} \quad \begin{bmatrix} a_n \\ b_n \end{bmatrix} = \begin{bmatrix} 0 & 1 \\ 1 & 2 \end{bmatrix} \begin{bmatrix} a_{n-1} \\ b_{n-1} \end{bmatrix}.$$

The eigenvalues of $\begin{bmatrix} 0 & 1 \\ 1 & 2 \end{bmatrix}$ are $1 \pm \sqrt{2}$ with corresponding eigenvectors $\begin{bmatrix} -1 \pm \sqrt{2} \\ 1 \end{bmatrix}$, respectively. The iterates of this linear system diverge in the direction of the eigenvector corresponding to the eigenvalue of largest magnitude, $1 + \sqrt{2}$; that is, they tend to infinity in the direction $\left[\begin{smallmatrix} -1+\sqrt{2} \\ 1 \end{smallmatrix}\right]$, and so $a_n/b_n \rightarrow -1 + \sqrt{2}$.

5.8 Here is such a procedure.

```
long max_value(const long* array, long nels) {
  long ans;

  ans = array[0];

  for (long k=1; k<nels; k++) {
    if (ans < array[k]) {
      ans = array[k];
    }
  }
  return ans;
}
```

Notice that the first argument is type const long*. The long* means that the argument is an array of long values. The const certifies that this procedure does not modify the values held in the array.

5.9 Here is a procedure to generate an array of Fibonacci numbers as a return value.

```
long* make_fibs(long n) {
  long* ans;
  ans = new long[n];
  ans[0] = 1;
  ans[1] = 1;
  for (int k=2; k<n; k++) ans[k] = ans[k-1] + ans[k-2];
  return ans;
}
```

Notice that the return type is long* because this procedure returns an array.

The problem with the main() given in the problem is that it suffers a memory leak. The array created by make_fibs is never released with a delete[] statement.

The use of new in this design is unavoidable. A better strategy is to design a procedure with type void make_fibs(long n, long* array);. It is then the responsibility of the programmer to allocate an array of the appropriate size before invoking this procedure:

```
long* fibs;
fibs = new long[20];
make_fibs(20,fibs);
// ...
delete[] fibs;
```

alternatively:

```
long fibs[20];
make_fibs(20,fibs);
//...
```

No `delete[]` is necessary in this second instance because `fibs` was not allocated with `new`.

5.10 Here is such a procedure.

```
long fibs(int n) {
  const int MAX_ARG = 40;
  static bool first_time = true;
  static long values[MAX_ARG+1];

  if (first_time) {
    values[0] = 1;
    values[1] = 1;
    for (int k=2; k<=MAX_ARG; k++) {
      values[k] = values[k-1] + values[k-2];
    }
    first_time = false;
  }

  if ((n<0) || (n>MAX_ARG)) return -1;
  return values[n];
}
```

The key idea is the use of `static` variables. The Boolean `first_time` is initialized to `true` so the code can detect when it is first run. The array `values` is then populated with Fibonacci numbers and `first_time` is set to `false` so the initialization is not repeated on subsequent calls. The array `values` is also declared to be `static` so that its contents persist between invocations.

5.11 Here is such a program.

```
#include <iostream>
using namespace std;

int main() {
  long size = 10000000; // ten million
  long* array;
  long count = 0;
  while (true) {
    array = new long[size];
    ++count;
    cout << "Success with array #" << count << endl;
  }
}
```

When this program is run on my computer, we see the following output.

```
Success with array #1
Success with array #2
Success with array #3
Success with array #4
...
Success with array #90
Success with array #91
*** malloc: vm_allocate(size=40001536) failed (error code=3)
```

```
*** malloc[4998]: error: Can't allocate region
Abort trap
```

5.12 Enter the digits (separated by spaces) into the On-Line Encyclopedia of Integer Sequences. It responds that this is sequence A059742 giving the decimal digits of $\pi + e$.

6.1 Here is a header file for the Line class.

```
#ifndef LINE_H
#define LINE_H

#include "Point.h"

class Line {
private:
  double a,b,c; // to specify the line ax+by+c=0

public:
  // constructors
  Line();
  Line(Point P, Point Q);
  Line(double aa, double bb, double cc);

  // get methods
  double getA() const;
  double getB() const;
  double getC() const;

  // reflection methods
  void reflectX();
  void reflectY();

  // check if a Point is on this Line
  bool incident(Point P) const;

  // generate a Point on this Line
  Point find_Point() const;

  // check if this Line equals another
  bool operator==(const Line& that) const;
};

// write a Line to an output stream
ostream& operator<<(ostream& os, const Line& L);

// find the distance between a Point and a Line
double dist(Line L, Point P);
double dist(Point P, Line L);

#endif
```

And here is the code file Line.cc.

```
#include "Line.h"
```

```
Line::Line() {
  a = b = c = 0.;
}

Line::Line(Point P, Point Q) {
  // get the coordinates of the points
  double x1 = P.getX();
  double y1 = P.getY();
  double x2 = Q.getX();
  double y2 = Q.getY();

  // If the points are the same, make the Line horizontal
  // through the one point.
  if (P==Q) {
    cerr << "Warning: Constructing Line from two equal Points"
         << endl;
    a = 0.;
    b = 1.;
    c = -y1;
    return;
  }

  // If the points are distinct, we continue starting
  // with the special case of a vertical line
  if (x1 == x2) {
    a = 1.;
    b = 0.;
    c = -x1;
    return;
  }

  // The points are distinct and not vertical
  a = y2 - y1;
  b = x1 - x2;
  c = x2*y1 - x1*y2;
}

Line::Line(double aa, double bb, double cc) {
  // Check if the data are valid
  if ((aa==0.) && (bb==0.)) {
    cerr << "Warning: Invalid call to Line(aa,bb,cc)" << endl;
    a = 0.;
    b = 1.;
    c = cc;
    return;
  }
  a = aa;
  b = bb;
  c = cc;
}

// get methods

double Line::getA() const { return a; }
double Line::getB() const { return b; }
```

```cpp
double Line::getC() const { return c; }

// reflection methods
void Line::reflectX() { b = -b; }
void Line::reflectY() { a = -a; }

// check if a Point is on this Line
bool Line::incident(Point P) const {
  return a*P.getX() + b*P.getY() + c == 0. ;
}

// generate a Point on a Line
Point Line::find_Point() const {
  // if it's a vertical line, return x-intercept
  if (b==0.) {
    return Point(-c/a, 0);
  }
  // otherwise, return the x-intercept
  return Point(0,-c/b);
}

bool Line::operator==(const Line& that) const {
  // Special case when a == 0
  if (a==0) {
    if (that.a != 0) return false;
    return c/b == that.c/that.b;
  }
  // When a is not 0
  if (that.a == 0) return false;
  return (b/a == that.b/that.a) && (c/a == that.c/that.a);
}

// for writing to an output stream
ostream& operator<<(ostream& os, const Line& L) {
  os << "[" << L.getA() << "," << L.getB() << ","
     << L.getC() << "]";
  return os;
}

// determine the distance from a Point to a Line
double dist(Line L, Point P) {
  // fetch a,b,c for the line
  double a = L.getA();
  double b = L.getB();
  double c = L.getC();

  // fetch the coordinates of the point
  double x = P.getX();
  double y = P.getY();

  // normalize by dividing by sqrt(a*a+b*b)
  double d = sqrt(a*a + b*b);
  a /= d;
  b /= d;
  c /= d;
```

```
    // project P onto the unit vector (a,b) and add c
    double ans = x*a + y*b + c;

    return fabs(ans);
}

double dist(Point P, Line L) { return dist(L,P); }
```

6.2 It is natural to suspect that something is wrong with the logic and algebra used to create the procedures, but that is not the case. If we examine the $[a,b,c]$ triples for the two lines, we see that the triple for L is a scalar multiple of the triple for M (the factor is 1.4), so these represent the same line.

The problem is roundoff. Indeed, if we print the distance from Y to M the computer responds 8.88178e-16. Minute errors in the divisions cause the problems; note that the slope of the line is $-5/7$ and the quotient cannot be held exactly in a double variable.

There is no perfect solution to this problem. A good strategy is to replace exact equality tests with $|x-y| < \varepsilon$ where ε is either built into the code or is set by the user.

6.3 The modification to Line.h is modest. The old private section would be replaced by this:

```
private:
   Point A;
   Point B;
```

Nothing else in Line.h would need to be modified.

Extensive repairs to Line.cc are now needed. For example, the get methods would need to be completely reworked.

However, code that uses the Line class would not require any modification at all!

6.4 A good way to do this is with a procedure declared like this:

```
bool intersect(Line L, Line M, Point& P);
```

The input Line objects are L and M. If they are parallel, then the procedure should return false (to mean *do not intersect*) and leave P unaffected. Otherwise, we set P to be the Point of intersection and return true.

6.5 It is not unreasonable to make the end points of a LineSegment object be public data elements because if a program modifies these data directly (and not through a set method), there is no possibility of creating an invalid object (we allow the two end points to be the same). By contrast, were we to allow open access to a Line object's data, an unwary user might set both a and b to 0 creating an invalid Line object.

In general, it is better to keep the data hidden and write simple `get` and `set` methods to manipulate those data.

Here is a header file `LineSegment.h`. Writing the `LineSegment.cc` file is left to you.

```
#ifndef LINE_SEGMENT_H
#define LINE_SEGMENT_H

#include <iostream>
#include "Point.h"
using namespace std;

class LineSegment {
private:
  // the end points of the segment
  Point A;
  Point B;

public:
  // constructors
  LineSegment();
  LineSegment(Point P, Point Q);
  LineSegment(double x1, double y1, double x2, double y2);

  // get & set methods
  Point getA() const;
  Point getB() const;
  void setA(Point P);
  void setB(Point Q);

  // Length and midpoint methods
  double length() const;
  Point midpoint() const;

  // Check for equality
  bool operator==(const Point& that);
};

// for output
ostream& operator<<(ostream& os, const LineSegment& LS);

#endif
```

6.6 Here are two solutions. In both cases, updating the fields `x` and `y` is straight-forward. The key issue is: how do we write the `return` statement at the end of the procedure.

In this first solution, we make a copy of the `Point` and return that:

```
Point Point::translate(double dx, double dy) {
  x += dx;
  y += dy;
  return Point(x,y);
}
```

This code is clear but suffers a minor inefficiency because it makes a copy of the `Point` and then that copy gets sent back as a return value. We can make this more efficient by returning the `Point` object itself. Here is how that is done.

```
Point Point::translate(double dx, double dy) {
    x += dx;
    y += dy;
    return *this;
}
```

The object `*this` is the `Point` object itself, so no extra copy is needed.

In this simple example, the inefficiency is minimal. However, if this method were to be used extensively then the change from `return Point(x,y);` to `return *this;` can make a noticeable difference.

6.7 This is a subtle problem. Suppose we define a procedure `proc` with one argument. There are two ways we can declare this procedure:

```
Point proc(Point Q);
Point proc(Point &Q);
```

If we use the first syntax, then a copy of the argument is sent to the procedure, but if we use the second syntax, then the argument itself is sent. That is, when we call `proc(W)` the first version cannot modify `W`, but the second version can.

Similarly, the normal behavior of `return x` is to send back a copy of `x`. However, if we want to return the object `x` itself, we need to add an `&` to the return type in the procedure declaration.

For the problem at hand, we need to change the return type of the `translate` method from `Point` to `Point&`. In the `public` section in `Point.h` we declare the procedure like this.

```
Point& translate(double dx, double dy);
```

In `Point.cc` we have the following code.

```
Point& Point::translate(double dx,double dy) {
    x += dx;
    y += dy;
    return *this;
}
```

Let's examine how the old versions (from the previous problem) and this new version behave for the code `(P.translate(1,2)).translate(10,10);`.

In either old version `P.translate(1,2)` returns a copy of `P`. (The first old version returns a copy of a copy of `P`!) So the second invocation of `translate` acts on a copy of `P` and not the object `P` itself. Therefore, `P` is not modified by `translate(10,10)`.

However, in the new version, the statement `return *this;` returns the object itself and not a copy. Therefore the second `translate(10,10)` acts on P and not on some unnamed copy of P.

One last point: In the solution to Exercise 6.6, both `return Point(x,y);` and `return *this;` have the intended effect of returning a copy of the `Point`. In this exercise, the code `return Point(x,y);` is inappropriate. We do not want to return a copy of the `Point`; we want to return the `Point` itself. So we must use `return *this;`.

7.1 Here is the file `Interval.h`. Note that the keyword `inline` is optional for the class methods but mandatory for the `operator<<` procedure. We require both `#include <iostream>` and `using namespace std;` to use `ostream` objects.

```cpp
#ifndef INTERVAL_H
#define INTERVAL_H
#include <iostream>
using namespace std;

class Interval {
private:
  double a,b; // end points of [a,b]

public:
  Interval() {
    a = 0.;
    b = 1.;
  }

  Interval(double x, double y) {
    if (x<y) {
      a = x;
      b = y;
    }
    else {
      a = y;
      b = x;
    }
  }

  double getA() const { return a; }
  double getB() const { return b; }

  bool operator==(const Interval& that) const {
    return (a==that.a) && (b==that.b);
  }

  bool operator!=(const Interval& that) const {
    return !(*this == that);
  }

  bool operator<(const Interval& that) const {
    if (a < that.a) return true;
    if (a > that.a) return false;
```

```
        return b < that.b;
      }

};

inline ostream& operator<<(ostream& os, const Interval& I) {
   os << "[" << I.getA() << "," << I.getB() << "]";
   return os;
}

#endif
```

7.2 The following program accomplishes the required task. It does its work by the following observation. Suppose the intervals are $I_1, I_2, \ldots, I_n$ where $I_j = [a_j, b_j]$ with $a_j < b_j$. Let $\alpha = \max_j a_j$ and $\beta = \min_j b_j$. Interval I_k meets all intervals if and only if $a_k \leq \beta$ and $b_k \geq \alpha$.

```
#include "Interval.h"
#include "uniform.h"
using namespace std;

double find_max_A(const Interval* list, long ni) {
   double ans = list[0].getA();
   for (long k=1; k<ni; k++) {
      if (ans < list[k].getA()) ans = list[k].getA();
   }
   return ans;
}

double find_min_B(const Interval* list, long ni) {
   double ans = list[0].getB();
   for (long k=1; k<ni; k++) {
      if (ans > list[k].getB()) ans = list[k].getB();
   }
   return ans;
}

bool one_meets_all(const Interval* list, long ni) {
   double alpha = find_max_A(list,ni);
   double beta  = find_min_B(list,ni);

   for (long k=0; k<ni; k++) {
      if ((list[k].getA() <= beta) && (list[k].getB() >= alpha)) {
         return true;
      }
   }
   return false;
}

int main() {
   seed();
   long ni;
   cout << "Enter number of intervals --> ";
   cin >> ni;
```

```
    long nreps;
    cout << "Enter number of repetitions --> ";
    cin >> nreps;

    Interval* list;
    list = new Interval[ni];

    long count = 0;

    for (long j=0; j<nreps; j++) {

      for (long k=0; k<ni; k++) {
        list[k] = Interval(unif(), unif());
      }
      if (one_meets_all(list,ni)) count++;
    }

    cout << "Success rate: " << 100*double(count)/double(nreps)
         << "%" << endl;

    delete[] list;
    return 0;
}
```

Here is a sample run of the program.

```
Enter number of intervals --> 100
Enter number of repetitions --> 100000
Success rate: 66.659%
```

It appears that about $\frac{2}{3}$ of the time there is an interval that meets all the others.
One might conjecture that as $n \to \infty$, the probability there is such an interval
approaches $\frac{2}{3}$ (and this is true). On the other hand, it is not hard to show
that for $n = 2$, the probability that the two intervals intersect is exactly $\frac{2}{3}$.
The surprise is that for any value of $n \geq 2$, the probability there is an interval
that intersects all the others is $\frac{2}{3}$. [Reference: J. Justicz, E. Scheinerman, and
P. Winkler, Random intervals, *American Mathematical Monthly* **97** (December
1990) 881–889.]

7.3 How can we sort the array and not modify it? Work with a copy. Here's the
code.

```
#include <iostream>
#include <algorithm>
using namespace std;

double median(const double* array, long nels) {
   if (nels < 0) return 0.;
   if (nels == 0) return array[0];

   // make a copy of the array
   double* copy_array;
   copy_array = new double[nels];
```

```
for (int k=0; k<nels; k++) copy_array[k] = array[k];

// sort the copy
sort(copy_array, copy_array+nels);

// extract the single middle element or the averages
// of two most central elements in the sorted list
double ans;
if (nels%2 == 1) {
  ans = copy_array[nels/2];
}
else {
  ans = (copy_array[nels/2] + copy_array[(nels/2)+1]) / 2.;
}

// free the copy and return the result
delete[] copy_array;
return ans;
}
```

7.4 We can modify Program 7.3 to do the work. We need to generate all primitive Pythagorean triples containing a leg or a hypotenuse whose length is at most 100. To do this, we run the constructor PTriple(m,n) for all $m, n \leq 100$. Note that if either m or n is greater than 100, then all three of $2mn$, $m^2 - n^2$, and $m^2 + n^2$ exceed 100 and may be ignored. Here's a program.

```
#include "PTriple.h"
#include <iostream>
#include <algorithm>
using namespace std;

/**
 * Count the number of triples with a given leg and
 * a given hypotenuse
 */

int main() {
  PTriple* table;        // table to hold the triples
  const int N = 100;     // maximum length we're counting

  int leg_count[N+1];    // tally leg lengths
  int hyp_count[N+1];    // tally hypotenuse lengths

  // Make sure counters are all zero
  for (int k=0; k<=N; k++) {
    leg_count[k] = 0;
    hyp_count[k] = 0;
  }

  // Allocate space for the table
  table = new PTriple[2*N*N];

  // Populate the table with all possible PTriples
  long idx = 0;              // index into the table
  for (long m=1; m<=N; m++) {
```

```
      for (long n=1; n<=N; n++) {
        PTriple P = PTriple(m,n);
        if (P.getA() <= N) {
          table[idx] = P;
          idx++;
        }
      }
    }

    // Sort the table
    sort(table, table+idx);

    // Process unique elements in the tables
    leg_count[table[0].getA()]++;
    leg_count[table[0].getB()]++;
    hyp_count[table[0].getC()]++;

    for (int k=1; k<idx; k++) {
      if (table[k] != table[k-1]) {
        long a,b,c;
        a = table[k].getA();
        b = table[k].getB();
        c = table[k].getC();

        if (a<=N) leg_count[a]++;
        if (b<=N) leg_count[b]++;
        if (c<=N) hyp_count[c]++;
      }
    }
    cout << "Length\tLeg\tHypotenuse" << endl;
    for (int k=0; k<=N; k++) {
      cout << k << "\t" << leg_count[k] << "\t"
           << hyp_count[k] << endl;
    }

    // Release memory held by the table
    delete[] table;
    return 0;
}
```

When the program is run, we see the following output.

Length	Leg	Hypotenuse
0	1	0
1	1	1
2	0	0
3	1	0
4	1	0
5	1	1
6	0	0
7	1	0
8	1	0
9	1	0
....		
90	0	0
91	2	0

92	2	0
93	2	0
94	0	0
95	2	0
96	2	0
97	1	1
98	0	0
99	2	0
100	2	0

For an explanation of these results, see: Eric W. Weisstein. "Pythagorean Triple." From *MathWorld*—A Wolfram Web Resource.

`http://mathworld.wolfram.com/PythagoreanTriple.html`

8.1 Declare a set of sets of integers as set< set<long> > x; Incidentally, the following does not work: set<set<long>> x; because the >> can be confused with the right-shift operator.

Here's code to build the set $\left\{\{1,2,3\},\{4,5\},\{6\}\right\}$.

```
#include <set>
using namespace std;

int main() {
  set<long> tmp;
  set< set<long> > bigset;

  tmp.insert(1);
  tmp.insert(2);
  tmp.insert(3);
  bigset.insert(tmp);

  tmp.clear();
  tmp.insert(4);
  tmp.insert(5);
  bigset.insert(tmp);

  tmp.clear();
  tmp.insert(6);
  bigset.insert(tmp);

  return 0;
}
```

8.2 The procedure should be declared as

```
void print_set(const set<int>& A);
```

Here is the program.

```
#include <set>
#include <iostream>
using namespace std;
```

```
void print_set(const set<int>& A) {
  long n = A.size();        // number of el'ts in the set
  long k;                   // for counting the set
  set<int>::iterator Ai;    // iterator into the set

  cout << "{";
  k = 0;
  for (Ai = A.begin(); Ai != A.end(); ++Ai) {
    cout << *Ai;
    ++k;
    if (k < n) cout << ",";  // comma if not the last element
  }
  cout << "}";
}
```

8.3 C++ set containers may only be used to house data types for which < and == are defined. The complex<double> type does not define <, so we cannot house these objects in a set.

However, C++ ordered pairs (pair<type1,type2>) can be held in a set provided < is defined for both type1 and type2.

Although we cannot put the value $2 + 3i$ into a set, we can put the ordered pair $(2,3)$ into a set. We use code that looks like this.

```
#include <complex>
#include <set>
using namespace std;

int main() {
  complex<double> z(2.,3.);
  set< pair<double,double> > A;
  A.insert( make_pair(z.real(),z.imag()) );
  return 0;
}
```

A more satisfactory solution is developed in Exercise 10.3.

8.4 We must show that a_n equals the number of ordered factorizations of n.

Proof. The proof is by induction with the basis case $a_1 = 1$ given. Suppose the result is true for all values less than n and we ask for the number of ordered factorizations of n. We condition on the first term in the factorization; say it's d. The number of ordered factorizations of a_n whose first factor is $d > 1$ is exactly the number of ordered factorizations of n/d (the balance of the factorization), and that equals $a_{n/d}$ by induction. Summing over all $d > 1$ gives

$$a_n = \sum_{d|n, d>1} a_{n/d}.$$

However, this is just a rearranged version of the sum

$$\sum_{d|n, d<n} a_d$$

and the result follows.   □

This sequence also arises in counting the number of perfect partitions of an integer. See the *On-Line Encyclopedia of Integer Sequences*, sequence A002033 and Eric W. Weisstein, "Perfect Partitions," from *Mathworld*—A Wolfram Web Resource.

```
http://mathworld.wolfram.com/PerfectPartition.html
```

8.5 Here is a file `Partition.h` that creates the class (with all methods and procedures written inline).

```cpp
#ifndef PARTITION_H
#define PARTITION_H

#include <set>
#include <iostream>
#include <vector>
using namespace std;

class Partition {
private:
  /// A multiset to hold the parts
  multiset<int> parts;
  /// An integer to hold the sum
  int sum;

public:
  /// Default constructor: create null partition of 0
  Partition() {
    sum = 0;
    parts.clear();
  }

  /// Add a new part to this partition
  void add_part(int n) {
    if (n <= 0) {
      cerr << "Cannot add nonpositive part to a partition"
           << endl;
      return;
    }
    parts.insert(n);
    sum += n;
  }

  /// What is the sum of the parts?
  int get_sum() const { return sum; }

  /// How many parts in this partition?
  int nparts() const { return parts.size(); }

  /// Get a vector containing the parts
  vector<int> get_parts() const {
    vector<int> ans;
    ans.resize(nparts());
```

```
      multiset<int>::iterator pi;
      int idx = 0;
      for (pi=parts.begin(); pi!=parts.end(); pi++) {
        ans[idx] = *pi;
        ++idx;
      }
      return ans;
    }

  /// Compare two parts
  bool operator<(const Partition& that) const {
    // first compare the number partitioned
    if (sum < that.sum) return true;
    if (sum > that.sum) return true;

    // same sum, so compare number of parts
    if (nparts() < that.nparts()) return true;
    if (nparts() > that.nparts()) return false;

    // last resort, compare element by element
    vector<int> my_parts = get_parts();
    vector<int> that_parts = that.get_parts();

    for (int k=0; k<nparts(); k++) {
      if (my_parts[k] < that_parts[k]) return true;
      if (my_parts[k] > that_parts[k]) return false;
    }

    // Only way to reach here is if the partitions are equal
    return false;
  }
};

inline ostream& operator<<(ostream& os, const Partition& P) {
  if (P.get_sum() == 0) {
    os << 0;
    return os;
  }

  vector<int> list = P.get_parts();
  int np = P.nparts();
  os << P.get_sum() << " = ";
  for (int i=np-1; i>=0; i--) {
    os << list[i];
    if (i > 0) os << "+" ;
  }
  return os;
}

#endif
```

8.6 Here is a program that uses a recursive approach conditioning on the largest part in the partition. The procedure make_partitions(n,mp) generates the set of all partitions of n whose largest part is at most mp.

```
#include "Partition.h"
```

```
set<Partition> make_partitions(int n, int mp) {
  set<Partition> ans;
  set<Partition> tmp;

  if (mp>n) mp = n;

  if (n==0) {
    ans.insert(Partition());
    return ans;
  }

  for (int k=1; k<=n; k++) {
    tmp.clear();
    tmp = make_partitions(n-k,k);
    set<Partition>::iterator sp;
    for (sp = tmp.begin(); sp != tmp.end(); sp++) {
      Partition P = *sp;
      P.add_part(k);
      ans.insert(P);
    }
  }
  return ans;
}

set<Partition> make_partitions(int n) {
  return make_partitions(n,n);
}

int main() {
  cout << "Enter n: ";
  int n;
  cin >> n;

  set<Partition> PS = make_partitions(n);
  set<Partition>::iterator pp;
  for (pp = PS.begin(); pp != PS.end(); pp++) {
    cout << *pp << endl;
  }
  cout << PS.size() << " partitions of " << n << endl;
  return 0;
}
```

Here is a sample run of the program.

```
Enter n: 6
6 = 6
6 = 5+1
6 = 4+2
6 = 3+3
6 = 4+1+1
6 = 3+2+1
6 = 2+2+2
6 = 3+1+1+1
6 = 2+2+1+1
6 = 2+1+1+1+1
```

```
6 = 1+1+1+1+1+1
11 partitions of 6
```

8.7 The solution is to use a `static` look-up table that takes *pairs* of `long` values as keys. To create such a table, be sure to `#include <map>` and give the following declaration.

```
static map< pair<long,long> , long > table;
```

Here is a program with a `binomial` procedure and a `main` to check that it works.

```cpp
#include <map>
#include <iostream>
using namespace std;

long binomial(long n, long k) {

   static map< pair<long,long> , long > table;

   if (k>n) return 0;
   if (k==0) return 1;
   if (k==n) return 1;

   pair<long, long> args = make_pair(n,k);

   if (table.count(args) > 0) {
     return table[args];
   }

   table[args] = binomial(n-1,k-1) + binomial(n-1,k);

   return table[args];
}

int main() {
   for (int n=0; n<10; n++) {
     for (int k=0; k<10; k++) {
       cout << binomial(n,k) << "\t";
     }
     cout << endl;
   }
   return 0;
}
```

The code gives Pascal's triangle as output.

1	0	0	0	0	0	0	0	0	0
1	1	0	0	0	0	0	0	0	0
1	2	1	0	0	0	0	0	0	0
1	3	3	1	0	0	0	0	0	0
1	4	6	4	1	0	0	0	0	0
1	5	10	10	5	1	0	0	0	0
1	6	15	20	15	6	1	0	0	0
1	7	21	35	35	21	7	1	0	0
1	8	28	56	70	56	28	8	1	0
1	9	36	84	126	126	84	36	9	1

8.8 The inputs to the procedure are the array and an integer specifying the length
of the array. The `vector` can either be the return value of the procedure or a
reference argument:

```
vector<long> array2vector(const long* list, long nels);

void array2vector(const long* list, long nels,
                  vector<long>& vlist);
```

The first is, perhaps, more natural, however, the second is more efficient. The
problem with the first approach is that when the `return` statement executes,
the vector<long> built in the procedure is copied to the receiving variable
in the calling procedure.

Here is code for the procedure using the second approach.

```
#include <vector>
using namespace std;

void array2vector(const long* list, long nels,
                  vector<long>& vlist) {
  vlist.resize(nels);
    for (int k=0; k<nels; k++) vlist[k] = list[k];
}
```

8.9 The solution is to use iterators that point to the first and one-past-the-last elements of the `vector` like this,

```
sort(values.begin(), values.end());
```

The following code illustrates this approach.

```
#include <vector>
#include <iostream>
using namespace std;

const int N = 10;

int main() {
  vector<long> values;
  values.resize(N);
  for (int k=0; k<N; k++) {
    values[k] = rand()%1000;
    cout << values[k] << " ";
  }
  cout << endl << endl;

  sort(values.begin(), values.end());

  for (int k=0; k<N; k++) cout << values[k] << " ";
  cout << endl;
  return 0;
}
```

Here is the output of this program.

```
807 249 73 658 930 272 544 878 923 709

73 249 272 544 658 709 807 878 923 930
```

8.10 This is dangerous. The iterator is now focused on a part of a set that no longer exists. Such an action renders the iterator invalid. Even worse, *all* iterators referring to the modified set are now invalid.

8.11 It is tempting (but wrong) to write code like this:

```
set<long>::iterator sp;
for (sp = A.begin(); sp != A.end(); ++sp) {
  if (*sp%2 == 1) A.erase(*sp);
}
```

The problem, as we discussed in the solution to Exercise 8.10, is that once we erase the element referred to by an iterator, the iterator becomes invalid. We need a different approach.

The technique we illustrate here is to step through the set and place a copy of the odd elements we find in a container (in the following code we use a stack, but other choices would work as well). Once we have accumulated copies of all the odd elements in the set, we run through the stack and delete the corresponding elements from the set. Here's the code.

```
#include <set>
#include <stack>
using namespace std;

void delete_odds(set<long>& A) {
  stack<long> eliminate;
  set<long>::const_iterator sp;
  for (sp=A.begin(); sp!=A.end(); ++sp) {
    if (*sp % 2 == 1) eliminate.push(*sp);
  }
  while (!eliminate.empty()) {
    A.erase(eliminate.top());
    eliminate.pop();
  }
}
```

8.12 Here is the code.

```
#include <set>
#include <algorithm>
#include <iostream>
using namespace std;

void print_element(long x) { cout << x << " "; }

void print_set(set<long>& A) {
  cout << "{ ";
  for_each(A.begin(), A.end(), print_element);
```

```
      cout << "}" << endl;
   }
```

Note that we first define the procedure `print_element`. This procedure is used as an argument to `for_each` in the `print_set` procedure.

9.2 Here is a complete solution. First we present the header file `Time.h` in which the short methods are defined inline. The private method `adjust()` is used to correct the variables so they fall in the proper ranges. For example, if the hour/minute/second variables have values $(5, 59, 65)$, then adjust would change these to $(6, 0, 5)$.

Notice the use of a `static` variable `ampm_style` that is modified via the `static` methods `ampm()` and `military()`.

```
#ifndef TIME_H
#define TIME_H

#include <iostream>
using namespace std;

class Time {
private:
  long hour, min, sec;
  static bool ampm_style;
  void adjust();

public:
  Time() { hour = min = sec = 0; }
  Time(int H, int M, int S);
  Time operator+(long n) const;
  Time operator-(long n) const;
  Time operator++();
  Time operator--();
  int get_hour() const { return hour; }
  int get_minute() const { return min; }
  int get_second() const { return sec; }
  static void ampm() { ampm_style = true; }
  static void military() { ampm_style = false; }
  static bool is_ampm() { return ampm_style; }
};

Time operator+(long n, const Time& T);
ostream& operator<<(ostream& os, const Time& T);

#endif
```

Next we give the code file `Time.cc`.

```
#include "Time.h"

bool Time::ampm_style = true;

void Time::adjust() {
  // adjust the seconds field first
```

```
  if (sec < 0) {
    long change = (-sec)/60 + 1;
    sec += change * 60;
    min -= change;
  }
  if (sec > 59) {
    long change = sec/60;
    sec -= change*60;
    min += change;
  }
  // adjust the min field next
  if (min < 0) {
    long change = (-min)/60 + 1;
    min += change*60;
    hour -= change;
  }
  if (min > 59) {
    long change = min/60;
    min -= change*60;
    hour += change;
  }
  // finally, adjust the hour
  if (hour < 0) {
    long change = (-hour/24) + 1;
    hour += change*24;
  }
  if (hour > 23) {
    hour %= 24;
  }
}

Time::Time(int H, int M, int S) {
  hour = H;
  min = M;
  sec = S;
  adjust();
}

Time Time::operator+(long n) const {
  Time T = *this;
  T.sec += n;
  T.adjust();
  return T;
}

Time operator+(long n, const Time& T) {
  return T+n;
}

Time Time::operator-(long n) const {
  Time T = *this;
  T.sec -= n;
  T.adjust();
  return T;
}
```

```
Time Time::operator++() {
  ++sec;
  adjust();
  return *this;
}

Time Time::operator--() {
  --sec;
  adjust();
  return *this;
}

ostream& operator<<(ostream& os, const Time& T) {
  long h = T.get_hour();
  long m = T.get_minute();
  long s = T.get_second();

  if (Time::is_ampm()) {     // am-pm style
    if (h==0) {
      os << 12;
    }
    else if (h>12) {
      os << h-12;
    }
    else {
      os << h;
    }
    os << ":";
    if (m < 10) os << 0;
    os << m;
    os << ":";
    if (s < 10) os << 0;
    os << s;

    if (h < 12) {
      os << " am";
    }
    else {
      os << " pm";
    }
  }

  else {                    // military time
    os << h << ":";
    if (m < 10) os << 0;
    os << m << ":";
    if (s < 10) os << 0;
    os << s ;
  }
  return os;
}
```

Finally, the following code shows how to extract the current local time on a UNIX computer. Unfortunately, it is difficult to understand. Fortunately, it is unusual for a program that solves a mathematics problem to need to deal with the current time of day.

```
#include <ctime>
#include <iostream>
#include "Time.h"
using namespace std;

Time now() {
  time_t clock = time(0);
  long h = localtime(&clock)->tm_hour;
  long m = localtime(&clock)->tm_min;
  long s = localtime(&clock)->tm_sec;

  return Time(h,m,s);
}

int main() {
  cout << "At the tone, the time will be " << now() << endl;
  return 0;
}
```

If you need to deal extensively with date and time matters, it is worth your while to download, build, and install a package created by someone else. For example, the Boost C++ package provides classes for working with date and time. (See Chapter 13 about working with packages you find on the Web including a brief description of Boost on page 286.)

9.3 Here is the header file `EuclideanVector.h`.

```
#ifndef EUCLIDEAN_VECTOR_H
#define EUCLIDEAN_VECTOR_H

#include <vector>
#include <iostream>
using namespace std;

class EuclideanVector {
private:
  static int DEFAULT_DIM;
  int dim;
  vector<double> coords;

public:
  EuclideanVector();
  EuclideanVector(int n);
  static int  get_default_dim() { return DEFAULT_DIM; }
  static void set_default_dim(int n);
  double get(int n) const;
  void set(int n, double x);
  int get_dim() const { return dim; }
  EuclideanVector operator+(const EuclideanVector& that) const;
  EuclideanVector operator*(double s) const;
  bool operator==(const EuclideanVector& that) const;
  bool operator!=(const EuclideanVector& that) const;
};
```

```
EuclideanVector operator*(double s, const EuclideanVector& v);
ostream& operator<<(ostream& os, const EuclideanVector& v);

#endif
```

Here is the code file EuclideanVector.cc.

```
#include "EuclideanVector.h"

int EuclideanVector::DEFAULT_DIM = 2;

EuclideanVector::EuclideanVector() {
  dim = DEFAULT_DIM;
  coords.resize(dim);
  for (int k=0; k<dim; k++) coords[k] = 0.;
}

EuclideanVector::EuclideanVector(int n) {
  if (n < 0) {
    cerr << "Cannot construct vector with negative dimension"
         << endl << "using zero instead" << endl;
    n = 0;
  }
  dim = n;
  coords.resize(n);
  for (int k=0; k<dim; k++) coords[k] = 0.;
}

void EuclideanVector::set_default_dim(int n) {
  if (n < 0) {
    cerr << "Cannot set default dimension to be negative"
         << endl << "using zero instead" << endl;
    n = 0;
  }
  DEFAULT_DIM = n;
}

double EuclideanVector::get(int n) const {
  n %= dim;
  if (n < 0) n += dim;
  return coords[n];
}

void EuclideanVector::set(int n, double x) {
  n %= dim;
  if (n < 0) n += dim;
  coords[n] = x;
}

EuclideanVector
EuclideanVector::operator+(const EuclideanVector& that) const {
  if (dim != that.dim) {
    cerr << "Attempt to add vectors of different dimensions"
         << endl;
    return EuclideanVector(0);
  }
  EuclideanVector ans(dim);
```

```
    for (int k=0; k<dim; k++) {
      ans.coords[k] = coords[k] + that.coords[k];
    }
    return ans;
}

EuclideanVector
EuclideanVector::operator*(double s) const {
  EuclideanVector ans(dim);
  for (int k=0; k<dim; k++) ans.coords[k] = s*coords[k];
  return ans;
}

bool
EuclideanVector::operator==(const EuclideanVector& that) const {
  if (dim != that.dim) return false;
  for (int k=0; k<dim; k++) {
    if (coords[k] != that.coords[k]) return false;
  }
  return true;
}

bool
EuclideanVector::operator!=(const EuclideanVector& that) const {
  return !( (*this) == that );
}

EuclideanVector operator*(double s, const EuclideanVector& v) {
  return v*s;
}

ostream& operator<<(ostream& os, const EuclideanVector& v) {
  os << "[ ";
  for (int k=0; k<v.get_dim(); k++) os << v.get(k) << " ";
  os << "]";
  return os;
}
```

9.4 Here are the files `s.h` and `s.cc` that implement the set S and its operation $*$.

```
#ifndef S_H
#define S_H
#include <iostream>
#include <cmath>
using namespace std;

class S {
private:
  long n;
public:
  S() { n = 0; }
  S(long a) {
    if (a<0) a = -a;
    n = a;
  }
  long getN() const { return n; }
```

```
    double value() const { return sqrt(n); }
    S operator*(const S& that) const { return S(n + that.n); }
};

ostream& operator<<(ostream& os, const S& s);

#endif
```

```
#include "S.h"

ostream& operator<<(ostream& os, const S& s) {
  os << "sqrt(" << s.getN() << ")";
  return os;
}
```

9.5 See the file quaternions.h on the CD-ROM that accompanies this book. This file includes embedded Doxygen comments and the CD-ROM includes the Web pages they generate.

10.1 The two classes are defined in the following .h file. There is no need for a .cc file.

```
#ifndef _RECTANGLE_
#define _RECTANGLE_

class Rectangle {
protected:
  double h,w;    // hold the height and width
public:
  Rectangle(double x=1.0, double y=1.0) {
    h = x;
    w = y;
  }

  double get_width() const  { return w; }
  double get_height() const { return h; }

  void set_width(double x)  { w = x; }
  void set_height(double y) { h = y; }

  double area() const { return h*w; }
  double perimeter() const { return 2.*(h+w); }
};

class Square : public Rectangle {
public:
  Square(double s = 1.) : Rectangle(s,s) { }
  void set_width(double x)  { w = x; h = x; }
  void set_height(double y) { w = y; h = y; }
};
```

```
#endif
```

10.2 Here is a file `parallelogram.h` that defines all three classes.

```cpp
#ifndef PARALLELOGRAM_H
#define PARALLELOGRAM_H
#include <cmath>

class Parallelogram {
protected:
  double a,b,c;
public:
  Parallelogram() { a = b = c = 0.; }
  Parallelogram(double x1, double x2, double y2) {
    if (x1 < 0) x1 = -x1;
    if (y2 < 0) y2 = -y2;
    a = x1;
    b = x2;
    c = y2;
  }
  double area() const { return c*a; }
  double perimeter() const {
    return 2*sqrt(b*b + c*c) + 2*a;
  }
};

class Rectangle : public Parallelogram {
public:
  Rectangle(double height, double width) :
    Parallelogram(width, 0., height) {}
  double perimeter() const { return 2*a + 2*c; }
};

class Rhombus : public Parallelogram {
public:
  Rhombus(double x, double y) :
    Parallelogram(sqrt(x*x)+sqrt(y*y), x, y) {}
  double perimeter() const { return 4*a; }
};

#endif
```

10.3 The class `mycomplex` can be defined in a header file, `mycomplex.h`, like this.

```cpp
#ifndef MY_COMPLEX_H
#define MY_COMPLEX_H

#include <complex>
using namespace std;

class mycomplex : public complex<double> {
public:
  mycomplex() : complex<double> (0.,0.) {}
  mycomplex(double r) : complex<double> (r,0.) {}
  mycomplex(double r, double i) : complex<double> (r,i) {}
```

```
    bool operator<(const mycomplex& that) const {
      if (real() < that.real()) return true;
      if (real() > that.real()) return false;
      if (imag() < that.imag()) return true;
      return false;
    }
};

#endif
```

We define the three constructors by passing arguments up to the constructor for complex<double>. No further action is required by these constructors and so the bodies of these three constructors are empty: {}.

Next the operator < is defined. Note that we use the complex<double> methods real() and imag() to access the data, and then use lexicographic ordering.

Here is a short main to test the mycomplex class.

```
#include "mycomplex.h"
#include <iostream>
#include <set>
using namespace std;

int main() {
  mycomplex z;
  mycomplex w(2);
  mycomplex v(-5,2);

  set<mycomplex> S;
  S.insert(z);
  S.insert(w);
  S.insert(v);

  set<mycomplex>::iterator si;
  for (si = S.begin(); si != S.end(); ++si) {
    cout << *si << " ";
  }
  cout << endl;

  return 0;
}
```

This produces the following output.

```
(-5,2) (0,0) (2,0)
```

10.4 The difficulty here is that each class depends on the other, leaving us in a quandary as to which to define first. In C++ one must declare classes and variables before they can be used. So, the solution is to give a preliminary declaration of the Segment class before giving a full declaration of Point. This is done with the statement class Segment; in the header file.

The operator+ in `Point` uses the `Segment(Point,Point)` constructor to convert a pair of points into a line segment. We cannot put the code for operator+ inline in the declaration of the `Point` class because the required `Segment(double,double)` constructor has not yet been declared. So we are compelled to write the code for operator+ in a separate `.cc` file.

Here are the files `pointseg.h` and `pointseg.cc`.

```
#ifndef POINTSEG_H
#define POINTSEG_H

#include <iostream>
using namespace std;

class Segment;

class Point {
private:
  double x,y;
public:
  Point() { x = y = 0; }
  Point(double xx, double yy) {
    x = xx;
    y = yy;
  }
  double getX() const { return x; }
  double getY() const { return y; }
  Segment operator+(const Point& that) const;
};

class Segment {
private:
  Point A,B;
public:
  Segment() { A = Point(0,0); B = Point(1,0); }
  Segment(Point X, Point Y) { A = X; B = Y; }
  Point getA() const { return A; }
  Point getB() const { return B; }
  Point midpoint() const;
};

ostream& operator<<(ostream& os, const Point& P);
ostream& operator<<(ostream& os, const Segment& S);

#endif
```

```
#include "pointseg.h"

Segment Point::operator+(const Point& that) const {
  return Segment(*this, that);
}

Point Segment::midpoint() const {
  double x = (A.getX() + B.getX()) / 2;
  double y = (A.getY() + B.getY()) / 2;
  return Point(x,y);
```

```
  }

  ostream& operator<<(ostream& os, const Point& P) {
    os << "(" << P.getX() << "," << P.getY() << ")";
    return os;
  }

  ostream& operator<<(ostream& os, const Segment& S) {
    os << "[" << S.getA() << "," << S.getB() << "]";
    return os;
  }
```

10.5 No. Suppose that class `Alpha` declares a data member of type `Beta` and vice versa. If this were allowed, we could create an infinite nesting: `Alpha` objects contain `Beta` objects which in turn contain `Alpha` objects, ad infinitum.

Examine this header file and the problem should become clear.

```
class Beta;

class Alpha {
private:
  Beta b;
};

class Beta {
private:
  Alpha a;
};
```

Trying to `#include` this header file into a program generates a compiler error message like this.

```
In file included from alphabeta.cc:1:
alphabeta.h:5: error: field 'b' has incomplete type
```

10.6 The key idea is to derive `Complexx` as a subclass of `complex<double>` and add two Boolean data fields to signal whether the object is infinite and whether the object is invalid. To save some typing, we `typedef` the symbol `c` to be shorthand for `complex<double>`.

To define the arithmetic operators, we rely on the corresponding operators from `complex`. To convert an object `z` of type `Complexx` to its parent class `c`, we use the expression `c(z)`. So, to multiple `Complexx` objects `w` and `z`, we first check for special cases where either is infinite or invalid, and then to get the required product we may use the expression `c(w)*c(z)`.

Here are the files `Complexx.h` and `Complexx.cc`. To make this class more useful, we should add the operators +=, -=, *=, and /=. We should also extend all the arithmetic operators so that one of the arguments can be a `double`.

```
#include <complex>
using namespace std;
```

```
typedef double R;
typedef complex<R> C;

class Complexx : public C {
private:
  bool infinite;
  bool invalid;
public:
  Complexx(C z) : C(z) {
    infinite = false;
    invalid = false;
  }
  Complexx(R a, R b) : C(a,b) {
    infinite = false;
    invalid = false;
  }

  bool isZero() const {
    return !invalid && !infinite && C(*this) == C(0);
  }
  bool isInfinite() const {
    return !invalid && infinite;
  }
  bool isInvalid() const { return invalid; }

  bool operator==(const Complexx& that) const {
    if (invalid || that.invalid) return false;
    if (infinite && that.infinite) return true;
    return C(*this) == C(that);
  }

  bool operator!=(const Complexx& that) const {
    if (invalid || that.invalid) return false;
    return !(*this == that);
  }

  Complexx operator+(const Complexx& that) const;
  Complexx operator-() const;
  Complexx operator-(const Complexx& that) const;
  Complexx operator*(const Complexx& that) const;
  Complexx operator/(const Complexx& that) const;
};

ostream& operator<<(ostream& os, const Complexx& z);
```

```
#include "Complexx.h"

Complexx Complexx::operator+(const Complexx& that) const {
  Complexx ans(C(*this) + C(that));
  if (invalid || that.invalid) {
    ans.invalid = true;
    return ans;
  }
  if (infinite && that.infinite) {
    ans.invalid = true;
```

```
      return ans;
    }
    if (infinite || that.infinite) {
      ans.infinite = true;
      return ans;
    }
    return ans;
  }

  Complexx Complexx::operator-() const {
    Complexx ans(-C(*this));
    if (invalid) {
      ans.invalid = true;
      return ans;
    }
    if (infinite) {
      ans.infinite = true;
      return ans;
    }
    return ans;
  }

  Complexx Complexx::operator-(const Complexx& that) const {
    return (*this) + (-that);
  }

  Complexx Complexx::operator*(const Complexx& that) const {
    Complexx ans(C(*this) * C(that));
    if (invalid || that.invalid) {
      ans.invalid = true;
      return ans;
    }
    if (infinite && that.isZero()) {
      ans.invalid = true;
      return ans;
    }
    if (isZero() && that.infinite) {
      ans.invalid = true;
      return ans;
    }
    if (infinite || that.infinite) {
      ans.infinite = true;
      return ans;
    }
    return ans;
  }

  Complexx Complexx::operator/(const Complexx& that) const {
    Complexx ans(C(*this) / C(that));
    if (invalid || that.invalid) {
      ans.invalid = true;
      return ans;
    }
    if (isZero() && that.isZero()) {
      ans.invalid = true;
      return ans;
```

```
    }
    if (infinite && that.infinite) {
      ans.invalid = true;
      return ans;
    }
    if (infinite) {
      ans.infinite = true;
      return ans;
    }
    if (that.infinite) {
      ans = Complexx(0.,0.);
      return ans;
    }
    if (that.isZero()) {
      ans.infinite = true;
      return ans;
    }
    return ans;
  }

  ostream& operator<<(ostream& os, const Complexx& z) {
    if (z.isInvalid()) {
      os << "INVALID";
      return os;
    }
    if (z.isInfinite()) {
      os << "Infinity";
      return os;
    }
    os << C(z);
    return os;
  }
```

11.1 Here is a hint. Suppose that π is written in disjoint cycle notation. The order of π is the least common multiple of lengths of the cycles. Also for positive integers a and b, the least common multiple of a and b is $ab/\gcd(a,b)$.

11.2 Here is the file Counted.h. Notice how we use the constructor just to increment the variable n_objects and the destructor to decrement it.

```
#ifndef COUNTED_H
#define COUNTED_H

class Counted {
private:
  static long n_objects;

public:
  Counted()  { n_objects++; }
  ~Counted() { n_objects--; }
  long static count() { return n_objects; }
};

#endif
```

This header file contains nearly everything except the code to initialize the static class variable n_objects. This is done in a separate source file, Counted.cc:

```
#include "Counted.h"
long Counted::n_objects = 0;
```

11.3 Here are the files Partition.h and Partition.cc.

```
#ifndef PARTITION_H
#define PARTITION_H
#include <iostream>
using namespace std;

class Partition {
private:
  int sum;
  int n_parts;
  int *parts;

public:
  Partition() {
    sum = n_parts = 0;
    parts = new int[1];
  }
  Partition(const Partition& that) {
    sum = that.sum;
    n_parts = that.n_parts;
    parts = new int[n_parts+1];
    for (int k=0; k<n_parts; k++) parts[k] = that.parts[k];
  }
  ~Partition() {
    delete[] parts;
  }
  Partition operator=(const Partition& that) {
    sum = that.sum;
    n_parts = that.n_parts;
    delete[] parts;
    parts = new int[n_parts+1];
    for (int k=0; k<n_parts; k++) parts[k] = that.parts[k];
    return *this;
  }
  void add_part(int n);
  int get_sum() const { return sum; }
  int nparts() const { return n_parts; }
  int operator[](int k) const;
  bool operator<(const Partition& that) const;
};

ostream& operator<<(ostream& os, const Partition& P);

#endif
```

```
#include "Partition.h"

void Partition::add_part(int n) {
```

```
    if (n <= 0) {
      cerr << "Cannot add a negative part" << endl;
      return;
    }
    sum += n;
    int* new_parts = new int[n_parts+1];
    for (int k=0; k<n_parts; k++) {
      new_parts[k] = parts[k];
    }
    new_parts[n_parts] = n;
    n_parts++;
    sort(new_parts, new_parts+n_parts);
    delete[] parts;
    parts = new_parts;
  }

  int Partition::operator[](int k) const {
    if ((k<0) || (k>=n_parts)) {
      cerr << "Index out of range" << endl;
      return -1;
    }
    return parts[k];
  }

  bool Partition::operator<(const Partition& that) const {
    if (sum < that.sum) return true;
    if (sum > that.sum) return false;
    if (n_parts < that.n_parts) return true;
    if (n_parts > that.n_parts) return false;
    for (int k=0; k<n_parts; k++) {
      if (parts[k] < that.parts[k]) return true;
      if (parts[k] > that.parts[k]) return false;
    }
    return false;
  }

  ostream& operator<<(ostream& os, const Partition& P) {
    if (P.nparts() < 2) {
      os << P.get_sum();
      return os;
    }
    for (int k=0; k<P.nparts(); k++) {
      os << P[k];
      if (k < P.nparts()-1) os << "+";
    }
    return os;
  }
```

11.5 Here is the file SmartArray.h that implements the class.

```
#ifndef SMART_ARRAY_H
#define SMART_ARRAY_H

class SmartArray {
private:
  long N;
```

```
    long* data;
    long adjust_index(long k) {
      k %= N;
      if (k<0) k += N;
      return k;
    }
public:
    SmartArray(long nels = 1) {
      N = (nels > 1) ? nels : 1;
      data = new long[N];
    }
    ~SmartArray() { delete[] data; }
    long get_N() const { return N; }
    long& operator[](long k) {
      long kk = adjust_index(k);
      return data[kk];
    }
};

#endif
```

Note that we define two `private` data elements: `N` (the size of the array) and `data` (a place to hold the data elements).

We also define a private method `adjust_index` that maps an array index to fall between `0` and `N-1` (inclusive).

The constructor, after ensuring that its argument is at least 1, allocates storage for the `data` array. The destructor's only job is to release the memory held by `data`. The `get_N` method is a convenient way to learn the size of the array. (Note: In this implementation we do not provide any way to resize an array.)

And now we come to the main point of this exercise: the `operator[]` method. Notice that we declared the return type of this method to be `long&`. If we had, instead, declared the return type to be `long`, then `return data[kk];` would simply return *a copy of the value* held in slot `kk` of the `data`. This is just like call by value, but in this case, it's return by value.

By declaring the return type to be `long&`, we switch from returning a copy of a value to returning the actual item itself; that is, we have implemented return by reference.

Suppose the following statements appear in a `main()`,

```
SmartArray X(10);
X[-1] = 4;
```

The first declares `X` to be a `SmartArray` housing 10 values. When the second statement is encountered, here's what happens. First `adjust_index(-1)` runs and returns the value 9 (and that's held in the temporary variable `kk`). Next the code executes `return data[-1];`. So the statement `X[-1] = 4;` is, effectively, transformed into `X.data[9] = 4;`.

11.6 The following code is a file named `LFT.h` that defines the class `LFT`. Notice that we defined `C` to be an abbreviation for `complex<double>`. We include get methods to extract the coefficients of the transformation and an `operator<<` to print the transformations to the screen.

```
#ifndef LFT_H
#define LFT_H
#include <complex>
#include <iostream>
using namespace std;
typedef complex<double> C;

class LFT {
private:
  C a,b,c,d;
public:
  LFT(C aa, C bb, C cc, C dd) {
    a = aa; b = bb; c = cc; d = dd;
  }
  LFT() {
    a = 1; b = 0; c = 0; d = 1;
  }
  C getA() const { return a; }
  C getB() const { return b; }
  C getC() const { return c; }
  C getD() const { return d; }
  C operator()(C z) const {
    return (a*z+b)/(c*z+d);
  }
  LFT operator*(const LFT& T) const {
    C aa,bb,cc,dd;
    aa = a*T.a + b*T.c;
    bb = a*T.b + b*T.d;
    cc = c*T.a + d*T.c;
    dd = c*T.b + d*T.d;
    return LFT(aa,bb,cc,dd);
  }
};

inline ostream& operator<<(ostream& os, const LFT& T) {
  os << "z |--> (" << T.getA() << "*z + " << T.getB()
     << ")/("      << T.getC() << "*z + " << T.getD()
     << ")";
  return os;
}

#endif
```

11.7 The files `Path.h` and `Path.cc` follow. Note that we used a `vector` container to house the points. This is easier than managing a `Path*` array. As a bonus, because we do not need to `delete[]` any arrays we allocated, we do not need a `~Path()` destructor.

Also note that we did not define an `operator+` for the case `Path+Point`. The C++ compiler knows how to convert a `Point` to a `Path` (thanks to the

single-argument constructor). However, a separate `Point+Path` procedure is
necessary.

```
#ifndef PATH_H
#define PATH_H
#include "Point.h"
#include <vector>

class Path {
private:
  vector<Point> pts;
  long npts;
public:
  Path() { npts = 0; }
  Path(const Point& P) {
    npts = 1;
    pts.resize(1);
    pts[0] = P;
  }
  long size() const { return npts; }
  Path operator+(const Path& that) const;
  Point operator[](int k) const;
};

Path operator+(const Point& X, const Path& P);
ostream& operator<<(ostream& os, const Path& P);

#endif
```

```
#include "Path.h"

Path Path::operator+(const Path& that) const {
  Path ans;
  ans.npts = npts + that.npts;
  ans.pts.resize(ans.npts);

  for (int k=0; k<npts; k++) ans.pts[k] = pts[k];
  for (int k=0; k<that.npts; k++) ans.pts[k+npts] = that.pts[k];

  return ans;
}

Point Path::operator[](int k) const {
  if (npts==0) return Point(0,0);
  k %= npts;
  if (k<0) k = -k;
  return pts[k];
}

Path operator+(const Point& X, const Path& P) {
  return Path(X) + P;
}

ostream& operator<<(ostream& os, const Path& P) {
  os << "[ ";
  for (int k=0; k<P.size(); k++) os << P[k] << " ";
```

```
      os << "]";
      return os;
  }
```

12.1 The code is a mildly edited version of the answer to Exercise 7.3; it works for
any type for which the < is defined. Here is the file `median.h`.

```
#ifndef MEDIAN_H
#define MEDIAN_H
#include <algorithm>
using namespace std;

template <class T>
T median(const T* array, long nels) {
  if (nels < 0) return 0.;
  if (nels == 0) return array[0];

  T* copy_array;
  copy_array = new T[nels];
  for (int k=0; k<nels; k++) copy_array[k] = array[k];

  sort(copy_array, copy_array+nels);

  T ans;
  ans = copy_array[nels/2];
  delete[] copy_array;
  return ans;
}
#endif
```

There is an interesting difference between the algorithm we use here and the
algorithm used in our solution to Exercise 7.3. In the first version, if the array
contains an even number of elements, we return the average to the two central
values. However, this new version might be applied to a type for which addition and division by 2 is not defined. Our solution is to return element number
$\lfloor n/2 \rfloor$ of (the sorted copy of) an array with n elements.

12.2 Here is the new version of `SmartArray.h`.

```
#ifndef SMART_ARRAY_H
#define SMART_ARRAY_H

template <class T>
class SmartArray {
private:
  long N;
  T* data;
  long adjust_index(long k) {
    k %= N;
    if (k<0) k += N;
    return k;
  }

public:
  SmartArray(long nels = 1) {
```

```
      N = (nels > 1) ? nels : 1;
      data = new T[N];
    }
    ~SmartArray() { delete[] data; }
    long get_N() const { return N; }
    T& operator[](long k) {
      long kk = adjust_index(k);
      return data[kk];
    }
};

#endif
```

12.3 Here is a file `derivative.h` that creates a template procedure `derivative`.

```
#ifndef DERIVATIVE_H
#define DERIVATIVE_H
#include "Polynomial.h"

template <class T>
Polynomial<T> derivative(const Polynomial<T>& P) {
  Polynomial<T> ans;

  for (long k=1; k<=P.deg(); k++) {
    ans.set(k-1,k*P[k]);
  }
  return ans;
}
#endif
```

Alternatively, we could add a `derivative()` method to the `Polynomial` class template.

12.4 Here is some advice on handling multiple roots. If $p(x)$ has multiple roots, then

$$q(x) = \frac{p(x)}{\gcd(p(x), p'(x))}$$

has the same roots as $p(x)$, but all the roots are simple.

12.5 Here is a file `triple.h` that defines the `triple` class and the `make_triple` procedure templates.

```
#ifndef TRIPLE_H
#define TRIPLE_H

template <class A, class B, class C>
class triple {
public:
  A first;
  B second;
  C third;

  triple() {};
  triple(const A& x, const B& y, const C& z) {
```

```
      first = x; second = y; third = z;
    }

  bool operator<(const triple& that) const {
    if (first < that.first) return true;
    if (first > that.first) return false;
    if (second < that.second) return true;
    if (second > that.second) return false;
    if (third < that. third) return true;
    return false;
  }
};

template <class A, class B, class C>
triple<A,B,C> make_triple(const A& x, const B& y, const C& z) {
  return triple<A,B,C>(x,y,z);
}

#endif
```

12.6 Use the `Polynomial` template created in this chapter as a basic building block. Your template should start like this:

```
template <class T>
class RationalFunction {
private:
   Polynomial<T> numerator;
   Polynomial<T> denominator;
public:
   ....
};
```

12.7 Notice that in the following program we do not use C++'s `set` containers. The `print_set` procedure takes an integer, examines each of its bits, and prints out a set element when it finds a nonzero bit.

```
#include <iostream>
using namespace std;

int LIMIT = 8*sizeof(unsigned long);

void print_set(unsigned long x) {
  cout << "{ ";
  for (int k=0; k<LIMIT; k++) {
    unsigned long mask = 1<<k;
    if ((x&mask) != 0) cout << k+1 << " ";
  }
  cout << "}" << endl;
}

int main() {
  cout << "Enter n (up to " << LIMIT-1 << ") --> ";
  int n;
  cin >> n;
```

```
      if ((n<0) || (n>=LIMIT)) {
        cerr << "Please use a positive value that is less than "
             << LIMIT << endl;
        return -1;
      }

      unsigned long top = (1<<n) - 1;

      for (unsigned long x = 0 ; x <= top; x++) {
        print_set(x);
      }
      return 0;
    }
```

Here is a sample run.

```
Enter n (up to 31) --> 3
{ }
{ 1 }
{ 2 }
{ 1 2 }
{ 3 }
{ 1 3 }
{ 2 3 }
{ 1 2 3 }
```

13.1 Of course, the "right" way to solve this problem is to see that there are 24 factors of 5 in 100!. Here is a solution that uses the GMP package.

```
#include <iostream>
#include "gmpxx.h"
using namespace std;

mpz_class factorial(long n) {
  mpz_class ans(1);
  for (int k=2; k<=n; k++) ans *= k;
  return ans;
}

int main() {
  mpz_class value = factorial(100);
  int count = 0;
  while (value % 10 == 0) {
    value /= 10;
    count++;
  }
  cout <<  count << endl;
  return 0;
}
```

13.2 Here is the program.

```
#include <iostream>
using namespace std;

const int NROWS = 21;
```

```
int main() {
  int table[NROWS][NROWS];

  table[0][0] = 1;
  for (int k=1; k<NROWS; k++) table[0][k] = 0;

  for (int n=1; n<NROWS; n++) {
    table[n][0] = 1;
    table[n][n] = 1;
    for (int k=1; k<n; k++)
      table[n][k] = table[n-1][k-1]+table[n-1][k];
    for (int k=n+1; k<NROWS; k++) table[n][k] = 0;
  }

  for (int n=0; n<NROWS; n++) {
    for (int k=0; k<NROWS; k++) {
      cout << table[n][k] << "\t";
    }
    cout << endl;
  }

  return 0;
}
```

To extend this to a 100×100 table, it is not enough to change the value of NROWS. The binomial coefficients grow too large to fit in `long` variables. Even `long long` variables are not big enough (assuming that these are 64 bits). Instead, we need to use arbitrary precision integers (e.g., from the GMP package) or settle for decimal approximations (i.e., use a `double` table).

14.1 Here are two solutions. First, we can convert the `string` into a `char*` array, and then use `atoi`:

```
int n;
n = atoi(s.c_str());
```

Alternatively, we can wrap the `string` inside an `istringstream` and then extract the integer value using the `>>` operator:

```
istringstream is(s);
int n;
is >> n;
```

14.2 `ifstream fin(file_name.cstr());`

14.3 The addition of a `char` and an `int` results in an `int`, and so instead of writing the letters a through z, instead we see the ASCII values for these letters.

The solution is to replace `'a'+k` with `char('a'+k)`.

14.4 Here are the header and code files, `Free.h` and `Free.cc`, that implement the Free class. Note that we include constructors for building Free objects from a single character, a `string`, and a `char*` array. The helper methods named reduce are used to cancel adjacent inverse elements. The `clear_check` method is used by the constructor to prevent the inclusion of improper characters.

```
#ifndef FREE_H
#define FREE_H
#include <iostream>
#include <cctype>
using namespace std;

class Free {
private:
  string word;
  bool reduce(int k);
  void reduce();
  void check_clear();
public:
  Free() { word.clear(); }
  Free(char ch) {
    word = string(" ");
    word[0]=ch;
    check_clear();
  }
  Free(const string& S) {
    word = S;
    check_clear();
  }
  Free(const char* str) {
    word = string(str);
    check_clear();
  }
  string toString() const { return word; }
  Free operator*(const Free& that) const {
    Free ans;
    ans.word = word + that.word;
    ans.reduce();
    return ans;
  }
  Free inverse() const;
};

inline
ostream& operator<<(ostream &os, const Free& f) {
  os << '(' << f.toString() << ')';
  return os;
}

#endif
```

```
#include "Free.h"

void Free::check_clear() {
  for (unsigned k=0; k<word.size(); k++) {
    while (!isalpha(word[k])) {
      word.erase(k,1);
    }
```

```
  }
  reduce();
}

bool Free::reduce(int k) {
  char c1 = word[k];
  char c2 = word[k+1];
  if ((islower(c1)&&isupper(c2))||(isupper(c1)&&islower(c2))){
    if (toupper(c1) == toupper(c2)) {
      word.erase(k,2);
      return true;
    }
  }
  return false;
}

void Free::reduce() {
  if (word.size() < 2) return;
  for (int k=0; k<int(word.size())-1; k++) {
    while (reduce(k)) {
      if (k>0) --k;
    }
  }
}

Free Free::inverse() const {
  Free ans;
  ans.word = word;
  int n = int(word.size());
  for (int k=0; k<n; k++) {
    char ch = word[k];
    ch = isupper(ch) ? tolower(ch) : toupper(ch);
    ans.word[n-1-k] = ch;
  }
  return ans;
}
```

14.5 The following program reads a text file (given as a command-line argument) and reports letter frequencies.

```
#include <iostream>
#include <fstream>
using namespace std;

int main(int argc, char** argv) {
  ifstream fin;
  if (argc < 2) {
    cerr << "Usage: " << argv[0] << " filename" << endl;
    return 1;
  }
  fin.open(argv[1]);
  if (fin.fail()) {
    cerr << "Unable to open " << argv[1] << " for reading"
         << endl;
    return 2;
  }
```

```
    string word;
    long count[26];
    long total = 0;
    for (int k=0; k<26; k++) count[k] = 0;

    while (true) {
      fin >> word;
      if (fin.fail()) break;
      for (int k=0; k<word.size(); k++) {
        char ch = tolower(word[k]);
        if (isalpha(ch)) {
          count[ch-'a']++;
          total++;
        }
      }
    }

    for (char ch='a'; ch <= 'z'; ch++) {
      int idx = ch - 'a';
      cout << ch << "\t" << 100*double(count[idx])/double(total)
          << endl;
    }

    return 0;
}
```

We ran this code on four 19th-century texts; three of these were in English and one in French. The results are in the accompanying table. It's not hard to distinguish one from the other three.

14.6 A complete program to count polygrams in a text file is included on the accompanying CD-ROM.

14.7 The following program uses the `Point` class from Chapter 6 (files `Point.h` and `Point.cc`) plus the files `Koch.h`, `Koch.cc`, and `main.cc` that we present here.

```
#ifndef KOCH_H
#define KOCH_H

#include <vector>
#include <cmath>
#include "Point.h"

using namespace std;

class Koch {

private:
  vector<Point> pts;

public:
  Koch(int n=0);
  int size() const { return pts.size(); }
```

Letter frequency chart to accompany solution to Exercise 14.5.

Letter	Frequency (percent)			
a	8.4	8.7	8.1	8.3
b	1.5	1.2	1.4	1.5
c	2.6	3.1	2.6	2.4
d	4.4	3.8	4.4	4.8
e	12.4	14.9	12.6	12.5
f	2.4	1.1	2.4	2.3
g	2.4	1.0	2.2	2.4
h	6.2	1.2	5.6	6.0
i	6.6	7.6	7.2	6.8
j	0.1	0.4	0.1	0.1
k	0.6	0.0	0.8	0.8
l	3.7	6.6	4.3	3.8
m	2.4	3.0	2.8	2.6
n	6.7	6.8	7.0	7.2
o	7.5	5.5	7.1	7.1
p	2.2	2.7	1.8	1.8
q	0.1	1.1	0.1	0.1
r	5.9	6.7	5.7	6.0
s	6.5	8.3	5.9	6.0
t	9.2	7.6	9.6	9.7
u	2.9	6.3	2.8	2.7
v	1.0	1.6	0.9	0.9
w	2.2	0.1	2.2	2.3
x	0.3	0.4	0.2	0.1
y	1.7	0.4	1.9	1.8
z	0.0	0.1	0.1	0.0

```
    Point get(int k) const { return pts[k]; }

};

void koch_step(const Point& P, const Point& Q,
               Point& X, Point& Y, Point& Z);

#endif
```

```
#include "Koch.h"

Koch::Koch(int n) {
  if (n<0) n=0;
  pts.push_back(Point(0,0));
  pts.push_back(Point(1,0));

  for (int k=0; k<n; k++) {
    vector<Point> tmp;
    for (unsigned j=0; j<pts.size()-1; j++) {
      Point P = pts[j];
      Point Q = pts[j+1];
```

```
        Point X,Y,Z;
        koch_step(P,Q,X,Y,Z);
        tmp.push_back(P);
        tmp.push_back(X);
        tmp.push_back(Y);
        tmp.push_back(Z);
      }
      tmp.push_back(Point(1,0));
      pts = tmp;
  }
}

const double theta = M_PI/3;

void koch_step(const Point& P, const Point& Q,
               Point& X, Point& Y, Point& Z) {
  double a = P.getX();
  double b = P.getY();
  double c = Q.getX();
  double d = Q.getY();

  X = Point( (2*a+c)/3, (2*b+d)/3 );
  Z = Point( (a+2*c)/3, (b+2*d)/3 );

  double x = (c-a)/3;
  double y = (d-b)/3;

  double xx = cos(theta)*x - sin(theta)*y;
  double yy = sin(theta)*x + cos(theta)*y;

  Y = Point( X.getX() + xx, X.getY() + yy);
}
```

```
#include "Koch.h"
#include "plotter.h"
#include <fstream>
using namespace std;

const char* OUTPUT_FILE = "koch.eps";

int main() {
  int n;
  cout << "Enter depth: ";
  cin >> n;
  Koch K(n);
  PlotterParams params;
  ofstream pout(OUTPUT_FILE);
  PSPlotter P(cin,pout,cerr,params);

  if (P.openpl() < 0) {
    cerr << "Unable to open plotter for file "
         << OUTPUT_FILE << endl;
    return 1;
  }

  P.fspace(0.,0.,1.,1.);
```

```
    for (int k=0; k<K.size()-1; k++) {
      double x1 = K.get(k).getX();
      double y1 = K.get(k).getY();
      double x2 = K.get(k+1).getX();
      double y2 = K.get(k+1).getY();
      P.fline(x1,y1,x2,y2);
    }

    P.closepl();
    return 0;
}
```

When run at depth 5, the following image is produced.

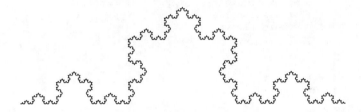

14.8 Here is a program to draw the Mandelbrot set.

```
#include <iostream>
#include <fstream>
#include <complex>
#include "plotter.h"
using namespace std;

const int MAX_ITS = 500;
const int PIXELS = 1000;
const char* FILE_NAME = "mandelbrot.gif";

double norm(complex<double> z) {
  double a = z.real();
  double b = z.imag();
  return a*a + b*b;
}

bool in_set(complex<double> c) {
  complex<double> z = 0.;
  for (int k=0; k<MAX_ITS; k++) {
    z = z*z + c;
    if (norm(z) > 4.) return false;
  }
  return true;
}

complex<double> pix_to_complex(int i, int j) {
  double a = (4.2*i)/PIXELS - 2.1;
  double b = (4.2*j)/PIXELS - 2.1;
  return complex<double>(a,b);
}
```

```
int main() {
  PlotterParams params;
  ofstream pout(FILE_NAME);
  GIFPlotter P(cin, pout, cerr, params);
  if (P.openpl() < 0) {
    cerr << "Unable to open plotter for file " << FILE_NAME
        << endl;
    return 1;
  }

  P.space(0,0,PIXELS,PIXELS);
  P.pencolorname("black");
  for (int i=0; i<PIXELS; i++) {
    for (int j=0; j<PIXELS; j++) {
      complex<double> z = pix_to_complex(i,j);
      if (in_set(z)) {
        P.point(i,j);
      }
    }
  }
  P.closepl();
  return 0;
}
```

It produces the following image.

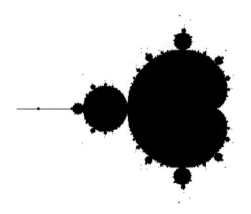

15.1 Here is `Card.h` with all methods and procedures inline.

```
#ifndef CARD_H
#define CARD_H

#include <vector>
#include <string>
#include <iostream>
using namespace std;

const int ACE = 1;
const int JACK = 11;
const int QUEEN = 12;
```

```
const int KING = 13;

const string value_names[] = { "nil", "ace", "two", "three",
                               "four", "five", "six", "seven",
                               "eight", "nine", "ten",
                               "jack", "queen", "king" };

const string suit_names[] = { "clubs", "diamonds",
                              "hearts", "spades" };

enum Suit { CLUBS, DIAMONDS, HEARTS, SPADES };

class Card {
private:
  int value;
  Suit suit;
  int card2int() const {
    return 13*suit + value;
  }

public:
  Card() { value = ACE; suit = SPADES; }
  Card(int spots, Suit family) {
    spots %= 13;
    if (spots <= 0) spots += 13;
    value = spots;
    suit = family;
  }
  bool operator<(const Card& that) const {
    return card2int() < that.card2int();
  }
  bool operator==(const Card& that) const {
    return card2int() == that.card2int();
  }
  bool operator!=(const Card& that) const {
    return card2int() != that.card2int();
  }
  friend ostream& operator<<(ostream& os, const Card& C);
};

inline ostream& operator<<(ostream& os, const Card& C) {
  os << value_names[C.value] << " of "
     << suit_names[C.suit];
  return os;
}

#endif
```

15.2 The statement `T* xp = new T(10);` creates a single new object of type `T` in which the value 10 is supplied to the class `T` constructor.

The statement `T* xp = new T[10];` creates a new array to hold 10 objects of type `T`.

15.3 Notice that in the following program we catch a `bad_alloc` that is thrown by new when there is not enough memory to allocate. (The prefix `std::` is not necessary because we declared `using namespace std;`.)

```
#include <iostream>
using namespace std;

int main() {
  long size = 10000000; // ten million
  long* array;
  long count = 0;
  while (true) {
    try {
      array = new long[size];
    }
    catch (bad_alloc X) {
      cerr << "Unable to allocate memory" << endl;
      exit(1);
    }
    ++count;
    cout << "Success with array #" << count << endl;
  }
}
```

15.4 +. x+y addition or `string` concatenation, x++ and ++x increment, and unary +x is the identity function (not mentioned elsewhere in this book).

−. a−b subtraction, −x unary minus, x−− and −−x decrement, and combined with > to give −> arrow operator.

∗. a∗b multiplication, ∗x pointer/iterator dereferencing, `type* var` pointer/array declaration, and combined with / to delimit comments: `/*` and `*/`.

/. a/b division and // single line comment. See also ∗.

=. a=b assignment, a==b equality testing, combined with < and > for ≤ and ≥, and combined with ! for inequality testing.

<. a<b for less than, a<<b for bit shift and stream output, and used with > for declaring templates and for `#include` of standard headers. See also =.

>. a>b for greater than, a>>b for bit shift and stream input. See also =, −, and >.

!. !a Boolean not. See also =.

&. a&b bitwise and, &a address of, `type& var` reference declaration, a&&b Boolean and.

|. a|b bitwise or and a||b Boolean or.

˜. ˜a bitwise not and `˜Class` destructor.

:. `class : public subclass` inheritance, `label:` statement; labeling, `case 1:` label in a `switch` statement,

```
Child(argument_list) : Parent(argument_sublist) { ... }
```

parent class constructor invocation, `Class::name`/`namespace::name` scope resolution, and `a?b:c` in the trigraph operator.

15.5 This is a tricky problem combining many of the concepts we have encountered in this book.

Each constructible number can be represented as an expression tree. For example, the golden mean $(1 + \sqrt{5})/2$ can be parsed like this.

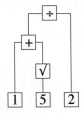

There are three types of nodes in this tree:

- Atomic nodes that contain integers.
- Unary operations nodes ($\sqrt{}$ and unary $-$) that have one child.
- Binary operation nodes ($+$, binary $-$, $\times$, and $\div$) that have a left and a right child.

Each child node is, itself, a constructible number.

Our solution begins by defining the data elements of the `Constructible` class. They are:

- An operation code op. This field indicates the type of node: an atomic node (i.e., an integer), a unary operation, or a binary operation. To this end we define an enum type called `Operation` that is one of NOP (for an atomic node), PLUS, MINUS, TIMES, DIVIDE, NEG, or SQRT.
- An integer `val` field. This is only used in the case where this is an atomic node.
- An integer `nargs` field. This is 0 for an atomic node, 1 for a unary operation, or 2 for a binary operation. It tell us how many direct children this node has.
- An array `args` containing the children. This array is dynamically allocated and its size is given in `nargs`.

The constructor creates an atomic node with a given integer value. The various `operator` methods create `Constructible` objects with one or two arguments (saved in the `args` array). An assignment operator is needed when copying a `Constructible` object so that a true independent copy is created. Because various operations allocate storage (via new), we need to be mindful to release that storage (e.g., in the destructor).

The code for the binary operator methods are remarkably similar, so we write a `private` helper method named `binary_op`. A similar `unary_op` is provided for the negation and square root.

The procedure to produce a decimal (complex) value for a `Constructible` object works recursively on the structure of the object. Note that the Standard Template Library does not define a square root method for `complex<double>` values, so we write our own.

The header `Constructible.h` and code files `Constructible.cc` follow.

```
#ifndef CONSTRUCTIBLE_H
#define CONSTRUCTIBLE_H
#include <iostream>
#include <complex>
using namespace std;

enum Operation { NOP, PLUS, MINUS, TIMES, DIVIDE, NEG, SQRT };

class Constructible {
private:
  Operation op;
  long val;
  int nargs;
  Constructible *args;
  Constructible binary_op(Operation verb,
                          const Constructible& that) const;
  Constructible unary_op(Operation verb) const;

public:
  Constructible(long x = 0) { op = NOP; val = x;  nargs = 0; }
  ~Constructible() {
    if (nargs > 0) {
      delete[] args;
    }
  }
  Constructible(const Constructible& X);
  Constructible& operator=(const Constructible& that);
  Constructible operator+(const Constructible& that) const {
    return binary_op(PLUS,that);
  }
  Constructible operator-() const { return unary_op(NEG); }
  Constructible operator-(const Constructible& that) const {
    return binary_op(MINUS,that);
  }
  Constructible operator*(const Constructible& that) const {
    return binary_op(TIMES,that);
  }
  Constructible operator/(const Constructible& that) const {
    return binary_op(DIVIDE, that);
  }
  Constructible sqrt() const { return unary_op(SQRT); }
  complex<double> value() const;
  friend ostream& operator<<(ostream& os,
                             const Constructible& X);
};
```

```
inline Constructible operator+(long a, Constructible X) {
  return Constructible(a) + X;
}
inline Constructible operator*(long a, Constructible X) {
  return Constructible(a) * X;
}
inline Constructible operator-(long a, Constructible X) {
  return Constructible(a) - X;
}
inline Constructible operator/(long a, Constructible X) {
  return Constructible(a) / X;
}

complex<double> complex_sqrt(const complex<double>& z);

inline Constructible sqrt(const Constructible& X) {
  return X.sqrt();
}
#endif
```

```
#include "Constructible.h"
#include <cmath>

Constructible::Constructible(const Constructible& X) {
  op = X.op;
  val = X.val;
  nargs = X.nargs;
  if (nargs > 0) {
    args = new Constructible[nargs];
    for (int k=0; k<nargs; k++) {
      args[k] = X.args[k];
    }
  }
}

Constructible&
Constructible::operator=(const Constructible& that) {
  op = that.op;
  val = that.val;
  if (nargs > 0) {
    delete[] args;
  }
  nargs = that.nargs;
  if (nargs > 0) {
    args = new Constructible[nargs];
    for (int k=0; k<nargs; k++) {
      args[k] = that.args[k];
    }
  }
  return *this;
}

Constructible
Constructible::binary_op(Operation verb,
                         const Constructible& that) const {
```

```
  Constructible ans;
  ans.op = verb;
  ans.nargs = 2;
  ans.args = new Constructible[2];
  ans.args[0] = *this;
  ans.args[1] = that;
  return ans;
}

Constructible
Constructible::unary_op(Operation verb) const {
  Constructible ans;
  ans.op = verb;
  ans.nargs = 1;
  ans.args = new Constructible[1];
  ans.args[0] = *this;
  return ans;
}

complex<double> Constructible::value() const {
  switch(op) {
  case NOP:
    return complex<double>(double(val),0);
    break;
  case PLUS:
    return args[0].value() + args[1].value();
    break;
  case MINUS:
    return args[0].value() - args[1].value();
    break;
  case NEG:
    return -args[0].value();
    break;
  case TIMES:
    return args[0].value() * args[1].value();
    break;
  case DIVIDE:
    return args[0].value() / args[1].value();
    break;
  case SQRT:
    return complex_sqrt(args[0].value());
    break;
  default:
    cerr << "This shouldn't happen!" << endl;
  }
  return complex<double>(0);
}

const char* OPEN = "\\left(";
const char* CLOSE = "\\right)";

ostream& operator<<(ostream& os, const Constructible& X) {
  switch(X.op) {
  case NOP:
    os << X.val;
    break;
```

```
    case PLUS:
      os << OPEN  << X.args[0] << " + " << X.args[1] << CLOSE;
      break;
    case MINUS:
      os << OPEN << X.args[0] << " - " << X.args[1] << CLOSE;
      break;
    case TIMES:
      os << OPEN << X.args[0] << " \\times " << X.args[1]
         << CLOSE;
      break;
    case DIVIDE:
      os << "\\frac{" << X.args[0] << "}{" << X.args[1] << "}";
      break;
    case NEG:
      os << "-" << OPEN << X.args[0] << CLOSE;
      break;
    case SQRT:
      os << "\\sqrt{" << X.args[0] << "}";
      break;
    default:
      os << "INVALID CONSTRUCTIBLE OBJECT";
    }
    return os;
}

complex<double> complex_sqrt(const complex<double>& z) {
  double x = z.real();
  double y = z.imag();
  if ((x==0.) && (y==0.)) return complex<double>(0.);
  if (y==0.) {
    if (x >= 0.) {
      return complex<double>(sqrt(x),0);
    }
    else {
      return complex<double>(0., sqrt(-x));
    }
  }
  double r = sqrt(hypot(x,y));
  double t = atan2(y,x)/2.;
  return complex<double> (r*cos(t), r*sin(t));
}
```

The following brief `main()` illustrates the use of the `Constructible` class.

```
#include "Constructible.h"
int main() {
  Constructible five(5);
  Constructible PHI = (1+sqrt(five))/2;
  cout << "PHI = " << PHI.value().real() << endl;
  cout << "TeX form: " << PHI << endl;
  return 0;
}
```

This gives the following output.

```
PHI = 1.61803
TeX form: \frac{\left(1 + \sqrt{5}\right)}{2}
```

The TeX code produces this: $\frac{\left(1+\sqrt{5}\right)}{2}$.

Showing that two `Constructible` objects represent the same complex number is tricky so we did not define an `==` operator. See E.R. Scheinerman, "When close enough is close enough," *American Mathematical Monthly* **107** no. 6 (June–July 2000) 489–499 for one approach.

15.6 A full solution is available on the accompanying CD-ROM. In this solution we use several classes.

- `Token`: Objects of this type are "atoms" of a Boolean expression. This may be a variable, a constant, a unary operation, or a binary operation.

- `Expression`: Objects of this type represent Boolean expressions.

- `LookupTable`: These are devices that associate Boolean values with variables.

- `ExpressionLoader`: This is a device to read input streams and produce `Expression` objects.

- `TautologyChecker`: This is a device that runs through all possible substitutions for the variables in an `Expression` to see if any yield FALSE. If not, then the expression is a tautology.

Index